AF479154

The IMA Volumes
in Mathematics
and Its Applications

Volume 4

Series Editors
George R. Sell   Hans Weinberger

Institute for Mathematics and Its Applications
IMA

The **Institute for Mathematics and Its Applications** was established by a grant from the National Science Foundation to the University of Minnesota in 1982. The IMA seeks to encourage the development and study of fresh mathematical concepts and questions of concern to the other sciences by bringing together mathematicians and scientists from diverse fields in an atmosphere that will stimulate discussion and collaboration.

The IMA Volumes are intended to involve the broader scientific community in this process.

Hans Weinberger, Director
George R. Sell, Associate Director

IMA Programs

1982–1983 Statistical and Continuum Approaches to Phase Transition

1983–1984 Mathematical Models for the Economics of Decentralized Resource Allocation

1984–1985 Continuum Physics and Partial Differential Equations

1985–1986 Stochastic Differential Equations and Their Applications

1986–1987 Scientific Computation

1987–1988 Applied Combinatorics

1988–1989 Nonlinear Waves

Springer Lecture Notes from the IMA

The Mathematics and Physics of Disordered Media
Editors: Barry Hughes and Barry Ninham
(Lecture Notes in Mathematics, Volume 1035, 1983)

Orienting Polymers
Editor: J. L. Ericksen
(Lecture Notes in Mathematics, Volume 1063, 1984)

New Perspectives in Thermodynamics
Editor: James Serrin
(Springer-Verlag, 1986)

Models of Economic Dynamics
Editor: Hugo Sonnenschein
(Lecture Notes in Economics, Volume 264, 1986)

Dynamical Problems in Continuum Physics

Edited by
J.L. Bona, Constantine Dafermos,
J.L. Ericksen, and David Kinderlehrer

With 74 Illustrations

Springer-Verlag
New York Berlin Heidelburg
London Paris Tokyo

J.L. Bona
Department of Mathematics, The Pennsylvania State University, University Park, PA
16802, U.S.A.

Constantine Dafermos
Division of Applied Mathematics, Brown University, Providence, RI 02912, U.S.A.

J.L. Ericksen
School of Mathematics and Department of Aerospace Engineering and Mechanics,
University of Minnesota, Minneapolis, MN 55455, U.S.A.

David Kinderlehrer
School of Mathematics, University of Minnesota, Minneapolis, MN 55455, U.S.A.

AMS Classification: 73-06, 73BXX, 76-06, 76AXX

AMS Classification: 73-06, 73BXX, 76-06, 76AXX

Library of Congress Cataloging-in-Publication Data
Dynamical problems in continuum physics.
  (The IMA volumes in mathematics and its applications ; v. 4)
  Proceedings of a workshop held during the
1984–85 IMA program on continuum physics and
partial differential equations.
  1. Field theory (Physics)—Congresses.
2. Nonlinear theories—Congresses.   I. Bona, J. L.
II. University of Minnesota. Institute for
Mathematics and Its Applications.   III. Series.
QC173.68.D96   1987   530.1′4   86-28035

Printed and bound by R.R. Donnelley & Sons, Harrisonburg, Virginia.
Printed in the United States of America.

9 8 7 6 5 4 3 2 1

ISBN 0-387-96463-0 Springer-Verlag New York Berlin Heidelberg
ISBN 3-540-96463-0 Springer-Verlag Berlin Heidelberg New York

The IMA Volumes in Mathematics and Its Applications

Current Volumes:

Volume 1: Homogenization and Effective Moduli of Material and Media
  Editors: J.L. Ericksen, David Kinderlehrer, Robert Kohn, and J.-L. Lions
Volume 2: Oscillation Theory, Computation, and Methods of Compensated Compactness
  Editors: Constantine Dafermos, J.L. Ericksen, David Kinderlehrer, and Marshall Slemrod
Volume 3: Metastability and Incompletely Posed Problems
  Editors: Stuart S. Antman, J.L. Ericksen, David Kinderlehrer, and Ingo Müller
Volume 4: Dynamical Problems in Continuum Physics
  Editors: J.L. Bona, Constantine Dafermos, J.L. Ericksen, and David Kinderlehrer

Forthcoming Volumes:

1984–1985: Continuum Physics and Partial Differential Equations

  Theory and Applications of Liquid Crystals
  Amorphous Polymers and Non-Newtonian Fluids

1985–1986: Stochastic Differential Equations and Their Applications

  Random Media
  Percolation Theory and Ergodic Theory of Infinite Particle Systems
  Hydrodynamic Behavior and Interacting Particle Systems and Applications
  Stochastic Differential Systems, Stochastic Control Theory and Applications

1986–1987: Scientific Computation

  Computational Fluid Dynamics and Reacting Gas Flows
  Numerical Algorithm for Modern Parallel Computer Architectures

CONTENTS

FOREWORD

This IMA Volume in Mathematics and its Applications

Dynamical Problems in Continuum Physics

represents the proceedings of a workshop which was an integral part of the
1984-85 IMA program on CONTINUUM PHYSICS AND PARTIAL DIFFERENTIAL EQUATIONS.
We are grateful to the Scientific Committee:

> J.L. Ericksen
>
> D. Kinderlehrer
>
> H. Brezis
>
> C. Dafermos

for their dedication and hard work in developing an imaginative, stimulating, and
productive year-long program.

> George R. Sell
>
> Hans Weinberger

PREFACE

The behavior of matter and waves in a dynamical setting offers many challenging problems to the mathematician and the materials scientist alike. Under review in this volume are a variety of nonlinear phenomena whose consideration entails new perspectives, not commonly found in the literature. Of particular note is the experimental aspect of many of the papers. In addition, attention has been given to the interaction of electromagnetic and mechanical properties of materials.

Questions arise which cannot now be answered. Attempts are made to describe and to understand phenomena which are far from equilibrium or which suffer abrupt changes in behavior. Some of this requires tentative physical or analytical assumptions. The bases for these hypotheses lie in the quest for a rational theory which agrees with experiment.

This Volume and Volume 2, Oscillation theory, compution, and methods of compensated compactness, offer different viewpoints of some of the dynamical studies considered during the 1984-1985 IMA program, Continuum Physics and Partial Differential Equations. Contents of the other volumes, found at the end of the book, contain other relevant titles.

The workshop brought together researchers in a number of areas of engineering, mathematics, and physics. The conference committee greatly appreciates the concerted efforts of the speakers and discussants to make their presentations intelligible to a mixed audience. Especial thanks are due to George Herrmann and Ivar Stakgold for their informal contributions.

The conference committee would like to take this opportunity to thank the staff of the IMA, Professors Weinberger and Sell, Mrs. Pat Kurth, and Mr. Robert Copeland for their assistance in arranging the workshop. Special thanks are due to Mrs. Debbie Bradley, Mrs. Patricia Brick and Mrs. Kaye Smith for their preparation of manuscripts. We gratefully acknowledge the support of the National Science Foundation.

J. Bona

C. Dafermos

J. L. Ericksen

D. Kinderlehrer

Conference committee

SOLITARY WATER-WAVES IN THE PRESENCE OF SURFACE TENSION

C.J. Amick

Department of Mathematics
University of Chicago
Chicago, Illinois 60680

and

K. Kirchgässner

Mathematische Institut A
Universitat Stuttgart
Pfaffenwaldring 57/vi   7 Stuttgart 80
West Germany

1.   Introduction

In this paper we consider steady two-dimensional surface waves under the
action of gravity on an ideal fluid, with surface tension, and contained in a channel
of infinite extent. Although a global theory exists for periodic and solitary
waves for fluids lacking surface tension [1], [3], [4], [10], the only rigorous
results for fluids with surface tension are for small-amplitude periodic waves
[5], [9], [16], [17], [18]. In this paper, we prove the existence of small-
amplitude solitary waves for fluids with surface tension. A global result along
with the proofs of some technical lemmas stated here will be given in [2].

We comment briefly on the problem for periodic waves, which may be refor-
mulated as a nonlinear eigenvalue problem [9]

$$\theta = G(\nu, \theta; \gamma, h, \lambda) \tag{1.1}$$

where γ is a given positive number proportional to the surface tension,
$h \in (0, \infty]$ is the given mean-depth of the fluid, and $\lambda \in (0, \infty)$ is the given
wavelength. Here the number $\nu > 0$ and the function $\theta(s)$, $s \in [-\lambda/2, \lambda/2]$, are
the unknowns, where ν is proportional to the square of the mean-speed, and $-\theta$
denotes the angle between the unknown free surface and the horizontal. Since γ,
h, and λ are fixed, one may just consider the equation $\theta = G(\nu, \theta)$. Now $G(\nu, 0)$
$= 0$ for all $\nu \in R$, and so one may seek non-trivial solutions by standard bifurcation theory. The case of bifurcation from a simple eigenvalue may be considered
in the usual way. For (λ, h) fixed, the eigenvalues for the linearized problem are

simple for all $\gamma > 0$ except for a sequence of γ's converging monotonically to zero. When γ equals one of these critical values, there is bifurcation from a double eigenvalue, the analysis of which has recently been provided by Jones and Toland [9]. In the case of a simple eigenvalue, the global theory due to Rabinowitz [12] ensures that a branch of solutions is either unbounded or returns to another eigenvalue. Numerical calculations [7], [8] suggest the former, but there are no rigorous results yet to show it.

2. The Mathematical Problem

Consider a wave moving from right to left on the surface of a two-dimensional incompressible fluid. We assume that the density ρ of the fluid is constant, say $\rho = 1$, that the flow is irrotational, and that the fluid is at rest at infinity where the asymptotic height of the wave is prescribed and equals $h > 0$. If the wave is moving without change of form and with (unknown) speed c, then in a moving reference frame the flow is steady, and the velocity field $q = (q_1, q_2)$ satisfies $q \to (c,0)$ at infinity. We shall be interested in this steady flow throughout the paper.

Bernoulli's equation ensures that

$$p(x,y) + \frac{1}{2} |q(x,y)|^2 + gy = \text{constant} \tag{2.1}$$

in the flow domain S, where p denotes the pressure and g is the acceleration of gravity. We assume that the unknown free surface Γ is given by a function $y = Y(x)$, so $\Gamma = \{(x,Y(x)): x \in R\}$ and the flow domain may be described by $S = \{(x,y): 0 < y < Y(x), x \in R\}$ where $R = (-\infty, \infty)$. We assume that the pressure is identically zero in the region above the fluid and that the jump in pressure across Γ is inversely proportional to the radius of curvature [11], [15]. The constant of proportionality is the surface tension $T > 0$, which is given. The use of this in (2.1) yields

$$\frac{1}{2} |q(x,Y(x))|^2 + gY(x) - TY''(x)/(1+Y'(x)^2)^{3/2} = \text{constant}, \ x \in R. \tag{2.2}$$

Note that the constant is just $\frac{1}{2} c^2 + gh$.

Since the fluid is incompressible and the flow is irrotational, we have $q = (q_1, q_2) = (\phi_x, \phi_y) = (\psi_y, -\psi_x)$. We define an analytic function, the complex potential, $w(z) = \phi(z) + i\psi(z)$, $z \in S$, where points in S are denoted by either (x,y) or $z = x + iy$ when convenient.

Since there is no flow through the solid bottom $\{y = 0\}$ or across the free surface Γ, the function ψ has constant values there. We set $\psi = 0$ on $\{y = 0\}$ and $\psi = Q$ on Γ. Note that $Q = ch$ since $\psi_y \to c$ at infinity in S and $Y(x) \to h$ as $|x| \to \infty$. The map $z \to w(z)/Q$ takes the <u>unknown</u> domain S onto the <u>fixed</u> domain $T = R \times (0,1)$. We seek w in the form

$$w'(z) = c \exp (\tau(z) + i\theta(z))$$

where $\tau + i\theta$ is analytic in S. Since the complex velocity $w'(z) = q_1(z) - iq_2(z)$, it follows that the speed $|q| = c \exp(\tau)$ and that $-\theta$ is the angle q makes to the horizontal. The condition that $q \to (c,0)$ at infinity in S ensures that θ and τ vanish at infinity.

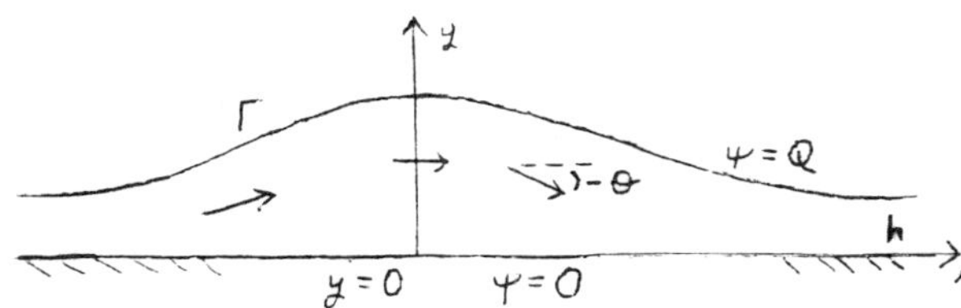

Figure 1. A solitary wave of elevation

We take coordinates $\zeta = \tilde{\phi} + i\tilde{\psi}$ in T, and let $m(\zeta)$ denote the inverse of $\frac{1}{Q} w$. Define $\tilde{\tau}(\zeta) + i\tilde{\theta}(\zeta) = \tau(m(\zeta)) + i\theta(m(\zeta))$, $\zeta \in T$. Then the boundary condition (2.2) on Γ becomes one on $\{\tilde{\psi} = 1\}$, and we merely state it here since it or analogous versions have been derived elsewhere [2], [5], [9]:

$$\frac{\nu}{2} \exp(2\tilde{\tau}(\tilde{\phi},1)) - \int_0^{\tilde{\phi}} \sin \tilde{\theta}(s,1) \exp(-\tilde{\tau}(s,1)) ds + \gamma \tilde{\theta}_{\tilde{\phi}}(\tilde{\phi},1) \exp(\tilde{\tau}(\tilde{\phi},1))$$

$$= \text{constant}, \quad \tilde{\phi} \in R, \tag{2.3}$$

where $\nu = c^2/gh$ and $\gamma = T/gh^2$. Note that $\gamma > 0$ is <u>fixed</u> while $\nu > 0$ is

4

<u>unknown</u>. If we can find a suitable solution of (2.3), then it is simple to reverse the arguments and find a solution to the physical problem. Since we will not be returning to the physical coordinates (x,y), we now simplify notation by writing (x,y) for $(\tilde{\phi}, \tilde{\psi})$ and (τ,θ) for $(\tilde{\tau}, \tilde{\theta})$.

Our mathematical problem is find a number $\nu > 0$ and functions θ and τ satisfying

$$\tau + i\theta \text{ is analytic in } T = R \times (0,1), \tag{2.4}$$

$$\tau(-x,y) = \tau(x,y), \quad \theta(-x,y) = -\theta(x,y), \quad (x,y) \in T, \tag{2.5}$$

$$\theta(x,0) = 0, \quad x \in R, \tag{2.6}$$

$$\theta(x,y), \tau(x,y) \to 0 \text{ as } |(x,y)| \to \infty \tag{2.7}$$

and

$$\nu\theta_y e^{2\tau} - \sin\theta\, e^{-\tau} + \gamma\theta_{xx} e^{\tau} + \gamma\theta_x\theta_y e^{\tau} = 0 \text{ on } \{y = 1\}. \tag{2.8}$$

We remark that (2.5) is due to our insistence on symmetric waves, (2.6) states that the flow is horizontal on the solid bottom $\{y = 0\}$, (2.7) is due to the form in which we seek w', and (2.8) follows from differentiating (2.3) in the horizontal direction. Since the conjugate harmonic function τ is uniquely determined by θ and by (2.7), one may consider (2.8) as an equation for ν and θ alone. For any value of ν, the function $\theta \equiv 0$ is a solution of (2.8); we are interested in non-trivial solutions.

It is natural to ask at what values of ν solutions might bifurcate from the branch of trivial solutions. An answer is suggested by the following formula, which holds under quite general assumptions on Y defining Γ:

$$\nu = 1 + \{\frac{3}{2h} \int_{-\infty}^{\infty}(Y - h)^2 + \frac{T}{gh} \int_{-\infty}^{\infty} (\sqrt{1 + (Y')^2} - 1)\} / \int_{-\infty}^{\infty} (Y-h). \tag{2.9}$$

This formula is proved directly from the Euler equations without any assumption that the wave is symmetric; a derivation may be found in [2] where arguments analogous to those in [3] are used. The same analysis which gives (2.9) also shows

that ν can never equal unity for a non-trivial solitary wave. If there is a
sequence of small-amplitude solutions converging to zero, then $(Y(x) - h)$ will
become small, whence (2.9) suggests that bifurcation should occur from $\nu = 1$.
For a wave of elevation (Figure 1) where $Y(x) \geqslant h$, we necesarily have $\nu > 1$ while
for waves of depression we have $\nu < 1$. Before proceeding to the existence of
solitary waves, we remark that (2.4) - (2.8) serve as the equations for periodic
waves if (2.7) is replaced by the condition that θ and τ be periodic in the x-
direction.

We rewrite (2.8) in the form

$$\theta_y - \theta + \gamma\theta_{xx} = f, \quad x \in R, \tag{2.10}$$

where

$$f(x) = (\sin \theta(x,1)\exp(-2\tau(x,1)) - \theta(x,1)) - \gamma\theta_x(x,1)\tau_x(x,1)$$

$$+ \tau_x(x,1)(1 - \exp(\tau(x,1))) + (1 - \nu)\tau_x(x,1)$$

$$+ (\nu - 1)\tau_x(x,1)(1 - \exp(\tau(x,1))).$$

Note that f is small compared to (θ,τ) when these quantities are small and ν
is close to unity. We proceed formally for a moment: Assume that $\theta(x,y)$ goes to
zero sufficiently fast at infinity in T. Taking the Fourier transform

$$\hat{\theta}(k,y) = \frac{1}{\sqrt{2\pi}} \int_{-\infty}^{\infty} e^{ikx} \theta(x,y)dx$$

of Laplace's equation and (2.6) yields

$$\left. \begin{array}{l} \dfrac{d^2}{dy^2} \hat{\theta}(k,y) - k^2\hat{\theta}(k,y) = 0, \ y \in (0,1), \\[2em] \hat{\theta}(k,0) = 0 \end{array} \right\} \ k \in R,$$

whence $\hat{\theta}(k,y) = A(k) \sinh ky/k$ with the usual meaning when $k = 0$. If we assume
that f is smooth and goes to zero rapidly enough at infinity, then (2.10) yields
$\frac{d}{dy} \hat{\theta}(k,1) - (1+\gamma k^2) \hat{\theta}(k,1) = \hat{f}(k), \ k \in R$, whence

$$\hat{\theta}(k,y) = \frac{\sinh ky}{kh(k;\gamma)} \hat{f}(k),$$

where $h(k;\gamma) = \cosh k - (1 + \gamma k^2) \sinh k/k$. Since the denominator will play a significant role in what is to follow, we state a lemma about the set $\sigma(\gamma) = \{k \in C: h(k;\gamma) = 0\}$. The proof is given in [2], and depends solely on Rouché's theorem and elementary calculus.

LEMMA 1. (a) _If_ $k \in \sigma(\gamma) - \{0\}$, _then_ k _is a simple root, and is either real or purely imaginary._

(b) _The set_ $\sigma(\gamma) \cap \{|\text{Imag } k| < \pi\}$ _consists of four roots counting multiplicity:_ $-r(\gamma), r(\gamma), 0, 0$. _If_ $\gamma \in (0, \frac{1}{3})$, _then_ $r(\gamma) > 0$ _and_ $r(\gamma) \to \infty$ _as_ $\gamma \to 0$ _and_ $r(\gamma) \to 0$ _as_ $\gamma \to \frac{1}{3}$. _If_ $\gamma = \frac{1}{3}$, _then_ $r(\gamma) = 0$. _If_ $\gamma > \frac{1}{3}$, _then_ $r(\gamma) = is(\gamma)$, _where_ $s(\gamma) > 0$ _and satisfies_ $s(\gamma) \to 0$ _as_ $\gamma \to \frac{1}{3}$ _and_ $s(\gamma) \to \pi$ _as_ $\gamma \to \infty$.

Since θ is harmonic in T and vanishes on $\{y = 0\}$, it is uniquely determined by its boundary values on $\{y = 1\}$, and so we may concentrate on the quantity

$$\hat{\theta}(k) \equiv \hat{\theta}(k,1) = \frac{\sinh k}{kh(k;\gamma)} \hat{f}(k) \ . \tag{2.11}$$

We expect that the purely imaginary roots of h will give some uniform exponetial decay to our solution, and that the zeros along the real-axis should be the most important. When $\gamma > \frac{1}{3}$, there are exactly two such roots (both at $k = 0$), while there are four if $\gamma \in (0,\frac{1}{3}]$. For the sake of brevity, _we shall consider the case of a fixed_ $\gamma > 1/3$ _for the time being_, and remark on the case $\gamma \in (0,\frac{1}{3}]$ later in section 5. In addition, we shall drop the dependence on γ in quantities like $h(k;\gamma)$.

Since $r(\gamma) = is(\gamma), s(\gamma) > 0$, there exists $d = d(\gamma)$ $(0, s(\gamma)/2)$ such that $1/h$ has only a double pole at $k = 0$ within the strip $\{|\text{Imag } k| < 2d\}$ and is otherwise regular there. The use of this observation in (2.11) yields

$$\hat{\theta}(k) = \hat{f}(k)/(\frac{1}{3} - \gamma)k^2 + H(k)\hat{f}(k) \tag{2.12}$$

where H is analytic in $\{|\text{Imag } k| < 2d\}$ and satisfies $|H(k)| <$ const.$/(1 + |k|^2)$ there. Define the inverse Fourier transform of H by

$$G(x) = \frac{1}{\sqrt{2\pi}} \lim_{\mu \to \infty} \int_{-\mu}^{\mu} e^{-ikx} H(k)\,dk.$$

The function G is even in x since H is even in k. If $x < 0$, we may move the contour up in the complex plane to find

$$G(x) = \frac{e^{2dx}}{\sqrt{2\pi}} \int_{-\infty}^{\infty} e^{-isx} H(s + 2id)\,ds$$

whence

$$G(x) = \frac{e^{-2d|x|}}{\sqrt{2\pi}} \int_{-\infty}^{\infty} e^{is|x|} H(s + 2id)\,ds, \quad x \in R. \tag{2.13}$$

The function G has some fixed exponential decay at infinity.

Define

$$A(x) = \frac{1}{\sqrt{2\pi}} \int_{-\infty}^{\infty} G(x-s)f(s)\,ds, \quad x \in R \tag{2.14}$$

and note that $\hat{A}(k) = \hat{G}(k)\hat{f}(k) = H(k)\hat{f}(k)$. This suggests we make the following decomposition in (2.12): $\theta(x) = Q(x) - A(x)$ where $(\frac{1}{3} - \gamma)k^2 \hat{Q}(k) = \hat{f}(k)$, or equivalently (at least formally),

$$Q''(x) = f(x)/(\gamma - \frac{1}{3}), \quad x \in R . \tag{2.15}$$

We now state a formula [13] which relates $\tau(x,1)$ to its conjugate function $\theta(x) = \theta(x,1)$:

$$\tau(x) \equiv \tau(x,1) = \beta + \frac{1}{2} \int_{-\infty}^{\infty} \theta(s)\, \{K(x-s) + K(s)\}\,ds, \quad x \in R \tag{2.16}$$

where $K(s) = \sinh s\pi/(-1 + \cosh s\pi)$ and the integral is taken as a principal value. Here $\beta \in R$ is a parameter to be chosen eventually so that $\tau(\pm\infty, 1) = 0$. We note that if θ is an odd function with sufficient decay at infinity, then the function

$$\frac{1}{2} \int_{-\infty}^{\infty} \theta(s)K(x-s)ds$$

vanishes at infinity and is the value on $\{y = 1\}$ of the function conjugate to θ. The difficulty with the formula is that it is not defined for elements of the function space we must use later.

3. Function Spaces and Properties of Mappings

We define $u \in R^2$ by $u = (u_1, u_2) = (Q,Q')$, and write Ω for $(u,A,\tau) = (u_1,u_2,A,\tau)$. The right-hand side f of (2.10) is a function of $\theta(x)$, $\theta'(x)$, $\tau(x)$, $\tau'(x)$, and $(1-\nu)$, and so setting $\theta(x) = Q(x) + A(x)$ yields $\theta_y - \theta + \gamma\theta_{xx} = f = g(u_1, u_2, A, A', \tau, \tau', \nu - 1)$, where $g : R^7 \to R$ is analytic and satisfies $g(0) = |\nabla g(0)| = 0$. Since we are only interested in small solutions to our problem (2.4) - (2.8) in this paper, there is no harm in introducing a truncation as follows. Let $U \in C^{\infty}(R \to R)$ be an odd function such that $U(x) = x$, $|x| \leqslant 1$, and $|U'(x)| \leqslant 1$, $|U(x)| \leqslant 2$ for all $x \in R$.

For each $\varepsilon > 0$, define

$$g_{\varepsilon}(u,A,\tau,\nu) \equiv g_{\varepsilon}(\Omega,\nu) = g(\varepsilon U(\tfrac{1}{\varepsilon} u_1),\ldots,\ \varepsilon U(\tfrac{1}{\varepsilon} \tau'),\ \varepsilon U(\tfrac{1}{\varepsilon}(\nu-1))) . \tag{3.1}$$

For each $\alpha = (\alpha_1, \alpha_2) \in R^2$ and $\beta \in R$ define

$$N_1(\Omega, \alpha, \beta, \nu; \varepsilon)(x) = \alpha_1 + \alpha_2 x + \frac{1}{\gamma - \frac{1}{3}} \int_0^x (x-s)g_{\varepsilon}(\Omega, \nu)(s)ds, \tag{3.2}$$

$$N_2(\Omega, \alpha, \beta, \nu; \varepsilon)(x) = \alpha_2 + \frac{1}{\gamma - \frac{1}{3}} \int_0^x g_{\varepsilon}(\Omega, \nu)(s)ds, \tag{3.3}$$

$$N_3(\Omega, \alpha, \beta, \nu; \varepsilon)(x) = \frac{1}{\sqrt{2\pi}} \int_{-\infty}^{\infty} G(x-s)\, g_{\varepsilon}(\Omega, \nu)(s)ds, \tag{3.4}$$

$$N_4(\Omega, \alpha, \beta, \nu; \varepsilon)(x) = \beta + \frac{1}{2} \int_{-\infty}^{\infty} (u_1(s) + A(s))\{K(x-s) + K(s)\}ds \tag{3.5}$$

for all $x \in R$. For $p \in R$, let

$$C_p = \{\phi \in C(R \to R): \sup_{\substack{x,y \in R \\ |x-y| \leqslant 1}} e^{-n|x|} \{|\phi(x)| + \frac{|\phi(x) - \phi(y)|}{\sqrt{|x-y|}}\} \equiv \|\phi\|_p < \infty\},$$

and let $C_p^1 = \{\phi \in C^1(R \to R): \phi, \phi' \in C_p\}$ with the norm $\|\phi\|_{p,1} = \|\phi\|_p + \|\phi'\|_p$. Let $d \in (0, s(\gamma)/2)$ be as in section 2, and set

$$X = C_d^1 \times C_d \times C_d^1 \times C_d^1$$

with norm

$$\|\Omega\|_X = \|u_1\|_{d,1} + \|u_2\|_d + \|A\|_{d,1} + \|\tau\|_{d,1} \, .$$

Our choice of spaces deserves some elaboration. The exponential increase in C_d or C_d^1 is preserved by the operators N_i. _If we sought a space in which solutions were forced to decay at infinity, then N_1 and N_2 might not map it into itself._ The Hölder norm in C_d^1 ensures that $u_1 + A$ and its first derivative is Hölder continuous. This is needed since the operator given by the right-hand side of (3.5) maps weighted $C^{1,1/2}$ spaces into themselves, but not weighted C^1 spaces into themselves.

For each $\varepsilon > 0$, define a map on $X \times R^4$ by

$$F(\Omega, \alpha, \beta, \nu; \varepsilon) = (u_1 - N_1, u_2 - N_2, A - N_3, \tau - N_4, \alpha, \beta, \nu). \qquad (3.6)$$

One can show that the range of F is contained in $X \times R^4$. Since $g(0) = |\nabla g(0)| = 0$, the right-hand side of (3.2) - (3.4) (ignoring the terms involving α) should be small when ε is small, and this is made precise in the following lemma [2]. Define $\tilde{F}: X \times R \to X$ by

$$\tilde{F}_1(\Omega, \nu; \varepsilon) = \frac{1}{\gamma - \frac{1}{3}} \int_0^x (x-s')\, g_\varepsilon(\Omega, \nu)(s)ds, \qquad (3,7a)$$

$$\tilde{F}_2(\Omega, \nu; \varepsilon) = \frac{1}{\gamma - \frac{1}{3}} \int_c^x g_\varepsilon(\Omega, \nu)(s)ds, \qquad (3.7b)$$

$$\tilde{F}_3(\Omega, \nu; \varepsilon) = \frac{1}{\sqrt{2\pi}} \int_{-\infty}^0 G(x-s)\, g_\varepsilon(\Omega, \nu)(s)ds, \qquad (3.7c)$$

$$\tilde{F}_4(\Omega, \nu; \varepsilon) = 0. \qquad (3,7d)$$

LEMMA 2. $\quad \|\tilde{F}(\Omega^{(1)}, v_1; \varepsilon) - \tilde{F}(\Omega^{(2)}, v_2; \varepsilon)\|_X$

$$(3.8)$$

$$\leq \text{const. } \varepsilon \, (\|\Omega^{(1)} - \Omega^{(2)}\|_X + |v_1 - v_2|)$$

<u>where the constant is independent of</u> $(\Omega^{(1)}, v_1)$, $(\Omega^{(2)}, v_2) \in X \times R$ <u>and</u> $\varepsilon \in (0,1)$.

A standard theorem [6; p. 147] shows that F is a homeomorphism of $X \times R^4$ onto itself if is small enough. Since F is Lipschitz, so is its inverse.

Given any $(\tilde{\alpha}, \tilde{\beta}, \tilde{v}) \in R^4$, we may solve the equations $F(\Omega, \alpha, \beta, v; \varepsilon) = (0,0,0,0, \tilde{\alpha}, \tilde{\beta}, \tilde{v})$ uniquely. It follows that $(\Omega, \alpha, \beta, v) = F^{-1}(0,0,0,0,\tilde{\alpha},\tilde{\beta},\tilde{v})$, whence $\Omega = J(\alpha, \beta, v; \varepsilon)$ and J is a Lipschitz map from R^4 into X. We state our result in the following

THEOREM 3. <u>There exists</u> $\bar{\varepsilon} > 0$ <u>such that for all</u> $\varepsilon \in (0,\bar{\varepsilon}\,]$, <u>the equation</u> $\Omega = N(\Omega, \alpha, \beta, v; \varepsilon)$ $(N = (N_1, \ldots, N_4)$ <u>given in</u> $(3.2) - 3.5))$ <u>has a unique</u> <u>solution</u> $\Omega = J(\alpha, \beta, v; \varepsilon)$. <u>Here</u> J <u>is a Lipschitz map from</u> R^4 <u>into</u> X <u>with</u> $J(0,0,0,v;\varepsilon) = 0$ <u>for all</u> v.

<u>From now on we fix</u> $\varepsilon = \bar{\varepsilon}$, restrict v $(1-\bar{\varepsilon}, 1+\bar{\varepsilon})$ and drop the dependence on ε from functions like F and J. The plan for solving our problem is to show the following: For each v near to unity, there is $(\alpha(v), \beta(v)) \in R^3$ such that (i) Ω is small compared to $\bar{\varepsilon}$ and (ii) $\Omega(x)$ goes to zero rapidly at infinity. Now (i) means that u_1, u_2, A, A', τ, and τ' have maximum values on R less than $\bar{\varepsilon}$ whence we may remove the subscript ε denoting truncation in $(3.2) - (3.4)$. Part (ii) means that these quantities should decay exponentially at infinity. When (i) and (ii) are satisfied, it is immediate that $(v,\theta,\tau) = (v, J_1(\alpha(v), \beta(v),v) + J_3(\alpha(v), \beta(v), v), J_4(\alpha(v), \beta(v),v))$ gives rise to a solution of $(2.4) - (2.8)$. The main step in implementing this program is a proof of a centre-manifold theorem which shows that the component $A = J_3$ in $\Omega = (u_1, u_2, A, \tau) = J$ is a pointwise function of u_1, u_2, and τ. In the following theorem, we write $u_1(\alpha, \beta, v)$ for $J_1(\alpha, \beta, v),\ldots,$ and $\tau(\alpha,\beta,v)$ for $J_4(\alpha, \beta, v)$ when convenient.

THEOREM 4. <u>For each</u> $(\alpha, \beta, \nu) \in R^4$ <u>let</u>

$$P(\alpha, \beta, \nu) = (J_1(\alpha, \beta, \nu), J_2(\alpha, \beta, \nu), J_4(\alpha, \beta, \nu)). \quad \underline{Then}$$

$$J(\alpha, \beta, \nu)(x+x_0) = J(P(\alpha, \beta, \nu)(x_0), \nu)(x) \tag{3.9}$$

<u>for all</u> $(\alpha, \beta, \nu) \in R^4$ <u>and all</u> $x, x_0 \in R$. <u>In particular</u>,

$$A(\alpha, \beta, \nu)(x) = J_3(u_1(\alpha, \beta, \nu)(x), u_2(\alpha, \beta, \nu)(x), \tau(\alpha, \beta, \nu)(x), \nu)(0)$$
$$\tag{3.10a}$$

<u>for all</u> $(\alpha, \beta, \nu) \in R^4$ <u>and all</u> $x \in R$.

<u>Proof</u>. Fix (α, β, ν) and x_0, and define $\Omega_1(x) = J(\alpha, \beta, \nu)(x+x_0)$ and $\Omega_2(x) = J(P(\alpha, \beta, \nu)(x_0), \nu)(x)$. Then a direct and simple calculation yields $F(\Omega_1, \alpha, \beta, \nu) = F(\Omega_2, \alpha, \beta, \nu)$. Since F is a homeomorphism, we get $\Omega_1 = \Omega_2$.

q.e.d.

<u>Remark</u> If we differentiate (3.9) with respect to x and set $x = 0$, then

$$\frac{d}{dx} A(\alpha, \beta, \nu)(x) = \frac{d}{dx} J_3(u_1, u_2, \tau, \nu)(0) . \tag{3.10b}$$

Since $J(0,0,0,\nu) = 0 \in X$ by Theorem 3, one can show that

$$\|J(\alpha, \beta, \nu)\|_X \leq \text{const.} \, (|\alpha| + |\beta|) \tag{3.11}$$

for $(\alpha, \beta, \nu) \in R^4$. Since $g(0) = |\nabla g(0)| = 0$, it is easy to check that at $\varepsilon = \bar{\varepsilon}$
$\|g_\varepsilon(\Omega, \nu)\|_d \leq \text{const.} \, \{|1-\nu| + \|\Omega\|_X\} \|\Omega\|_X$ for all $\Omega \in X$. The use of this inequality with (3.4) and (3.11) yields

$$\|J_3(\alpha, \beta, \nu)\|_{d,1} \leq \text{const.} \, \{|1-\nu| + \|J(\alpha, \beta, \nu)\|_X\} \|J(\alpha, \beta, \nu)\|_X \tag{3.12}$$

$$\leq \text{const.} \, \{|1-\nu| + |\alpha| + |\beta|\} \, (|\alpha| + |\beta|).$$

It then follows from (3.10) that

$$|A(\alpha, \beta, \nu)(x)| + \left|\frac{d}{dx} A(\alpha, \beta, \nu)(x)\right| \leq \{|1-\nu| + |u_1| + |u_2| + |\tau|\}(|u_1| + |u_2| + |\tau|)$$
$$\tag{3.13}$$

for all $x \in R$ whenever the quantities on the right-hand side are sufficiently small. Here $u_1 = u_1(\alpha, \beta, \nu)(x)$ and similarly for u_2 and τ. We emphasize that the constant in (3.13) is an absolute one.

We write Q for u_1 and Q' for u_2; more precisely, $Q(\alpha, \beta, \nu)(x) = u_1(\alpha, \beta, \nu)(x)$. Then (3.2) ensures that

$$Q'' = \frac{1}{\gamma - \frac{1}{3}} \, g_{\bar{\varepsilon}}(Q, Q', A, \tau, \nu), \qquad (3.14)$$

$$Q(0) = \alpha_1 \quad , \quad Q'(0) = \alpha_2 .$$

We are now going to drop the $\bar{\varepsilon}$ subscript by assuming that the functions we are working with always satisfy

$$|\nu-1|, \; |Q(x)|, \; |Q'(x)|, \; |\tau(x)|, \; |\tau'(x)| \; < \bar{\varepsilon}, \; x \in R . \qquad (3.15)$$

The substitution of (3.15) into (3.13) and the choice of a smaller $\bar{\varepsilon}$ (if needed) yield

$$|A(x)|, \; |A'(x)| < \bar{\varepsilon}, \; x \in R . \qquad (3.16)$$

At this stage, it is not obvious that there are non-trivial solutions $J(\alpha, \beta, \nu) = (Q, Q', A, \tau)$ for which (3.15) - (3.16) are satisfied. We shall show that the solutions we construct in section 4 have this property.

The use of (3.15) - (3.16) with (3.1) in (3.14) yields

$$Q''(x) = \frac{1}{\gamma - \frac{1}{3}} \, g(Q(x), Q'(x), A(x), A'(x), \tau(x), \tau'(x), \nu - 1), \quad x \quad R. (3.17)$$

If we use (3.10) in the right-hand side we obtain

$$Q''(x) = \frac{1}{\gamma - \frac{1}{3}} \, \tilde{g}(Q(x), Q'(x), \tau(x), \tau'(x), \nu - 1), \; x \quad R . \qquad (3.18)$$

For our next assumption, we return to (2.5) and demand that

$$Q \text{ is an odd function and } \tau \text{ is an even function (of } x) . \qquad (3.19)$$

It follows from (3.4) that A is an odd function. We next satisfy (2.7) by demanding that

$$|Q(x)|, \ |Q'(x)|, \ |\tau(x)|, \ |\tau'(x)| \leq \text{const.} \ e^{-\delta|x|}, \ x \in R \qquad (3.20)$$

where the constant and $\delta > 0$ may depend on Q and τ.

4. The Rescaled Equations

We take $\alpha_1 = Q(0) = 0$ to satisfy (3.19). As remarked after (2.9), the parameter ν never equals unity except for trivial flows. Since $|\nu-1|$ is to be small, we set $\mu = \sqrt{1-\nu}$. (The choice $\mu = \sqrt{\nu-1}$ would lead to the trivial solution in the calculations to follow.) Since $\nu \in (1-\bar{\epsilon}, 1+\bar{\epsilon})$ in section 3, we have the restriction, $\mu \in [0, \sqrt{\bar{\epsilon}}\ ]$. We define rescaled variables by $Q(x) = \mu^3 \tilde{Q}(\mu x)$, $\tau(x) = \mu^2 \tilde{\tau}(\mu x)$, $\alpha_2 = \mu^4 \tilde{\alpha}_2$, and $\beta = \mu^2 \tilde{\beta}$; more precisely

$$\tilde{Q}(\tilde{\alpha}_2, \ \tilde{\beta}, \ \mu)(x) = \frac{1}{\mu^3} \ Q(0, \mu^4 \tilde{\alpha}_2, \ \mu^2 \tilde{\beta}, \ 1-\mu^2)(\frac{x}{\mu}), \qquad x \in R.$$

If we use this in (3.18) and recall that $\tilde{g}$ is related to g, which is itself the right-hand side of (2.10), then a simple calculation yields

$$\tilde{Q}'' = \frac{1}{\gamma - \frac{1}{3}} \ \{-2\tilde{Q}\tilde{\tau} - \tilde{\tau} \ \tilde{\tau}' + \tilde{\tau}'\} + j(\tilde{Q}, \ \tilde{Q}', \ \tilde{\tau}, \ \tilde{\tau}', \ \mu), \qquad (4.1)$$

$$\tilde{Q}(0) = 0, \ \tilde{Q}'(0) = \tilde{\alpha}_2 \qquad (4.2)$$

where j is a map vanishing when $\mu = 0$. The function j is odd if $\tilde{Q}$ is odd and $\tilde{\tau}$ is even. The equation for $\tilde{\tau}$ is

$$\tilde{\tau}(x) = \tilde{\beta} + \frac{1}{2} \ \int_{-\infty}^{\infty} \{\tilde{Q}(s) + k(\tilde{Q}(s), \ \tilde{Q}'(s), \tilde{\tau}(s), \mu)\}(K(\frac{x-s}{\mu}) + K(\frac{s}{\mu}))ds,$$

$$(4.3)$$

where $K(x) = \sinh x\pi/(-1+\cosh x\pi)$ and k has properties similar to j. In deriving (4.3), we have used the fact that $\tau = N_4$ in (3.5), the representation (3.10a) to rewrite A in terms of Q, Q', and τ, and the estimate (3.13).

Let us formally put $\mu = 0$ in (4.1) - (4.3):

$$\tilde{Q}'' = \frac{1}{\gamma - \frac{1}{3}}(-2\tilde{Q}\tilde{\tau} - \tilde{\tau} \ \tilde{\tau}' + \tilde{\tau}') \ , \qquad (4.4)$$

14

$$\tilde{Q}(0) = 0, \quad \tilde{Q}'(0) = \tilde{\alpha}_2, \tag{4.5}$$

$$\tilde{\tau}(x) = \tilde{\beta} + \frac{1}{2} \int_{-\infty}^{\infty} \tilde{Q}(s)\{H(x-s) + H(s)\}ds, \tag{4.6}$$

where $H(s)$ equals 1 if $s > 0$ and -1 if $s < 0$. If we restrict attention to odd functions $\tilde{Q}$ with a suitable decay rate at infinity, then we may choose

$$\tilde{\beta} = -\frac{1}{2} \int_{-\infty}^{\infty} \tilde{Q}(s)H(s)ds = -\int_{0}^{\infty} \tilde{Q}(s)ds \tag{4.7}$$

so that

$$\tilde{\tau}(x) = \frac{1}{2} \int_{-\infty}^{\infty} \tilde{Q}(s)H(x-s)ds = -\int_{x}^{\infty} \tilde{Q}(s)ds .$$

The use of this in (4.4) yields $\tilde{Q}'' = \{-3\tilde{\tau}\tilde{\tau}' + \tilde{\tau}'\}/(\gamma - \frac{1}{3})$, whence

$$\tilde{Q}' = \tilde{\tau}'' = \frac{1}{\gamma - \frac{1}{3}} \{-\frac{3}{2}\tilde{\tau}^2 + \tilde{\tau}\}. \tag{4.8}$$

This equation has the solution $\tilde{\tau}(x) = \operatorname{sech}^2 (x/2\sqrt{\gamma-1/3})$. Returning to the function Q, we have $Q(x) = \mu^2 \frac{d}{dx} \operatorname{sech}^2 (\mu x/2\sqrt{\gamma-1/3})$. Since $\theta(x) = Q(x) + A(x)$ and A is small compared to Q, we have $\theta(x) \simeq Q(x)$ for our formal arguments. The physical wave profile is given by $\{(s,Y(s)): s \in R\}$, where $Y'(s) \simeq -\theta(s/h)$. Since $Y(\pm \infty) = h$, we arrive at

$$Y(x) \simeq h + h \int_{x/h}^{\infty} Q(s)ds \simeq h - h\mu^2 \operatorname{sech}^2 (\frac{\mu x}{2h\sqrt{\gamma-1/3}}) ,$$

which is the approximate solution found by Korteweg and de Vries and which represents a small-amplitude solitary wave of depression when $\gamma > \frac{1}{3}$.

We now make the previous formal arguments rigorous. For suitable functions ϕ, define

$$\|\phi\|_3 = \sup_{\substack{x,y \in R \\ |x-y| \leqslant 1}} \exp(|x|/2\sqrt{-1/3}) \{|\phi(x)| + |\phi(x)-\phi(y)|/|x-y|\},$$

$$\|\phi\|_2 = \|\phi\|_3 + \|\phi'\|_3 \ , \quad \|\phi\|_1 = \|\phi\|_3 + \|\phi'\|_3 + \|\phi''|_3 \ .$$

Let

$$X_1 = \{\phi \in C^2(R \to R): \ \phi \text{ is an odd function of } x \text{ and } \|\phi\|_1 < \infty\},$$

$$X_2 = \{\phi \in C^1(R \to R): \ \phi \text{ is an even function of } x \text{ and } \|\phi\|_2 < \infty\},$$

and

$$X_3 = \{\phi \in C(R \to R): \ \phi \text{ is an odd function of } x \text{ and } \|\phi\|_3 < \infty\}.$$

Define $M: \ X_1 \times X_2 \times (-\sqrt{\bar{\epsilon}}, \sqrt{\bar{\epsilon}}) \to X_3 \times X_2$ by

$$M(\tilde{Q}, \tilde{\tau}, \mu) = (\tilde{Q}'' - \frac{1}{\gamma - \frac{1}{3}} \{-2\tilde{Q}\tilde{\tau} - \tilde{\tau}\tilde{\tau}' + \tilde{\tau}'\} - j(\tilde{Q}, \tilde{Q}', \tilde{\tau}, \tilde{\tau}', \mu),$$

$$\tag{4.9a}$$

$$\tilde{\tau} - \frac{1}{2} \int_{-\infty}^{\infty} (\tilde{Q} + k(\tilde{Q}, \tilde{Q}', \tilde{\tau}, \mu) K(\frac{x-s}{\mu}) ds) \quad \text{when} \quad \mu > 0$$

and by

$$M(\tilde{Q}, \tilde{\tau}, \mu) = (\tilde{Q}'' - \frac{1}{\gamma - \frac{1}{3}} \{-2 \tilde{Q}\tilde{\tau} - \tilde{\tau}\tilde{\tau}' + \tilde{\tau}'\}, \ \tilde{\tau} + \int_x^\infty \tilde{Q}(s) ds)$$

$$\text{when} \quad \mu < 0 \ . \tag{4.9b}$$

We have the known solution to $M(\tilde{Q}_0, \tilde{\tau}_0, 0) = 0$, where $\tilde{\tau}_0(x) = \text{sech}^2 (x/2\sqrt{\gamma - \frac{1}{3}}$ and $\tilde{Q}_0(x) = \tilde{\tau}_0'(x)$. We shall show that M and $M_{(\tilde{Q}, \tilde{\tau})}$, its partial derivative with respect to $(\tilde{Q}, \tilde{\tau})$, are continuous in a neighborhood of $(\tilde{Q}_0, \tilde{\tau}_0, 0)$, and that $M_{(\tilde{Q}, \tilde{\tau})}(\tilde{Q}_0, \tilde{\tau}_0, 0)$ is a linear homeomorphism. The implicit function theorem then gives us the totality of solutions of $M(\tilde{Q}, \tilde{\tau}, \mu) = 0$ in a neighborhood of $(\tilde{Q}_0, \tilde{\tau}_0, 0)$ as $\tilde{Q} = \tilde{Q}(\mu)$ and $\tilde{\tau} = \tilde{\tau}(\mu)$. We may then define

$$\tilde{\beta} = -\frac{1}{2} \int_{-\infty}^{\infty} \{\tilde{Q} + k(\tilde{Q}, \tilde{Q}', \tilde{\tau}, \mu)\} K(\frac{s}{\mu}) ds$$

so that (4.3) is satisfied. It is clear that upon choosing μ sufficiently small, we will have satisfied (3.15), (3.16), and (3.20).

All of the hard work in (4.9) is needed to estimate the effect of $K(\frac{x-s}{\mu})$ as $\mu \downarrow 0$. This is a tedious, but straightforward affair [2] and we summarize the results in the following

LEMMA 5. <u>There exists an open neighborhood of</u> $(\tilde{Q}_0,\ \tilde{\tau}_0,\ 0)$ <u>in</u> $X_1 \times X_2 \times R$ <u>such that</u> M <u>and</u> $M_{(\tilde{Q},\tilde{\tau})}$ <u>are continuous in this neighborhood. In addition, the</u> <u>map</u> $M_{(\tilde{Q},\tilde{\tau})}(\tilde{Q}_0,\ \tilde{\tau}_0,\ 0): X_1 \times X_2 \rightarrow X_3 \times X_2$ <u>is a homeomorphism.</u>

To show that the partial derivative is non-singular, note that

$$L(\tilde{Q},\tilde{\tau}) \equiv M_{(\tilde{Q},\tilde{\tau})}\ (\tilde{Q}_0,\ \tilde{\tau}_0,\ 0)(\tilde{Q},\tilde{\tau})$$

$$= (\tilde{Q}'' - \frac{1}{\gamma - \frac{1}{3}}\ \{-2\tilde{Q}_0\tilde{\tau} - 2\tilde{Q}\ \tilde{\tau}_0 - \tilde{\tau}\ \tilde{\tau}_0' - \tilde{\tau}'\tilde{\tau}_0 + \tilde{\tau}'\},\ \tilde{\tau} + \int_x^\infty \tilde{Q}).$$

To show that L is injective, assume that $L(\tilde{Q},\tilde{\tau}) = 0$. Then $\tilde{\tau}' = \tilde{Q}$, whence

$$\tilde{Q}'' = \frac{1}{\gamma - \frac{1}{3}}\ \{-3\tilde{Q}_0\tilde{\tau} - 3\tilde{Q}\ \tilde{\tau}_0 + \tilde{\tau}'\} = \frac{1}{\gamma - \frac{1}{3}}\ \frac{d}{dx}\ \{-3\tilde{\tau}_0\tilde{\tau} + \tilde{\tau}\}$$

and so

$$\tilde{\tau}'' = \tilde{Q}' = \frac{1}{\gamma - \frac{1}{3}}\ \{-3\tilde{\tau}_0\ \tilde{\tau} + \tilde{\tau}\}. \tag{4.10}$$

Assume (4.10) has a non-trivial solution $\tilde{\tau} \in X_2$. Upon differentiating (4.8) with respect to x, we see that $\tilde{\tau}_0'$ satisfies the differential equation (4.10). However, this function is an odd function of x, and we are assuming (4.10) also has an even solution, that is, an element of X_2. It follows that any solution of (4.10) must decay exponentially at $\pm\infty$. However, the function $\tilde{\tau}_0$ goes to zero at infinity, and it is easy to choose initial data for (4.10) at some large value of x such that the solution is unbounded at infinity. This contradiction ensures that L is injective.

To show that L is surjective, let $(g,h) \in X_3 \times X_2$, and consider the equation $L(\tilde{Q},\tilde{\tau}) = (g,h)$. A bit of manipulation shows that this equation is equivalent to

$$\Gamma'' - \Gamma/(\gamma - \tfrac{1}{3}) + 3\tilde{\tau}_0 \Gamma/(\gamma - \tfrac{1}{3}) = -h(1-3\tau_0)/(\gamma - \tfrac{1}{3}) + \int_x^\infty (g + 2\tilde{\tau}_0 h'/(\gamma - \tfrac{1}{3})),$$

where $\Gamma(x) = \int_x^\infty \tilde{Q}(s)ds$. The operator $\Gamma'' - \Gamma/((\gamma - \tfrac{1}{3}))$ may be inverted, and the problem becomes $\Gamma - S\Gamma = p$, where p is known and S is a compact operator (due to the exponential decay of $\tilde{\tau}_0$). Since unity is not an eigenvalue of S, the map $I - S$ is surjective, and so we may solve for Γ which then yields $\tilde{Q}$ and $\tilde{\tau}$.

We summarize our results in the following

THEOREM 6. <u>Let</u> $\gamma > \tfrac{1}{3}$. <u>There exists</u> $\bar{\mu} = \bar{\mu}(\gamma)$ <u>such that for each</u> $\mu \in (0,\bar{\mu}]$, <u>the problem</u> (2.4) - (2.8) <u>has a solution</u> (ν,θ,τ) <u>with</u> $\nu = 1-\mu^2$ <u>and</u>

$$\theta(x; \mu) = \theta(x,y = 1;\mu) = \mu^2 \frac{d}{dx} \operatorname{sech}^2 \frac{\mu x}{2\sqrt{\gamma - \tfrac{1}{3}}} + O(\mu^4 \exp(-\frac{\mu|x|}{2\sqrt{\gamma - \tfrac{1}{3}}}))$$

<u>and</u>

$$\tau(x;\mu) = \tau(x,y = 1;\mu) = \mu^2 \operatorname{sech}^2 \frac{\mu x}{2\sqrt{\gamma - \tfrac{1}{3}}} + O(\mu^3 \exp(-\frac{\mu|x|}{2\sqrt{\gamma - \tfrac{1}{3}}})).$$

<u>The physical wave-profile is given by</u> $\{(x,Y(x)): x \in R\}$, <u>where</u>

$$Y(x) = h - h\mu^2 \operatorname{sech}^2 \frac{\mu x}{2h\sqrt{\gamma - \tfrac{1}{3}}} + O(\mu^3 \exp(-\frac{\mu|x|}{2h\sqrt{\gamma - \tfrac{1}{3}}})).$$

<u>Remarks</u>

(i) As noted earlier, this wave represents a solitary wave of depression in the presence of surface tension. Since the term to order μ^2 for Y agrees with that predicted by Kortweg and de Vries, our rigorous results justify their model for small-amplitude solitary waves.

(ii) If we fix some $\tilde{\mu} \in (0,\bar{\mu})$, then we have a solution of (2.4)-(2.8) corresponding to some fixed $\tilde{\nu} = 1-\tilde{\mu}^2 < 1$ with θ and τ decaying exponentially at infinity, say,

$$|\theta(x)|, \ |\tau(x)| \leq \text{const.} \ e^{-\delta|x|}, \ x \in R \tag{4.11}$$

where $\delta > 0$. Equation (2.10) may be rewritten as $\nu\theta_y - \theta + \gamma\theta_{xx} = \tilde{f}$ where this quantity is quadratic in θ and τ when θ and τ are small.

If we restrict attention to $\nu < \tilde{\nu}$ and to functions with decay as in (4.11), then we may take Fourier transforms as in section 2 to obtain

$$\hat{\theta}(k) = \hat{\theta}(k,1) = \frac{\sinh k}{k} \frac{\hat{\tilde{f}}(k)}{h(k;\gamma,\nu)} \tag{4.12}$$

where $h(k;\gamma,\nu) = \nu \cosh k - (1 + \gamma k^2) \sinh k/k$. When $\gamma > \frac{1}{3}$ (the case under consideration) and $\nu < 1$, then $h(k;\gamma,\nu) < 0$, $k \in R$. If we restrict $\nu \in (0,\tilde{\nu}]$, then the distance of the zeros of $h(\cdot, \gamma, \nu)$ to the real-axis is bounded, independently of ν. (We have taken $\nu > 0$ since it is easy to show that (2.4)-(2.8) has only the trivial solution when $\nu = 0$.) Taking the inverse transform of (4.12) yields

$$\theta(x) = \theta(x,1) = \int_{-\infty}^{\infty} W(x-s; \nu) \tilde{f}(\theta, \theta', \tau, \tau',\nu)(s)ds, \; x \in R, \tag{4.13}$$

where W has exponential decay at infinity (cf.(2.13)). The equation relating τ to θ is (cf. (2.16) and the remarks thereafter)

$$\tau(x) = \frac{1}{2} \int_{-\infty}^{\infty} \theta(s) \, K(x-s)ds, \; x \in R.$$

The use of this equation in (4.13) yields $\theta = N(\nu, \theta)$ for $\nu \in (0,\tilde{\nu}]$. It is easy to check that N is compact for functions with decay as in (4.11). The Leray-Schauder degree of $\theta - N(\nu, \theta)$ may be calculated at the known solution where $\nu = \nu$, and it equals -1. These ideas are developed in [2] to give a global branch of solutions emanating from the bifurcation point $(\lambda = 1, \theta = 0)$.

5. The Case $\gamma \in (0, \frac{1}{3}]$

Lemma 1 shows that $h(\cdot; \gamma)$ has four real zeros (counting multiplicity) when $\gamma \in (0, \frac{1}{3}]$, and only two when $\gamma > \frac{1}{3}$. Since we have considered the case $\gamma > \frac{1}{3}$ in sections 2-4, we shall restrict attention here to $\gamma \in (0, \frac{1}{3})$. (The case $\gamma = \frac{1}{3}$ is covered by most of the arguments below but calls for a different scaling.) One is led to an equation like (2.12):

$$\hat{\theta}(k) = \frac{B(\gamma)\hat{f}(k)}{k^2(k^2-r^2(\gamma))} + \frac{D(\gamma)\hat{f}(k)}{k^2} + H(k;\gamma)\,\hat{f}(k)$$

where H has properties similar to those of its namesake in (2.12) and where $B(\gamma) < 0$, $\gamma \in (0,\frac{1}{3})$ and $B - Dr^2 = -r^2/(\frac{1}{3} - \gamma)$. One seeks $\theta(x) = Q(x) + E(x) + A(x)$, where

$$Q'''' + r^2(\gamma)Q'' = B(\gamma)f, \tag{5.1}$$

$$E'' = -D(\gamma)f,$$

and

$$A(x) = \frac{1}{\sqrt{2}\pi} \int_{-\infty}^{\infty} G(x-s;\gamma)f(s)ds.$$

It follows immediately that $E = -D(\gamma)\{Q'' + r^2Q\}/B(\gamma)$. The analysis proceeds as in section 3, and one finds A as a pointwise function of $Q,\ldots,Q'''$, and τ. The use of this representation in (4.1) yields

$$Q'''' + r^2(\gamma)Q'' = B(\gamma)g(Q,\ldots,Q''', \tau, \tau', \nu-1),$$

$$Q(0) = \alpha_1,\ldots, Q'''(0) = \alpha_4,$$

$$\tau(x) = \beta + \frac{1}{2} \int_{-\infty}^{\infty}(Q(s) + E(s) + A(s))\{K(x-s) + K(s)\}ds.$$

One introduces rescaled variables in a manner similar to that of section 4 (with the only change being $\mu = \sqrt{\nu - 1}$):

$$\mu^2\tilde{Q}'''' + r^2(\gamma)\tilde{Q}'' = B(\gamma)\{-2V\tilde{Q}\tilde{\tau} - \tilde{\tau}\tilde{\tau}' - \tilde{\tau}'\} + j(\tilde{Q},\ldots,\tilde{Q}''',\tilde{\tau},\tilde{\tau}',\mu) \tag{5.2}$$

where $V = 1 - Dr^2/B$ and

$$\tilde{\tau}(x) = \frac{1}{2} \int_{-\infty}^{\infty} (V\tilde{Q}(s) + k(\tilde{Q},\ldots,\tilde{Q}''', \tilde{\tau}, \mu))\, K(\frac{x-s}{\mu})ds.$$

If we formally set $\mu=0$, then we can solve the equation as before; the resulting $\tilde{Q}$ is now positive, and the physical wave represents a solitary wave of elevation. The formula agrees with that found by Korteweg and de Vries, and has the same form as the expansion for Y in Theorem 6, but with a plus sign in front of the sech^2 term. In setting $\mu = 0$, we have assumed that for small μ (i) the function $\tilde{\tau}$ is well-approximated by $\tilde{\tau}(x) = -V\int_x^\infty \tilde{Q}(s)ds$ and (ii) the term $\mu^2 \tilde{Q}''''$ can be ignored in (5.2). The estimates used in proving Lemma 5 show that (i) is a very reasonable assumption. However, (ii) is suspect since μ^2 multiplies the highest derivative in the equation.

A reasonable first approach to study (5.2) is to set $\tilde{\tau}(x) = - \int_x^\infty V\tilde{Q}(s)ds$ and to drop the term involving j. Then

$$\mu^2 \tilde{Q}'''' + r^2(\gamma)\tilde{Q}'' = B(\gamma)\frac{d}{dx} \left\{ -\frac{3}{2} \tilde{\tau}^2 - \tilde{\tau} \right\},$$

whence

$$\frac{\mu^2}{r^2} \tilde{\tau}'''' + \tilde{\tau}'' + \frac{1}{\gamma - \frac{1}{3}} \tilde{\tau} = -\frac{3}{2} \frac{1}{\gamma - \frac{1}{3}} \tilde{\tau}^2 .$$

After rescaling the independent and dependent variables and the parameter μ, we arrive at an equation of the form

$$\mu^2 C'''' + C'' - C - C^2 = 0, \tag{5.3}$$

$$C'(0) = C'''(0) = 0, \ C(x) \to 0 \text{ as } |x| \to \infty.$$

One might ask whether (5.3) has a solution depending on μ near to the solution $C_0(x) = -\frac{3}{2} \mathrm{sech}^2 \frac{x}{2}$ when $\mu = 0$. The result is not obvious since there is only a one-dimensional stable manifold and yet two boundary conditions are imposed at $x = 0$. (If we were allowed to replace μ^2 by $-\mu^2$ in (5.3), then the stable manifold would be two-dimensional, and the desired result easy to prove.)

Unfortunately, finding a non-trivial solution of (5.3) with zero on the right-hand side is not very informative because the physical problem (5.2) or a reasonable approximation to it demands that we be able to solve equations like (5.3) when the right-hand side is $O(\mu)$, say $\mu \ell(C,..,C''',\mu)$. One can write down explicit functions of this form for which (5.3) has only the trivial solution. All of this suggests that one must use the particular form of the nonlinear term or think of a completely different approach to the problem.

The conclusive results when $\gamma > \frac{1}{3}$ and the inconclusive ones when $\gamma < \frac{1}{3}$ are not unexpected. Although the model equations of Korteweg and deVries do not differ drastically in these two cases (the solutions merely change sign), recent numerical calculations of Vanden-Broeck [14] suggest there is a difference. Vanden-Broeck solved the periodic-wave problem numerically, and then let the wavelength $\lambda \to \infty$. (In the absence of surface tersion, the case $\gamma = 0$, it is known rigorously that periodic waves converge on compact sets to solitary waves [4].) When $\gamma > \frac{1}{3}$, Vanden-Broeck found the expected convergence to solitary waves of depression, but the calculations did not converge when $\gamma < \frac{1}{3}$. Hence, the existence of small-amplitude solitary waves when $\gamma < \frac{1}{3}$ remains open, and if they do exist, techniques will be needed that are far more subtle than the implicit function theorem used when $\gamma > \frac{1}{3}$.

References

[1] Amick, C.J. Fraenkel, L.E., and Toland, J.F., On the Stokes conjecture for the wave of extreme form. Acta Math., 148 (1982), 193-214.

[2] Amick, C.J. and Kirchgässner, K., A global theory of solitary water-waves in the presence of surface tension, to appear.

[3] Amick, C.J. and Toland, J.F., On solitary water-waves of finite amplitude. Arch. Rat. Mech. Anal., 76 (1981), 9-95.

[4] ——— On periodic water-waves and their convergence to solitary waves in the long-wave limit. Phil. Trans. Roy. Soc. Lond., A. 303 (1981), 633-669.

[5] Beale, J.T., The existence of cnoidal water waves with surface tension. J. Diff. Eqn., 31 (1979), 230-263.

[6] Deimling, K., Nonlinear Functional Analysis, Springer-Verlag (1985).

[7] Hogan, S.J., Some effects of surface tension on steep water waves, Part 1, J. Fluid Mech., 91 (1979), 167-180.

[8] ——————— , Some effects of surface tension on steep water waves, Part 2, J. Fluid Mech., 96 (1980), 417-445.

[9] Jones, M.C.W. and Toland, J.F., The bifurcation and secondary bifurcation of capillary-gravity waves. Proc. Roy. Soc. Lond., A 399 (1985), 391-417.

[10] Keady, G. and Norbury, J., On the existence theory for irrotational water waves, Math. Proc. Comb. Phil. Soc., 83 (1978), 137-157.

[11] Milne-Thompson, L.N., Theoretical Hydrodynamics, Macmillan (1977).

[12] Rabinowitz, P.H., Some global results for nonlinear eigenvalue problems. J. Func. Anal., 7 (1971), 487-513.

[13] Titchmarsh, E.C., Introduction to the Theory of Fourier Integrals, Oxford (1948).

[14] Vanden-Broeck, J.M., Solitary and periodic gravity-capillary waves of finite amplitude. J. Fluid Mech., 134 (1983), 205-219.

[15] Wehausen, J.V. and Laitone, E.V., Surface waves. In Handbuch der Physik (ed. C. Truesdell), vol. 9, Springer-Verlag (1960), 446-776.

[16] Zeidler, E., Existenzbeweis für cnoidal waves unter Berücksichtigung der Oberflächenspannung. Arch. Rat. Mech. Anal. 41 (1971), 81-107.

[17] ————— , Beiträge zur Theorie und Praxis freier Randwertaufgaben, Akademie-Verlag (1971).

[18] ————— , Bifurcation theory and permanent waves. In Applications of Bifurcation Theory (ed. P. Rabinowitz), Academic Press (1977), 203-223.

PRESENCE AND ABSENCE OF WEAK SINGULARITIES IN NONLINEAR WAVES

Michael Beals[*]

Department of Mathematics
Rutgers University
New Brunswick, NJ 08903

Introduction

This paper presents a survey of some of the recent results obtained and methods
used in the study of the propagation of smoothness and development of singularities in solutions to nonlinear, strictly hyperbolic equations. The results are
local in character, and consequently the solutions will be assumed to exist on
an open subset $\mathcal{O}$ of R^n, or of $R \times R^n$ (with coordinates $(t,x_1,\ldots,x_n)$) when
the principal part of the operator is the usual D'Alembertian $\Box = \partial_t^2 - \sum_{i=1}^{n} \partial_{x_i}^2$.
The general setting will be problems of the form

$$(1) \qquad p_m(x,D)u = f(x,u,\ldots,D^\beta u,\ldots)_{|\beta|<m} \qquad \text{(semilinear)}$$

or

$$(2) \quad \sum_{|\alpha|=m} a_\alpha(x,u,\ldots,D^{\alpha'}u,\ldots)_{|\alpha'|<m} D^\alpha u = f(x,u,\ldots,D^\beta u,\ldots)_{|\beta|<m} \qquad \text{(quasilinear)}$$

with initial values specified on some space-like hypersurface, or with the solution u given on some open set in the past. The operators on the left side of
(1) or (2) are assumed to be strictly hyperbolic on $\mathcal{O}$. In the quasilinear case
this means we assume that a solution u has been given, and that
$\sum a_\alpha(x,u(x),\ldots,D^{\alpha'}u(x),\ldots)\xi^\alpha$ is strictly hyperbolic in the usual sense as a
function of (x,ξ). The functions f on the right-hand sides are assumed to
depend smoothly on their arguments, but no additional conditions are made. At
this level of generality there will be no special conserved quantities, nor special
bounds on the growth or decay of f, either locally or globally. Consequently,
there will not be available the special techniques which apply to certain classes
of equations, allowing for example conclusions about long-time existence of solu-

---

[*] Research partially supported by NSF Grant #DMS-8201281

tions, or existence of solutions of very low regularity.

The singularities to be considered are weak in that it is assumed that the solution u of (1) or (2) exists in $H_s^{loc}(\mathcal{O})$ for s large enough that all of the functions $D^{\alpha}u$ appearing n a nonlinear fashion are continuous. (Of course, the greatest interest would be in relaxing this condition. Some recent investigations along this line will be mentioned in the last section.) For these values of s, solutions will exist at least locally in time by well-known arguments. Our considerations here are focused upon u during the interval in which it exists as a classical solution, e.g. before the onset of shocks. The singularities under consideration will be due to the presence of singularities in the (derivatives of the) data initially, that is, points where the data is H^s but not C^{∞}. Such singularities will propagate in the linear case, while for the nonlinear problem we shall see that additional singularities will in general appear. These "nonlinear singularities" are in a certain sense due both to the interaction of the linear ones with each other, and the interaction of a linear singularity with itself (in more than one space-dimension).

2. Propagation of Smoothness in Linear Equations

For the solution of a linear, strictly hyperbolic equation, the singular support may be described in a simple fashion. Let u satisfy

(3) $\qquad p(x,D)u = f,$ f   smooth and $p_m(x,\xi)$ (the principal part of p)

$\qquad\qquad\qquad$ strictly hyperbolic on $R^n \times R^n$.

If data is specified on a space-like hypersurface, and S is the singular support of the data, let $\tilde{S}$ denote the union of all the characteristic curves for p_m which intersect S. (Recall that the curve $(x(t),\zeta(t))$ is called a null bicharateristic for $p_m(x,\xi)$ if $\dot{x} = \nabla_\xi p_m(x,\xi),$ $\dot{\xi} = -\nabla_x p_m(x,\xi),$ $p_m(x,\xi) = 0.$ the characteristics are the projections $x(t)$ of the null bicharacteristics.) Then the singular support of u is contained in $\tilde{S}$. See Figure 1 for an example with the ordinary wave equation and data singular at two points. The surfaces of the two light cones contain the singular support of the solution in the set $t > 0$.

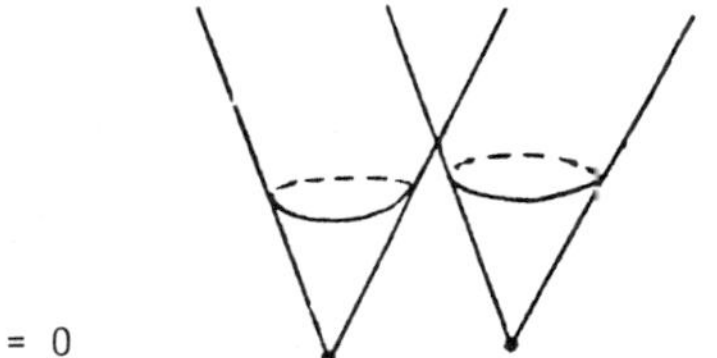

Singular support of a
solution to $\Box\, u = 0$ with
appropriate data singular
at two points.

$t = 0$

Figure 1.

More precise information about the singularities of the solution is obtained
by examining not only the set where u is non-smooth, but also the frequencies
which contribute to the non-smoothness at a point. Recall that the wave front set
of u, $WF(u)$, is defined as the complement in $R^n \times R^n \setminus \{0\}$ of
$\{(x_0, \xi_0): \exists\, \phi \in C_0^\infty(R^n)$ with $\phi(x_0) \neq 0$ and K a conic neighborhood of ξ_0 in
$R^n \setminus \{0\}$ with $\chi_K(\xi)\widehat{\phi u}(\xi)$ rapidly decreasing$\}$. Here $\,\widehat{}\,$ denotes the Fourier
transform, and χ_K the characteristic function of K. (See Hörmander [18].) The
wave front set refines the notion of singular support; the projection onto the
first factor of $WF(u)$ is exactly sing supp(u). The theory of pseudodifferntial
operators easily allows the conclusion that for $p(x,D)u = f$ with f smooth,
$WF(u)$ is contained in the characteristic set $\{(x,\xi): p_m(x,\xi) = 0\}$. Hörmander's
theorem [19] completely characterizes the wave front set of a solution to a linear
strictly hyperbolic equation.

Theorem 1. Let u satisfy (3). If $(x_0, \xi_0) \in WF(u)$ and Γ is the null
bicharacteristic through (x_0, ξ_0), then $\Gamma \subset WF(u)$.

This result is a statement about the propagation of a singularity, but
equivalently, if (x_0, ξ_0) is a characteristic point and $(x_0, \xi_0) \in WF(u)$, then
$\Gamma \cap WF(u) = \emptyset$. A quantitative version of this statement describes the regularity

along a null bicharacteristic in terms of "microlocal" Sobolev estimates. Say that $u \in H^r(x_0, \xi_0)$ if $\chi_K(\xi)(1 + |\xi|^2)^{r/2} \widehat{\phi u}(\xi) \in L^2(R)$, with ϕ and K as in the definition of the complement of the wave front set. And for $\Gamma \subset R^n \times R^n \setminus \{0\}$, let $u \in H^r(\Gamma)$ mean that $u \in H^r(x_0, \xi_0)$ for each $(x_0, \xi_0) \in \Gamma$. The quantitative version of Theorem 1 is as follows.

Theorem 1'. Let u satsisfy (3), and let Γ be the null bicharacteristic for p_m through (x_0, ξ_0). Then

$$(*) \qquad u \in H^r(x_0, \xi_0) \Rightarrow u \in H^r(\Gamma).$$

3. Regularity in Nonlinear Equations and Nonlinear Singularities

Two important points involving Theorem 1' lead to a particularly rich theory for nonlinear equations. First, let u satisfy (1) or (2) on $\mathcal{O}$, $u \in H^s_{loc}(\mathcal{O})$ for s sufficiently large, and let Γ now denote the component containing (x_0, ξ_0) of the null bicharacteristic intersected with $\mathcal{O} \times R^n \setminus \{0\}$. Then $(*)$ remains true <u>for certain values of</u> $r \geqslant s$. Precisely, the following holds.

<u>Theorem 2.</u> Let $u \in H^s_{loc}(\mathcal{O})$ satisfy (1) or (2), with $s > \frac{n}{2} + m - 1$ in the semilinear case, $s > \frac{n}{2} + m$ in the quasilinear case. If $u \in H^r(x_0, \xi_0)$, then $u \in H^r(\Gamma)$, for $s \leqslant r \leqslant 2s - \frac{n}{2} - m + 1$ in the semilinear case.

$$s \leqslant r \leqslant 2s - \frac{n}{2} - m \qquad \text{in the quasilinear case.}$$

It follows easily (using elliptic arguments microlocally away from char p_m) that if x_0 is a point for which every null bicharacteristic through (x_0, ξ) has projection which intersects the initial surface at points where the data are in H^r_{loc}, then $u \in H^r_{loc}$ near x_0. For example (see Figure 2) if the initial data are singular at only one point and the equation is $\square u = f(t, x, u, Du)$ on $R \times R^n$, then u has regularity of at least type $H^{2s - (n+1/2) - 1}$ on the interior of the light cone over the origin.

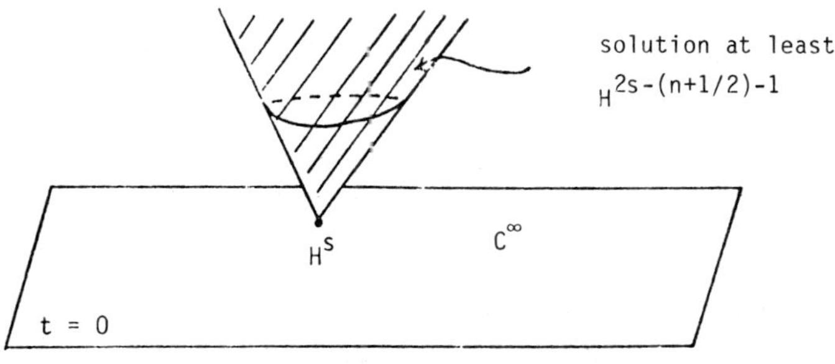

Figure 2.

Theorem 2 was proved in the general case by Bony [12], who developed the powerful theory of paradifferential operators to linearize the problem. This method involves quantizing operators in a fashion different from the usual one for differential or pseudodifferential equations, by taking a Littlewood-Paley decomposition of the functions involved. (The upper endpoint results, $r = 2s - \frac{n}{2} - m + 1$ or $r = 2s - \frac{n}{2} - m$, follow from Meyer [23].) At the same time, Rauch [27] proved Theorem 2 in the special semilinear case $p_m(x,D)u = f(x,u,\ldots,D^\beta u,\ldots)|_{|\beta| \leqslant m-2}$. There the hypotheses are $u \in H^s_{loc}(\Theta)$, $s > \frac{n}{2} + m - 2$, and the conclusion holds for $s < r \leqslant 2s - \frac{n}{2} - m + 3$. It is shown by Beals and Reed [9], [10], that Rauch's idea may be combined with a calculus of pseudodifferential operators, modified to include nonsmooth coefficients, to recover the general case of Theorem 2.

The central idea in either method of proof of the result is the following observation. Since, for example, $\hat{u}^2 = \widehat{u*u}$, it is clear that different directions ξ in the wave front set of u will interact when a nonlinear function is applied to u. But it turns out that singularities of type H^{s_1} and H^{s_2} concentrated microlocally near (x_0,ξ_1) and (x_0,ξ_2) yield singularities in $f(u)$ of strength no greater than $H^{s_1 + s_2 - (n/2)}$ near (x_0,ξ) for ξ away from conic neighborhoods of both ξ_1 and ξ_2. Thus the nonlinear interaction is weaker than the original nonsmoothness, and regularity up to this weaker order can be propagated.

On the other hand, the nonlinear theory does involve new phenomena, because (*) <u>does</u> <u>not</u> <u>remain</u> <u>true</u> (in general) for all $r \geq s$. Indeed, the first examples (in one space dimension) are given in Rauch and Reed [28]. For a third order semi-linear equation with principal symbol $\partial_t(\partial_t - \partial_x)(\partial_t + \partial_x)$, solutions are given with behavior as illustrated in Figure 3. With data singular at two points, if Thoerem 1' held for such a nonlinear equation, the solution could be singular only on the three characteristics coming from each point. But, for appropriately chosen data, u is constructed with singularities on the additional characteristics issuing from the crossing points in $t > 0$ of the original characteristics.

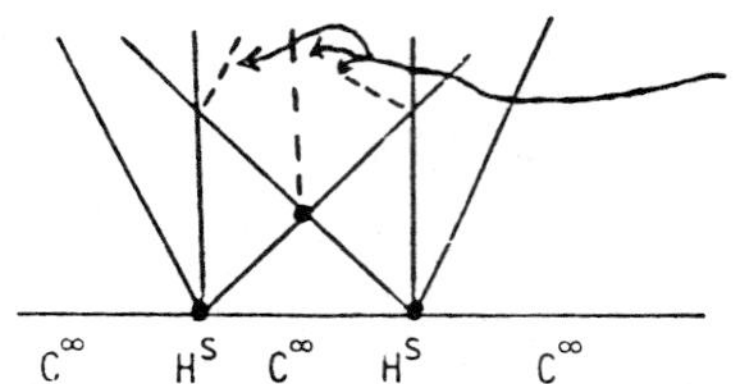

Figure 3.

In fact, these and "later generation crossings" are the only source of singularities present in the nonlinear case in one space dimension, and the new singularities propagate only along characteristics from the crossing points (and accumulation points of such). For semilinear strictly hyperbolic equations and systems, these results appear in Rauch and Reed [29], [30]. The picture in one space dimension is thus quite well understood. Further results include those on quasilinear equations (Messer [22], Godin [17]), and non-strictly hyperbolic systems (Micheli [24]).

For a second order semilinear equation in one space dimension, the results of [28] imply in particular that no nonlinear singularities appear for a second order equation, since when two characteristics cross, the only characteristics issuing from the crossing point are those already present. In higher dimensions, addi-

tional characteristics are present for a second order equation, and nonlinear
singularities due to the crossing of singularity-bearing characteristics do occur.
Examples in $R \times R^n$ for a semilinear equation with principal part $\Box$ are
constructed in Lascar [20] (for $n > 3$) and Beals [3] (for $n > 2$). Time slices
of such singularities are shown in Figure 4.

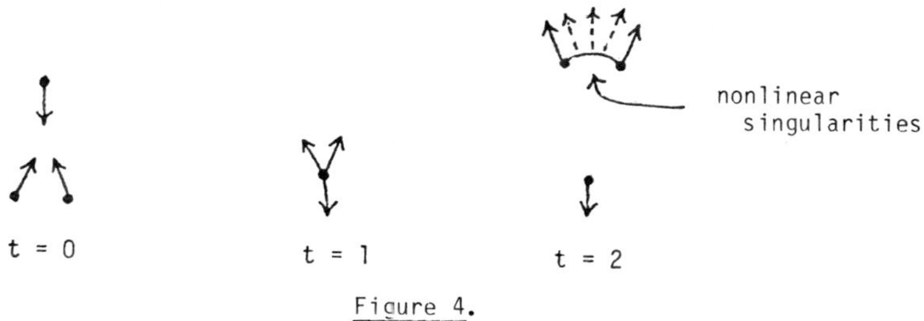

Figure 4.

4. Spreading of Singularities in Higher Dimensions

Theorem 2 yields limits on the strength of singularities in solutions to
equations like (1) or (2). And finite propagation speed for such equations gives
a limit to the size of the possible set of singularities when the location of the
nonsmooth points of the data is known. For example, in Figure 5a), the solution of
the equation $\Box u = f(u,Du)$ must be smooth outside the shaded set, and have
regularity of order at least $H^{2s + (n+1/2) - 1}$ on the lightly shaded set. But
as noted above, in one space dimension, the regularity for the solution of such an
equation is actually as shown in Figure 5b; singularities caused by nonlinearity
are completely absent. And for higher order equations, those which appear are
generally located on a much smaller set than that guaranteed merely by finite propagation speed. (E.g., refer to Figure 3.) A natural question is whether an analogous statement can be made in higher dimensions.

In fact, the situation in higher dimensions is in complete contrast with the
case of one dimension. No condition on only the location of the nonsmooth points
of the data will restrict the singularities of the solution to a set smaller than
that guaranteed by finite propagation speec. The following result is proved in [4].

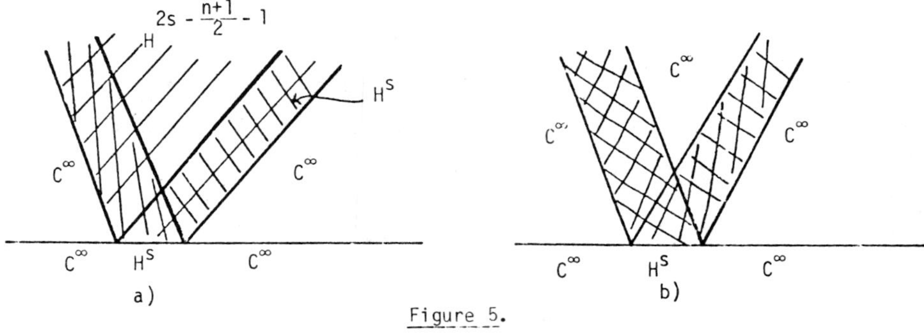

Figure 5.

Theorem 3. On $\mathcal{O} = (0,1) \times R^n$, $n > 1$, there is a choice of smooth function $\beta(t,x)$, and of compactly supported data singular only at $x = 0$, such that the solution $u \in H^s(\mathcal{O})$ of $\Box u = \beta(t,x)u^3$ with the given data has sing supp$(u) = \{(t,x): |x| \leq t\}$.

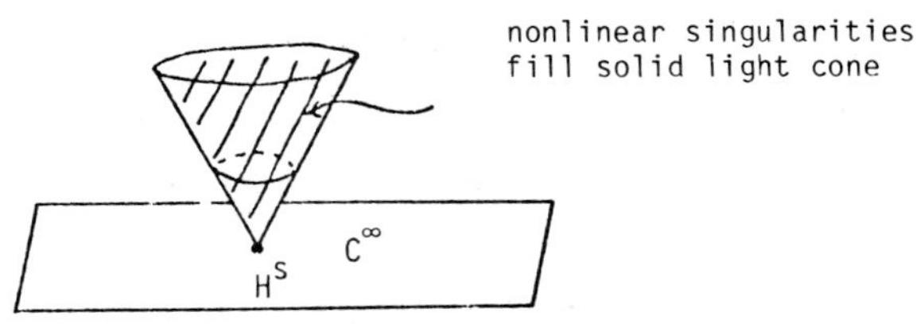

Figure 6.

It is natural to ask what makes the difference in higher dimensions, and whether there is any control at all in this case over the presence of nonlinear singularities. It will be seen in the next section that for many natural kinds of data, these singularities will be absent or sharply restricted in appearance. But more than just H^s regularity needs to be required initially. Such regularity will be present, for example, in the characteristic function of a half-space, or in the Dirac distribution, sufficiently smoothed to belong to an appropriate H^s space.

It is instructive to mention a few points from the proof of Theorem 3, which give an idea of how to detect the presence of nonlinear singularities in higher dimensions, and of conditions which would be necessary to prevent their appearance. First, let E be the solution operator for the linear, forward problem from time zero: Ew satisfies $\Box\, Ew = w$ and Ew has zero Cauchy data at time zero. Let v solve the linear problem $\Box\, v = 0$, with Cauchy data to be chosen as for Theorem 3. Then the solution of the nonlinear problem with that data is

$$(4) \qquad u = v + E(\beta u^3) = v + E(\beta v^3) + E[3v^2(E\beta u^3) + 3v(E\beta u^3)^2 + E(\beta u^3)^3]$$

The first term on the right-hand side of (4) exhibits only linear singularities. The third term turns out to be strictly smoother in the nonlinear microlocal directions than the middle term. This property is to be expected since E smooths by one order, and so from the results of Rauch and Bony mentioned previously, a term like $3v^2(E\beta u^3)$ should have smoothness of one order better than that of βv^3 in the directions due to interactions. Thus the linear solution v is a first approximation to u, but or the level of singularities $v + E\beta v^3$ is a better one, good enough to distinguish nonlinear singularities from a remainder term. In particular, for a problem of the form (1), let v be a solution of the corresponding linear problem $p_m(x,D)v = 0$ with specified data. The microlocal singularities of $f(x,v,\ldots,D^\beta v,\ldots)_{|\beta|<m}$ should give an indication of the type of spreading of singularities in the solution of the nonlinear problem. (The forward fundamental solution for p_m can take singularities spread on the Fourier transform side and propagate them along null bicharacteristics.)

In the proof of Theorem 3, a solution v to the linear problem which is singular along a single characteristic is constructed which interacts with itself in the nonlinear problem to produce singularities along other characteristics. This interaction is possible because a v which is, singular on a characteristic through (t_0,x_0), can have wave front set containing the entire line $\{(t_0,x_0,r,rw): r \in R, w \in S^{n-1} \text{ fixed}\}$. The properties of convolution and the definition of wave front then imply that $v^2 = v*v$ will be non-rapidly-decreasing on a set larger than the line through $(1,w)$, and similarly for $f(v)$ in general. (In one space dimension, the fact that v is a solution of the linear problem prevents this spreading.) Thus in higher dimensions a finer regularity notion than merely that of the classical wave front set, defined by decay in cones, is needed if the self-spreading of singularities is to be ruled out.

The type of smoothness to be required needs to be preserved under nonlinear maps, and must be respected by the forward fundamental solution of the corresponding linear operator, that is, respected by the principal part of the equation. In one space dimension, the operator may be factored into a product of vector fields. Vector fields automatically interact nicely with nonlinear maps, by the chain rule and Schauder's Lemma. For example, let f be smooth, $s > \frac{n}{2}$, and let L be a smooth, first order differential operator. Then

$$(5) \qquad u, Lu \ \epsilon \ H^s_{loc}(\Theta) \Rightarrow f(u), \ Lf(u) = f'(u)Lu \ \epsilon \ H^s_{loc}(\Theta).$$

And those vector fields which are factors of the hyperbolic operator interact correctly with the equation. As a result, those vector fields can be used to measure smoothness, as in [28], [30]. In higher dimensions, the operator can only be factored into a product of pseudodifferential operators. Unfortunately, if L in (5) is pseudodifferential, then the conclusion of (5) no longer holds in general. There is a weak substitute for second order equations: the operator itself measures a certain degree of smoothness that tempers self-spreading. Thus in this situation, even though nonlinear singularities form (Theorem 3), their strength is not given correctly by Theorem 2. Rather, for the second order case the conclusions of Theorem 2 hold up to order $r < 3s - n - 2$ (the semilinear problem) or $r < 3s - n - 4$ (the quasilinear case). See Beals [6].

5. Smoothness Conditions Limiting the Presence of Nonlinear Singularities in Higher Dimensions

It has been demonstrated by Bony [13], [14] that "conormal regularity" is an appropriate property to consider for solutions to semilinear, strictly hyperbolic equations. As a simple case consider $u_0 = x_1^r$, where r is a sufficiently large non-integer that $u_0 \ \epsilon \ H^s_{loc}$. Then

$$(6) \qquad (x_1 \partial_{x_1})^{\alpha_1} \partial_{x_2}^{\alpha_2} \dots \partial_{x_n}^{\alpha_n} u_0 \ \epsilon \ H^s_{loc} \qquad \text{for any multi-index } \alpha.$$

Note that (6) implies that $WF(u_0) \subset \{(0,x',\xi_1,0)\} = N^*(\{x_1 = 0\})$, but is in fact a much stronger statement. If $\phi \in C_0^\infty$, then $\phi u_0(\xi)$ has a strictly controlled rate of decay as ξ moves away from $\{\xi' = 0\}$) which is not implied merely by the statement about $WF(u_0)$.

Now, for any distribution u_0 satisfying (6), the same holds for $f(u_0)$ with f smooth, as long as $s > \frac{n}{2}$. Such a distribution is called "conormal with respect to $\{x_1 = 0\}$". Consider for example a semilinear equation with principal part $\square$. The two characteristic surfaces through $\{x_1 = 0\}$ are $\{t = x_1\}$ and $\{t = -x_1\}$. Let $\mathcal{M}$ denote the collection of smooth vector fields tangent to both. Generators of $\mathcal{M}$ over C_2^∞ are given by
$\{(t - x_1)(\partial_t - \partial_{x_1}),(t + x_1)(\partial_t + \partial_{x_1}),\partial_{x_2},\ldots,\partial_{x_n}\}$. Let
$N^{s,k} = \{u: M_I u \in H_{loc}^s(\mathcal{O}), 0 \leqslant |I| \leqslant k, M_1 = M_I \ldots M_{|I|}, M_i \in \mathcal{M}\}$. Then

$$(7) \qquad u \in N^{s,k} \Rightarrow f(u) \in N^{s,k} \quad \text{for } f \text{ smooth, } s > \frac{n+1}{2}.$$

Moreover, the commutators $[\square,M]$ for $M \in \mathcal{M}$ are in the span of $\square$ (over C^∞) and M (over operators of order one). Thus for instance

$$(8) \qquad \text{if } \square u = f(x,u), \text{ then } \square M_I u = \sum_{|J| \leqslant |I|} T_J M_J u + \sum_{|J'| \leqslant |I|} M_{J'} g_{J'}(u)$$

where each T_J has order at most one and $M_i, M_j, M_{j'} \in \mathcal{M}$.

In other words, the smoothness measured by $\mathcal{M}$ has the appropriate behavior with respect to the operator $\square$. It is not difficult to show that (7), (8), induction, the usual H^s estimates for solutions of the linear wave equation, and the assumption (6) on the initial data (for $|\alpha| \leqslant k$) yield the conclusion that $u \in N^{s,k}$. In particular, if $k = \infty$ then u is C^∞ away from the two characteristic surfaces, the same conclusion as in the linear case.

The general results of Bony [13], [14], apply to the interaction of singularities when a solution $u \in H^s(\mathcal{O})$ to the semilinear equation (1) is assumed to be conormal with respect to two characteristic hypersurfaces Σ_1, Σ_2 in the "past" $\mathcal{O}^-$. If these two hypersurfaces intersect in the "future" $\mathcal{O}^+$ (the domain of dependence for $\mathcal{O}^+$ is assumed to be contained in $\mathcal{O}^-$), let $\Sigma_3, \ldots, \Sigma_m$ denote the remaining, smooth, forward characteristic hypersurfaces for p_m issuing from $\Sigma_1 \cap \Sigma_2$. Away from the intersection conormal regularity is defined as in the flat case (6) above by allowing differentiation with respect to all vector fields tangent to the surface under consideration. Near the intersection, if $m > 2$, the vector field definition as in $\mathcal{M}$ above will not suffice. Instead, $\mathcal{M}$ is taken to be the collection of classical, first order, pseudodifferential operators whose principal symbols vanish on the intersection of the conormal sets $N^*(\Sigma_j)$, for each j, and on $N^*(\Sigma_1 \cap \Sigma_2)$. The space $N^{s,k}$ corresponding to this $\mathcal{M}$ is defined as above. Then the analogues of (7) and (8) hold; note that the presence of pseudodifferential operators makes (7) more difficult to verify, as remarked previously after (5). But Bony shows that microlocally one can reduce the study of the differentiation of a product to the case where vector fields can be used. The following result then holds (see Figure 7).

Theorem 4. With $0^{\pm}, \Sigma_1, \ldots, \Sigma_m$, and $N^{s,k}$ as above, let $u \in N^{s,k}(0^-)$, $s > \dfrac{n}{2} + m - 1$ satisfy (1) on $\mathcal{O}$. Then $u \in N^{s,k}(0)$. In particular, if $k = \infty$, u is smooth off of the union of the m forward characteristic hypersurfaces from $\Sigma_1 \cap \Sigma_2$.

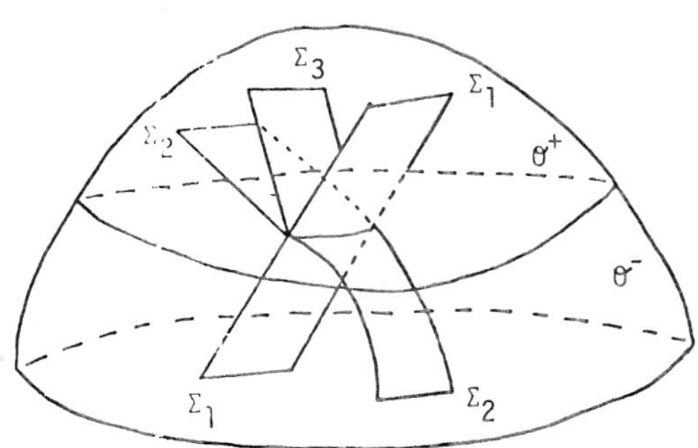

Figure 7.

Thus when two characteristic hypersurfaces carrying conormal singularities intersect, the only nonlinear singularities produced are on the remaining m - 2 hypersurfaces. If m = 2, no nonlinear singularities appear. The analysis of the situation for the intersection of three characteristic hypersurfaces is considerably more complex. An example exhibiting a nonlinear singularity in two space dimensions for the semilinear equation $\Box\, u = f(u,Du)$ is constructed in Rauch and Reed [31]. The new singularities lie on the surface of the light cone over the point of simultaneous intersection of the three surfaces. (Time slices are shown in Figure 8.) These are the only possible nonlinear singularities, under the assumption that u is conormal with respect to the three surfaces in the past, as proved in Melrose and Ritter [21] and Bony [15]. The analysis involved is considerably more difficult than in the case of Theorem 4. The family of operators measuring smoothness, corresponding to $\mathcal{M}$ above, does not satisfy the analogue of (8) near the point of triple intersection. Instead, the corresponding problem in polar coordinates near that point must be examined.

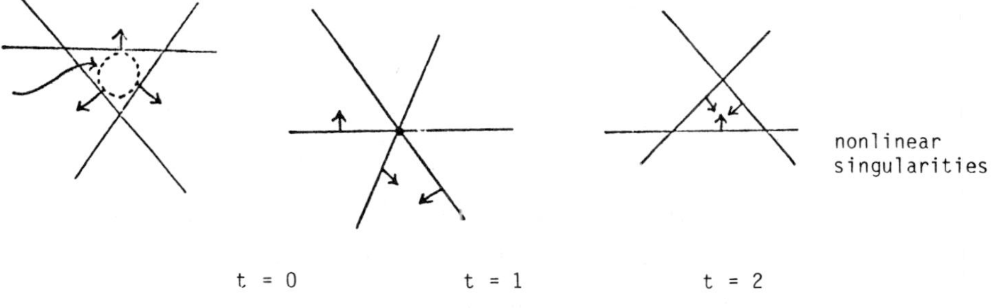

Figure 8.

Conormal assumptions in the past (or at time zero) restrict the type of singularities considered to those across a collection of hypersurfaces. The results in one space dimension do not require this restriction; certain analogues remain true in higher dimensions. The proofs combine the notion of smoothness measured by a family of vector fields and of some sort of redution to the lower dimension case. An example is the case of "angularly smooth" solutions considered in Beals [6]. Let M denote the family of vector fields generated by

$\{x_i \partial_{x_j} - x_j \partial_{x_i}\}_{i,j=1}^n$. The generators commute with $\Box$. Away from the line $\{x = 0\}$, a change to polar coordinates allows the reduction of a problem involving functions u such that

(9)
$$M_I u \in H_{loc}, \quad \text{all} \quad M_I = M_1 \cdots M_{|I|}, \quad M_i \in \;,$$

to a problem involving only the time and radial coordinates (t,r). Define a nonlinear domain of singular dependence C_p at the point $p = (t,x)$, $t > 0$, as follows: C_p is the union of the surface of the backward light cone L_p from p and the surface of the backward light cone N_p from L_p $\{x = 0\}$ if that set is nonempty. (See Figure 9.)

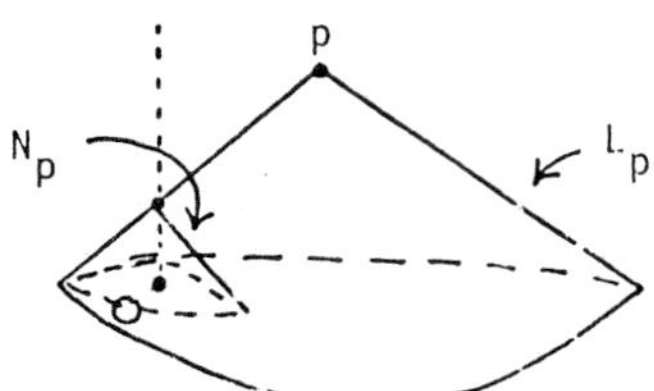

Figure 9.

Theorem 9. Let $\Box u = f(t,x,u,Du)$ on $(0,T) \times R^n$. If the Cauchy data $u(0,x)$, $\frac{\partial u}{\partial t}(0,x)$ satisfy (9) (with H_{loc}^s replaced by H_{loc}^{s-1} for $\frac{\partial u}{\partial t}$) with $s > \frac{n}{2} + 2$, then u satisfies (9) on $(0,T) \times R^n$. If $p \in (o,T) \times R^n$ and if the data are smooth on a neighborhood of C_p $\{t = 0\}$, then $u \in C^\infty$ near p.

It follows in particular that if the data are angularly smooth and singular only at $\{x = 0\}$ (for instance, $|x|^r$), then the solution is singular only on the surface of the light cone $\{t = |x|\}$, in contrast to Theorem 3. On the other hand, certain nonlinear singularities can appear for other such data: note that the set corresponding to C_p in the linear case is just L_p. An exmaple of a solution with spread singularities as in Figure 10 is constructed in [6].

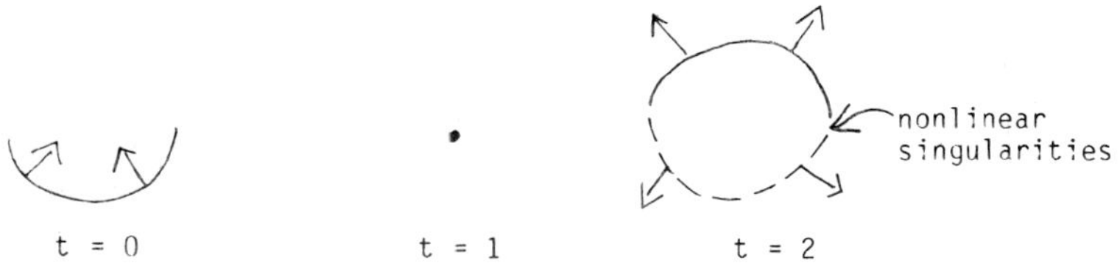

nonlinear
singularities

t = 0   t = 1   t = 2

Figure 10.

Similar arguments apply to the case of "striated data" considered by
Rauch and Reed [32]. It is assumed that the initial, space-like hypersurface is
foliated into leaves, and that the data are smooth with respect to differentiation
in the directions tangential to the foliation. Let the semilinear equation be of
second order, or more generally consider a "two-speed system" (see [32]). It is
proved that smoothness with respect to the families of pairs of characteristic
hypersurfaces through the leaves is maintained, as long as those hypersurfaces
remain smooth. (The angularly smooth case above is an example where the surfaces
degenerate.) For example, if in Figure 5b an additional space dimension is added
and the data is striated with respect to the new direction, the corresponding picture for smoothness will hold.

6. Mixed Initial-Boundary Problems and Further Topics

In addition to strictly hyperbolic problems on an open domain, certain
results have been obtained for domains with boundary. The problem in one space
dimension is considered in Berning and Reed [11] (the second order case), and is ana-
lyzed in great detail in Oberguggenberger [25], [26]. Again the nonlinear singularities appear in a very controlled fashion, as opposed to the situation that
obtains in higher dimensions. The multi-dimensional case is considered, away from
grazing rays, in Sablé and Tougeron [35], [36] where the result corresponding to
Theorem 2 for a general nonlinear strictly hyperbolic operator and well-posed
boundary conditions (the propagation of smoothness up to order roughly $2s - \frac{n}{2}$)

is given. The proof uses the analogue of Bony's paradifferential operators in the setting of a calculus tangential to the boundary. A treatment of a simple, semilinear problem along the lines of [27] is given in Shuxing [37], with the more general case handled in David [16].

Conormal solutions in the setting of a problem with boundary are considered in Beals and Metivier [7], [8]. A solution u to a semilinear problem of order m with well-posed boundary conditions is assumed to be conormal with respect to a characteristic hypersurface in the interior of the domain in the past. If the surface intersects the boundary transversally, and the boundary data are conormal with respect to the intersection, it is proved that u is conormal with respect to the family of reflected characteristic hypersurfaces in the future. (In particular, u is smooth away from these hypersurfaces, as in the linear case.) See Figure 11. If the family of reflected hypersurfaces has cardinality two, the analysis can be carried out using vector fields as in the second order case dealt with in Theorem 4. The case of a second order equation is particularly simple: on R^n, coordinates can be chosen to simultaneously flatten the boundary and characteristic hypersurfaces into $x_n = 0$, $x_1 - x_n = 0$, and $x_1 + x_n = 0$. The appropriate family of vector fields to be considered is then generated by

$$M = \{x_n(x_1 - x_n)(\partial_{x_1} - \partial_{x_n}), x_n(x_1 + x_n)(\partial_{x_1} + \partial_{x_n}), x_1\partial_{x_1} + x_n\partial_{x_n}, \partial_{x_2}, \ldots, \partial_{x_{n-1}}\}.$$

These operators are tangential to both hypersurfaces and to the boundary. They have appropriate commutation properties with the hyperbolic operator, which after division by a non-zero factor is necessarily of the form

$$p_2(x,D) = \partial_{x_n}^2 - \partial_{x_1}^2 + (x_1\partial_{x_1} + x_n\partial_{x_n})r_1(x,D) + \tilde{p}_2(x,D)$$

with $\partial_{x_2}, \ldots,$ or $\partial_{x_{n-1}}$ appearing in each term of $\tilde{p}_2(x,D)$. When a larger number of characteristic hypersurfaces is present in the reflected family, it is again necessary (as in the proof of Theorem 4) to use pseudodifferential operators to separate the analysis near the differential conormals. The presence of the boundary adds complications due to the types of pseudodifferential operators which must be used.

boundary of region

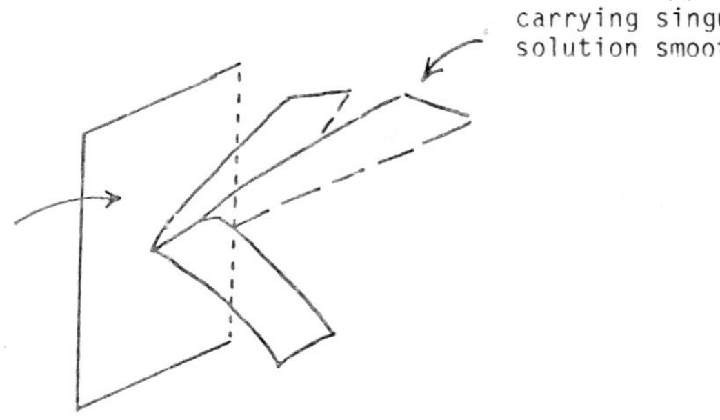

Figure 11.

Research into the phenomena considered above continues along various lines. Singularities stronger than those of type H^s, $s > \frac{n}{2}$, are of natural interest; such regularity is not required in one space dimension ([28]). For the existence of less regular solutions in higher dimensions in second-order problems, see Rauch and Reed [33]. In the conormal setting, existence theorems for lower values of s are proved in Ritter [34]. Almost all conormal results have been proved only for semilinear equations, since the characteristic hypersurfaces are in that case smooth (away from caustics). The study for general nonlinear equations has begun with the work of Alinhac [1], [2], where conormal regularity with respect to a single characteristic hypersurface is treated. The proof involves replacing the composition of non-smooth functions (as in changes of variables to flatten hyper-surfaces) with a "paracomposition", extending the methods of Bony. Additional topics of current study include the behavior of nonlinear singularities near characteristic hypersurfaces which form caustics, and near grazing directions in boundary value problems, as well as solutions of lower regularity to more general problems.

References

[1] S. Alinhac, Paracomposition et application aus equations non-lineaires, Sem. Bony-Sjöstrnd-Meyer, exp. no. 11 (1984-85).

[2] __________, Paracomposition et operateurs paradifferentiels, preprint.

[3] M. Beals, Spreading of singularities for a semilinear wave equation, Duke Math. J. 49 (1982), 275-286.

[4] ________, Self-spreading and strength of singularities for solutions to semilinear wave equations, Ann. of Math. 118 (1983), 187-214.

[5] ________, Nonlinear wave equations with data singular at one point, Contemp. Math. 27 (1984), 83-95.

[6] ________, Propagation of smoothness for nonlinear second order strictly hyperbolic differential equations, Proc. Symp. Pure Math. (1985), to appear.

[7] M. Beals and G. Métivier, Progressing wave solutions to certain nonlinear mixed problems, preprint.

[8] ________, Reflection of transversal progressing waves in nonlinear strictly hyperbolic mixed problems, in preparation.

[9] M. Beals and M. Reed, Propagation of singularities for hyperbolic pseudodifferential operators with nonsmooth coefficients, Comm. Pure Appl. Math. 35 (1982), 169-184.

[10] ________, Microlocal regularity theorems for nonsmooth pseudodifferential operators and applications to nonlinear problems, Trans. Am. Math. Soc. 285 (1984), 159-184.

[11] J. Berning and M. Reed, Reflection of singularities of one-dimensional semilinear wave equations at boundaries, J. Math. Anal. Appl. 72 (1979), 635-653.

[12] J.-M. Bony, Calcul symbolique et propagation des singularitiés pour les équations aux dérivéés partielles nonlineaires, Ann. Scien. de l'Ecole Norm. Sup. 14 (1981), 209-246.

[13] ________, Interaction des singularités pour les équations aux derivéés partielles nonlineaires, Sem. Goulaouic-Meyer-Schwartz, exp. no. 22 (1979-80).

[14] ________, Interaction des singularités pour les équations aux derivéés partielles nonlineaires, Sem. Goulaouic-Meyer-Schwartz, exp. no. 2 (1981-82).

[15] ________, Interaction des singularités pour les équations de Klein-Gordon non lineaires, Sem. Goulaouic-Meyer-Schwartz, exp. no. 10 (1983-84).

[16] F. David, Ph.D. thesis, Courant Institute of Mathematical Sciences (1984).

[17] P. Godin, Propagation of C^∞ regularity for fully nonlinear second order strictly hyperbolic equations in two variables, Trans. Am. Math. Soc., to appear.

[18] L. Hörmander, Fourier integral operators I, Acta Math. 127 (1971), 79-183.

[19] ________, Linear differential operators, Actes Congr. Inter. Math. Nice 1 (1970), 121-133.

[20] B. Lascar, Singularités des solutions d'equations aux derivéés partielles nonlineaires, C.R. Acad. Sci. Paris 287 (1978), 521-529.

[21] R. Melrose and N. Ritter, Interaction of nonlinear progressing waves, Ann. of Math. 121 (1985), 187-213.

[22] J. Messer, The propagation and creation of singularities of solutions of quasilinear, strictly hyperbolic systems in one space dimension, Ph.D. thesis, Duke University (1984).

[23] Y. Meyer, Régularite des solutions des équations aux derivées partielles non
lineaires, Sem. Bourbaki, no. 560 (1979-80).

[24] L. Micheli, Propagation of singularities for non-strictly hyperbolic semilinear systems in one space dimension, preprint.

[25] M. Oberguggenberger, Semilinear mixed hyperbolic systems in two variables:
reflection and density of singularities, preprint.

[26] _________________, Propagation and reflection of regularity for semilinear
hyperbolic (2×2) systems in one space dimension, preprint.

[27] J. Rauch, Singularities of solutions to semilinear wave equations, J. Math.
Pures et Appl. 58 (1979), 299-308.

[28] J. Rauch and M. Reed, Propagation of singularities for semilinear hyperbolic
equations in oe space variable, Ann. of Math. 111 91980), 531-552.

[29] _________________, Jump discontinuities of semilinear strictly hyperbolic
systems in two variables: creation and propagation, Comm. Math. Phys. 81
(1981), 203-227.

[30] _________________, Nonlinear microlocal analysis of semilinear hyperbolic
systems in one space dimension, Duke Math. J. 49 (1982), 379-475.

[31] _________________, Singularities produced by the nonlinear interaction of
three progressing waves: examples, Comm. P.D.E. 7 (1982), 1117-1133.

[32] _________________, Striated solutions of semilinear, two speed wave
equations, Indiana U. Math. J., to appear.

[33] _________________, Discontinuous progressing waves for semilinear systems,
preprint.

[34] N. Ritter, Progressing wave solutions to nonlinear hyperbolic Cauchy
Problems, Ph.D. thesis, M.I.T. (1984).

[35] N. Sablé-Tougeron, Paralinearisation de problèmes aux limites non lineaires,
C.R. Acad. Sci. Paris 299 (1984), 169-171.

[36] _________________, Regularité microlocale pour des problèmes aux limites non
lineaires, preprint.

[37] C. Shuxing, The reflection and interaction of the singularities of solutions
to semilinear wave equation in higher space dimension, Nonlinear Anal. 8
(1984), 1167-1179.

SOME DYNAMICAL PROBLEMS
IN CONTINUUM PHYSICS

Millard F. Beatty

Department of Engineering Mechanics
University of Kentucky
Lexington, KY 40506

1. Introduction.

This is a tutorial lecture[1] about some one and two dimensional dynamical problems in continuum physics. Three problems concerning the physical behavior of fluid and solid materials in a dynamical setting will be discussed. The first example is an old, but fascinating problem in the fluid mechanics of jet instability whose solution I will relate to a rather remarkable technological achievement by American industry. This illustration also will demonstrate the gap in the time between creation of important physical and mathematical results and their conversion to useful technology.

Of course, not every important scientific or mathematical discovery in the physics of continua has the same potential for technological application. Nevertheless, basic results seem always to find relevant applications that bear on the advancement of technology by providing useful supportive information about the physical nature and mechanical behavior of materials. The other two problems that I will discuss fall in the latter category. These involve the dynamical response of highly elastic solid continua.

An interesting phenomenon produced in the transverse vibration of a rubber cord stretched over its entire limit of extensibility will be described both by experiment and theory. This analysis involves the superposition of small amplitude oscillations on a finitely deformed equilibrium configuration of the string. The results will be related to the molecular structure of the rubber materials tested. Then we shall explore the finite amplitude, free longitudinal vibration of a body supported by a rubber spring. The exact solution of this nonlin-

----

[1] A somewhat similar address was broadcast live on UNITE television from the University of Minnesota at Minneapolis on May 28, 1985.

ear problem will be outlined, and its physical implications will be illustrated. Finally, the problem of small amplitude, free longitudinal oscillations superimposed on a finitely deformed axial stretch of the rubber support studied in the previous problem will be related to experimental data, some of which has application in human biology.

2. Applied Mathematics and Technology.

If one turns on a water faucet and then gradually reduces the flow rate from the spigot so that the flow is slow but steady, one usually will observe that the stream becomes unstable and disintegrates randomly into drops of irregular size, shape and spacing. Sometimes one may find that by whistling at a constant pitch near the stream, the induced vibration will enhance this readily observed jet breakup phenomenon and produce somewhat improved regularity in the drops. The general effect is shown in the series of stroboscopic photographs[2] in Figs. 1 to 4. The Figs. 2 and 3 are enlargements of the regions between the arrows indicated in Figs. 1 and 2, respectively; and Fig. 4 illustrates separately the ultimate jet disintegration.

2.1. Theory of the Fluid Jet Instability Phenomenon

This interesting instability behavior of liquid jets was first studied by several 19th century researchers. It was described in experiments by Bidone in 1830, Savart in 1833, and Magnus in 1859. In 1873, Plateau showed theoretically and experimentally that because of surface tension, the principal cause of this instability phenomenon, the cylindrical shape of a jet of fluid in a steady flow is an unstable configuration of equilibrium whenever its length λ exceeds its circumference πd_j, as shown[3] in Fig. 5. Thus, *Plateau's criterion for*

---

[2] These photographs were provided by Professor Charles F. Knapp, Department of Mechanical Engineering, University of Kentucky. I thank Dr. Knapp for permission to reproduce them in this report.

[3] Several of the illustrations in this presentation are adaptations of figures that have appeared in publications of the International Business Machines Corporation described in my acknowledgements. I thank Mr. W. L. Buehner of the Office Products Division of IBM for permission to reproduce items from his work in the references [3,4,7] and for his special assistance with arrangements for their use here.

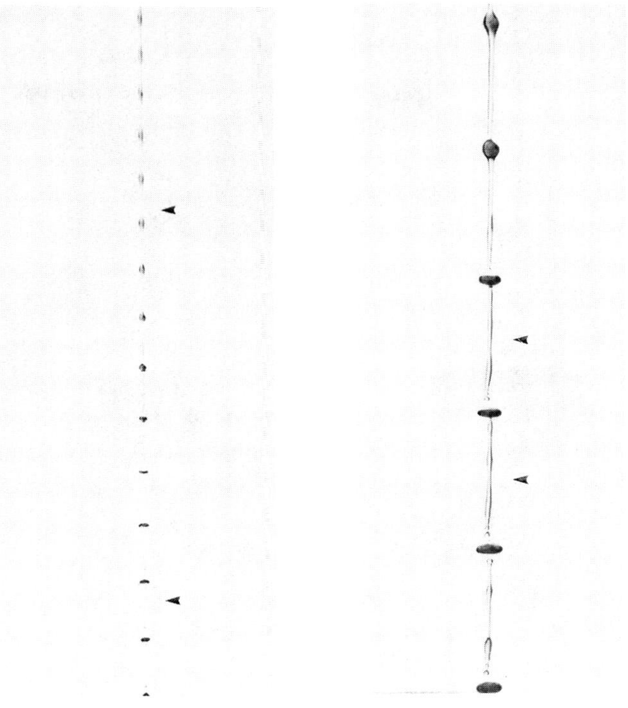

Fig. 1: Instability of a Fluid Jet

Fig.2: Formation of the Stream into Drops

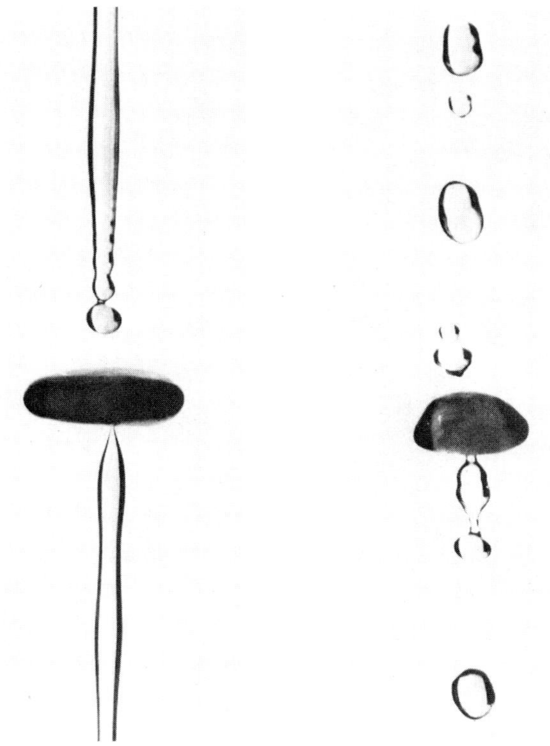

Fig. 3: Closeup of the Jet
Disintegration into a
Drop and a Possible
Satellite Droplet
Behind it.

Fig. 4: Disintegration of the
Liquid Jet into Drops of
Irregular Size, Shape
and Spacing; and
Appearance of Some
Satellite Droplets.

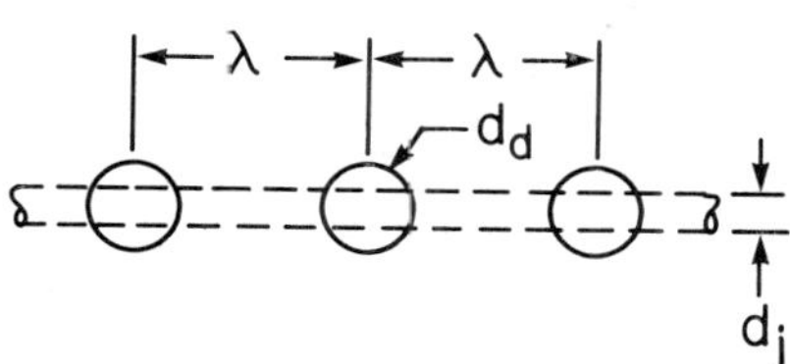

Fig. 5: Jet stream geometry and its uniform breakup. The droplets in a uniform stream are spaced a distance λ apart.
Each droplet of diameter d_d is composed of fluid previously contained in a fluid cylinder of length λ and
diameter equal to the mean jet diameter d_j. Adapted
from [3], courtesy of International Business Machines
Corporation.

the jet instability is that $\lambda > \pi d_j$. However, it was 1878 when Lord
Rayleigh [1-2] provided a particularly elegant mathematical analysis
of the phenomenon. A periodic pressure variation acting on the fluid
jet produces undulation of its lateral surface. Surface tension
effects act to induce jet instability where the jet stream is narrow-
est. Since the surface tension becomes large where the curvature is
high, it acts to further reduce the stream diameter until the stream
breaks into separate drops, as described above. Rayleigh showed that
a superimposed small sinusoidal diametral disturbance grows exponentially in time, as shown in Fig. 6, until eventually the stream breaks
into droplets at a rate corresponding to the imposed disturbance. He
found that the growth of the stream perturbation δd_j is given by

$$\delta d_j = \delta d_{j0} \, \exp\left[I(z) \left[\frac{8\sigma}{\rho d_j^3} \right]^{1/2} t \right], \qquad (2.1)$$

wherein d_j is the jet diameter, σ is the surface tension, ρ denotes
the fluid density, δd_{j0} is the initial stream disturbance, and

Rayleigh's Analysis (1878)

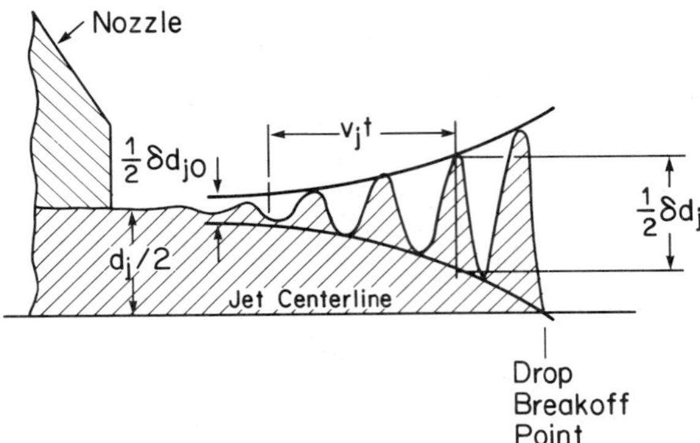

Fig. 6: Stream instability of the steady flow of an
ideal fluid jet. Adapted from [3], courtesy
of International Business Machines Corporation.

$$I(z) \equiv \left[\frac{z\ I_0'(z)}{I_0(z)}\ (1 - z^2) \right]^{1/2} \quad \text{with } z \equiv \frac{\pi d_j}{\lambda}. \qquad (2.2)$$

In (1.2) $I_0(z)$ denotes the modified Bessel function of the first kind.
The function $I(z)$, therefore, is a measure of the ease and rapidity
with which a stream breaks up into uniformly spaced droplets, so it
has been named the *instability factor* [4].   This measure is highly
dependent upon the *stream factor* k $\equiv \lambda/d_j$, the ratio of the drop
spacing in the axial direction to the jet stream diameter described in
Fig. 5.   A plot of the instability factor as a function of the stream
factor is shown in Fig. 7. For a stream factor less than π, we have z
$>$ 1 in (2.2). This is interpreted theoretically as implying that only
random stream breakup may occur for k $<$ π. Therefore, before uniform
breakup can occur, the length λ must exceed the stream circumference
πd_j.   Hence, Plateau's criterion for jet breakup falls out naturally
from Rayleigh's analysis. But there is more. As shown in Fig. 7, *the*
maximum instability occurs at a stream factor of 4.51, after which the
instability gradually decreases. This extreme is the value of the
stream factor at which the jet disintegrates with greatest ease into
droplets of uniform size and spacing; it characterizes the optimum
condition for jet instability according to Rayleigh's calculation.
Experiments [3] confirm that the best operating range for stable,

satellite-free drop formation occurs between factors of 4.5 and 6.5, as shown in Fig. 7. Hence, the stream factor introduced by Rayleigh turns out to have critical importance in the formation of stable, satellite-free drops. We shall see later on why this is significant.

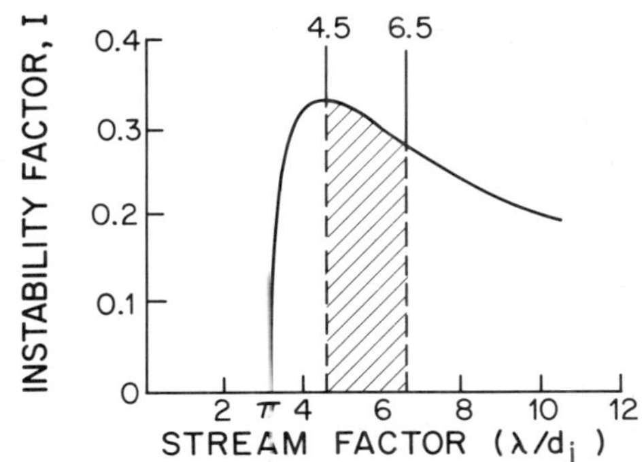

Fig. 7: Rayleigh's instability curve showing that λ must exceed the stream circumference before uniform breakup can occur. The optimum jet instability occurs at the stream factor $\lambda/d_j = 4.5$. Adapted from [3], courtesy of International Business Machines Corporation.

2.2. A Technological Application of Rayleigh's Analysis

In 1965, Richard G. Sweet of Stanford University showed that an electrical charge could be impressed on the drops that form out of a jet of electrically conducting ink [5]. He discovered that even when the drops were generated at the rate of 100,000 or more per second, the charge on each drop could be determined independently so that the trajectory followed by each could be controlled by passing the drops through a uniform electric field. His interest was in the construction of a high speed device for recording rapidly changing electrical signals on a moving strip chart. Hence, Sweet had in mind the application of Rayleigh's theory of jet instability to electronically controlled graphics. In modern terms, this would be identified as a form of *computer graphics*.

2.2A. A Technological Achievement

The same technique has found use in other applications, such as the sorting of cells in blood specimens, the atomization of fuels, and in the printing of alphabetic characters, which Sweet himself was among the first to investigate [6]. To get a feel for the performance speed in Sweet's printer application, let us estimate that the drops are formed at a rate of 100,000 per second and that an alphabetic or other character is composed of 100 drops. Then the maximum printing rate in Sweet's device would be roughly 1,000 characters per second. Of course, actual rates would be much slower because not every drop in the jet stream can be used. Nevertheless, it seems reasonable that rates of nearly 100 characters per second could be obtained in a practical printer without the use of high speed, mechanical impact printing elements, well-known for their inherent problems of wear, fatigue, and acoustical noise generation. A business office printer of this kind was first developed by the A.B. Dick Company, but I know nothing of its description, nor when it was first produced.

In 1976, however, these theoretical and experimental developments culminated in a remarkable modern technological achievement when the Office Products Division of IBM at Lexington, Kentucky, introduced an ink jet printer capable of producing character images of typewriter quality [7]. The ink jet printer, illustrated schematically in Fig. 8A, produces an image from tiny spherical droplets of electrically charged ink, approximately 30μ m ($\sim 1/1000$ in) in diameter, fired from a drop generator orifice the size of a human hair. Ink is pumped from its reservoir into the drop generator wherein a synchronizing signal induced by the vibrations of a piezoelectric crystal creates drops uniform in size, velocity, and spacing, at the synchronizing frequency rate of 117,000 drops per second. The periodic pressure variation applied by the crystal produces undulation of the lateral surface of the jet stream, and surface tension effects cause the stream to break up into droplets in accordance with Rayleigh's theory. The disintegrating ink jet then passes between charging electrodes where the conducting drops are selectively charged electrostatically following instructions from programmed electronic control circuits that describe the image characters in terms of charge-no charge language. Moving at about 40 mph (730 in/sec) initially, the charged droplets pass through a constant electric field that directs them onto the paper. While the vertical scanning occurs, a control mechanism moves the printer carriage across the paper at a constant speed of 7.7 in/sec, i.e. fast enough to traverse a standard page width in about one second. In this way, the ink jet printer composes about 80 characters each second, that is to say, a full line of type across an 8 ^{1}/$_2$ inch page in about one second.

2.2B. *The Working Principle of the Ink Jet Printer*

To perceive the working principle of the ink jet printing operation, we may consider a fluid droplet P of mass m and charge q given an initial velocity $\underset{\sim}{v}_0 = v_0 \underset{\sim}{i}$ relative to the printer carriage, as illustrated in Fig. 9. To simplify the analysis, we may ignore aerodynamic drag and wake effects and the influence of electric repulsive forces between droplets. Then the total force acting on the drop is the constant force due to its weight $\underset{\sim}{W} = -mg\underset{\sim}{j}$ and the applied

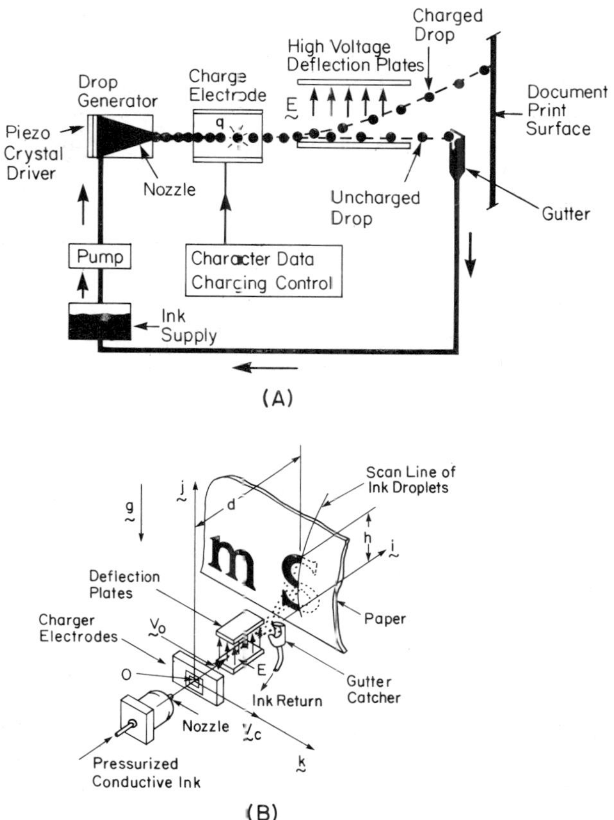

Fig. 8: Schema of the IBM ink jet printing process. Adapted from works [4] and [7], courtesy of International Business Machines Corporation.

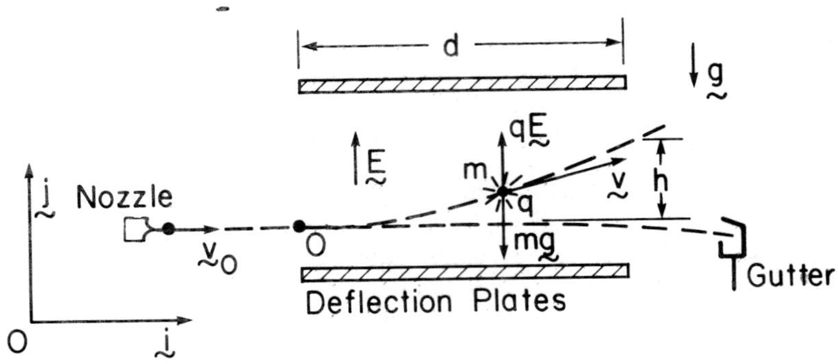

Fig. 9: Working principle of the ink jet printer.

electric field force $\underset{\sim}{F}_e = q\underset{\sim}{E} = qE\underset{\sim}{j}$. Thus, Newton's equation of motion $\underset{\sim}{W} + \underset{\sim}{F}_e = m\underset{\sim}{a}(P,t)$ yields the following motion for the trajectory of P:

$$\underset{\sim}{x}(P,t) = \frac{1}{2}(cE - g)t^2 \underset{\sim}{j} + v_0 t \underset{\sim}{i}, \qquad (2.3)$$

where $c \equiv q/m$. That is, the droplet path is the parabola

$$y(x) = \frac{1}{2v_0^2}(cE - g)x^2. \qquad (2.4)$$

Let us suppose for simplicity that the deflection plates of length d extend from the origin at the charger to the paper surface, as suggested in Figs. 8B and 9. Then (2.4) holds for $0 \leq x \leq d$. Hence, at $x = d$, the droplet deflection or scan height $h = y(d)$ at the paper surface is determined by

$$h = \frac{d^2}{2v_0^2}(cE - g). \qquad (2.5)$$

This result shows that when an electrostatically charged drop enters the uniform electric field, the electric force alters its free fall trajectory and deflects it vertically by an amount proportional to its charge. An uncharged drop passes straight through the field in free fall and is captured by a gutter catcher that returns the unused ink to its reservoir, as illustrated in Figs. 8 and 9. A charged drop strikes the paper. Alphabetic or any other characters are formed by directing the ink drops onto the paper in proper, overlapping patterns determined by the printer electronics. An example [4] is shown schematically in Fig. 8B. The decision to charge or not to charge is made automatically 117,000 times each second. The formula (2.5) shows that the final drop height, hence the character height, is inversely proportional to the square of the stream speed v_0 which is controlled by the pump pressure. The printer electronics controls the drop height automatically through its pump control circuit. And some interesting character style effects may be produced by varying the printer carriage rate. In this way, the ink jet printer is able to rapidly generate various characters of typewriter quality.

A remarkable stroboscopic microphotograph of drops of ink emerging from an ink jet printer is reproduced[4] in Fig. 10. A jet of ink that originated in the drop generator to the right has dissociated into equally spaced drops. The lower line of drops were not charged, so these are moving toward the ink gutter to the left. The larger gaps between the uncharged drops are the vacated positions formerly occupied by the field deflected, charged drops that are travelling on trajectories above.

The Mead Corporation has built a high speed printer that uses some 600 nozzles lined up across the width of a page, and paper is drawn continuously under these [6]. (See Fig. 11.) An entire line of drops is formed at one time, and, in consequence, the Mead printer is capable of printing 45,000 lines per minute, but with low resolution. We have all seen some printed matter produced in this way; indeed, this kind of high speed printing has received wide application in the printing of junk mail. I understand that a high resolution device of this sort has been investigated by IBM. I don't believe the purpose is to improve the printed quality of our junk mail. Rather, with improved resolution, it may be possible eventually to print newspapers, magazines, and standard forms, such as tax forms, by ink jet printer graphics.

2.2C. *An Open Mathematical Problem in Ink Jet Technology*

One may ask if there are any open mathematical problems associated with this technology. I believe that possibly there are many, but there is one in particular that I will mention briefly that may

----

[4] This extraordinary photograph by Mr. Carl Lindberg may be seen in color on the cover of the Number 1 issue of the 1977 IBM Journal of Research and Development. See [4]. I have enhanced the intensity of the droplets for greater clarity in the illustration used here.

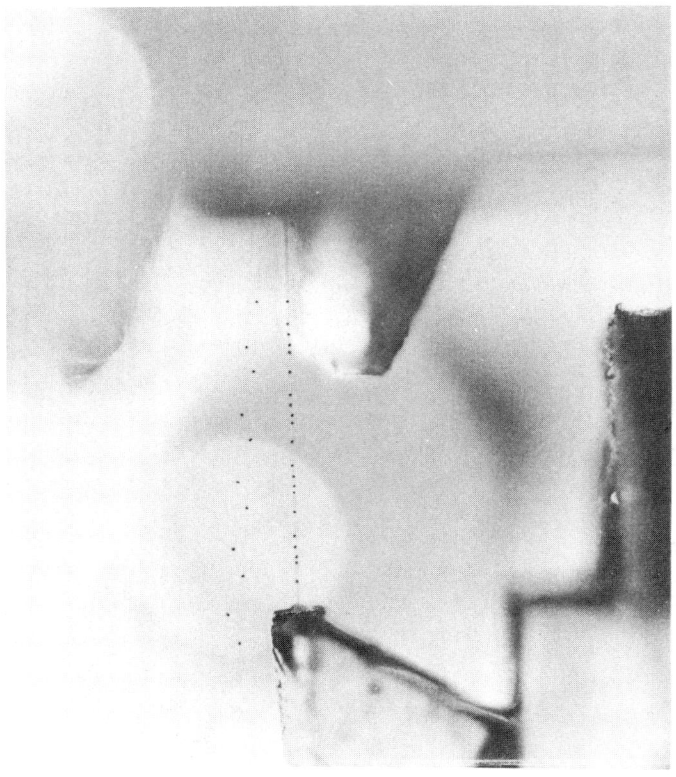

Fig. 10: Stroboscopic microphotograph of ink drops in a jet printer.
Copyright 1977 by International Business Machines Corporation;
reprinted with permission.

Fig. 11: Magnification of multiple streams of ink drops
emerging from ink jet nozzles at the left. Note
the uniformity in the drop size, spacing, and
speed. Reproduced from [6], courtesy of the IBM
Thomas J. Watson Research Center.

still constitute an unsolved problem and which would present a challenge for an applied mathematician or engineering scientist. This
concerns the formation of smaller, satellite droplets between the desired drops [3, 4, 6], as shown in Fig. 12A. These are troublesome
because they can contaminate the deflection apparatus and cause misplaced marks on the printed document. The linearized stability analyses by Rayleigh and others do not predict satellite droplet formation.
A nonlinear instability study [8], on the other hand, has predicted
the formation of satellites in conjunction with the breakup into
drops. Unfortunately, the nonlinear analysis predicts simultaneous
satellite separation at both ends of a main drop, and this stands in
contradiction to experimental observations described schematically in
Fig. 12B. A successful analysis of this problem could lead to a better
understanding of drop generation and improved drop generator design.

Let us now turn to a class of interesting problems in solid
continuum mechanics. The first will relate the molecular structure of
rubber to the vibrations of a rubber cord. The second will treat the
finite amplitude oscillations of a mass suspended by a rubber spring.
A final example will relate some of these ideas to a small amplitude
oscillation study that has an application in life science.

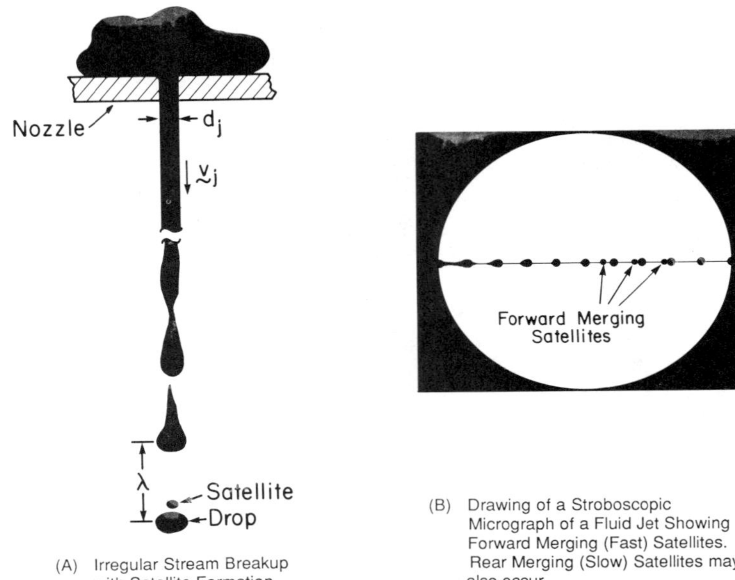

(A) Irregular Stream Breakup
with Satellite Formation

(B) Drawing of a Stroboscopic
Micrograph of a Fluid Jet Showing
Forward Merging (Fast) Satellites.
Rear Merging (Slow) Satellites may
also occur.

Fig. 12: Jet stream breakup showing the formation of satellite droplets which may contaminate the deflection apparatus and cause misplaced marks on the printed document. See [3,4,8]. Courtesy of International Business Machines Corporation.

3. MOLECULES AND THE VIBRATIONS OF A RUBBER CORD.

When a rubber band is stretched sufficiently between the fingers of both hands, held close to the ear and plucked like a string of a musical instrument, it is observed that the pitch of the sound will vary only slightly as the stretch is increased. The first experimental investigations of this phenomenon were reported independently at the turn of this century by two scientists, T.J. Baker in Great Britain [9] and V. von Lang in Germany [10]. But neither provided a consistent theoretical explanation of the phenomenon. Of course, at that time, the mathematical theory of rubber elasticity had not yet been developed. Since the early 1940's, however, theoretical understanding of the mechanical properties of rubber has grown substantially, and we are now able to evaluate more accurately the substance of the Baker-von Lang observations. Let us begin by recalling Taylor's basic frequency formula.

3.1. The Transverse Vibrational Frequency of a String

Let us consider a homogeneous and perfectly flexible string that is stretched and fixed at both ends. If the string is plucked in the middle with a small displacement so that the transverse motion is planar, then the frequency ν of the fundamental mode of the transverse vibration of the string is given by *Taylor's universal formula* [11]:

$$\nu = \frac{1}{2\ell} \left[\frac{T}{m} \right]^{1/2}, \qquad (3.1)$$

where ℓ is the stretched length between the clamped   ends,   T   is   the constant string tension, and m is the mass per unit length of the stretched string. Since the string is homogeneous, its total mass   is M = mℓ. If ℓ_o is the initial unstretched length of the string,   then $\ell = \ell_o\lambda$, as shown in Fig. 13. Hence, (3.1) may be rewritten as a function of the amount of stretch λ as follows:

$$\nu(\lambda) = \frac{1}{2\ell_o} \left[\frac{T(\lambda)}{m_o\lambda} \right]^{1/2}, \qquad (3.2)$$

in which m_o = M/ℓ_o is the mass per unit length of the unstretched string, and the function $T(\lambda)$ is determined by the constitutive   equation of the material. Of course, the stretch $\lambda = \ell/\ell_o$, as indicated in Fig. 13, is a measure of the deformation of the cord.

a) A Stretched Rubber Cord with
$\lambda = \ell/\ell_0$

b) Small Amplitude on a Finite Stretch λ

Fig. 13: Transverse vibration of a string
stretched an amount $\lambda = \ell/\ell_o$.

3.2. The Experimental Results

A low intensity laser beam apparatus was devised to determine the elastic properties of rubber by measurement of the small transverse vibrations of a rubber cord subjected to stretch varying to the maximum extension [12]. The apparatus is shown in Fig. 14, and the experimental results are shown in Fig. 15. The Baker-von Lang phenomenon is evident for all of the rubber materials studied. It is seen that the frequency rises rapidly at first, but after a stretch of roughly 2, the curve flattens considerably so that subsequent growth in the frequency is small. This is consistent with our earlier description of the response. But what is the source of this unusual phenomenon?

3.3. The Theoretical Results

It is now well-known that rubber consists of molecules connected in tangled chains. This chain structure is what gives rubber its highly elastic property, and several theoretical models that account for this so-called hyperelastic behavior have been developed [13]. Beatty and Chow [12] found that the nearly constant pitch of a sufficiently stretched rubber band is predicted in slightly different ways by three theoretical models studied. Only two of these models will be described here. One of them shows directly how the number of links in the molecular chain structure affects the extensibility of rubber. These results are presented next.

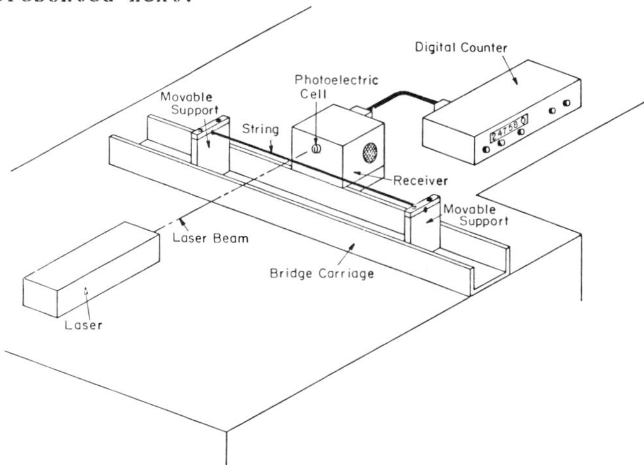

Fig. 14: Experimental apparatus for the
frequency test. See [12].

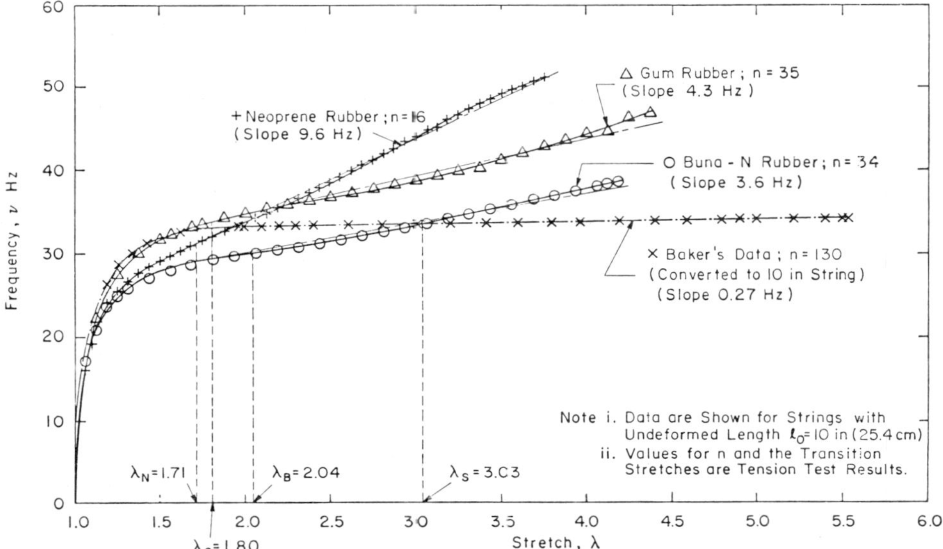

Fig. 15: Experimental values of the transverse frequency
vs. stretch averaged from four tests on each of
three rubber materials and from estimates of
Baker's data, all for strings having an initial
length ℓ_o = 10 in (25.4 cm). See [12].

3.3A. *The Neo-Hookean Model of Rubber Elasticity*

A neo-Hookean material model is an incompressible, isotropic
elastic material for which

$$T(\lambda) = \frac{A_o E}{3} \left(\lambda - \frac{1}{\lambda^2} \right). \tag{3.3}$$

Here A_o denotes the undeformed cross sectional area of the cord and E
is Young's modulus. This is the simplest model of rubberlike behavior;
it derives from the Gaussian statistical mechanics of a molecular network [13].

Use of (3.3) in (3.2) yields the theoretical formula for the
transverse vibrational frequency of a neo-Hookean rubber band,

$$\nu(\lambda) = \nu_\infty \left[1 - \frac{1}{\lambda^3} \right]^{1/2} \quad \text{with} \quad \nu_\infty = \left[A_o E / 12 M \ell_o \right]^{1/2}. \tag{3.4}$$

The graph in Fig. 16 of the frequency ratio ν/ν_∞ as a function of stretch shows clearly that the frequency rises rapidly as the cord length increases to about two or threefold, but with further extension the frequency remains almost unchanged [12]. Moreover, the phenomenon is a finite extensibility effect and may be attributed to the almost linearly elastic behavior of the material at large values of the stretch. Indeed, (3.4) gives $\nu/\nu_\infty > 0.98$ for $\lambda \geq 3$; and in this case the tension (3.3) is within 11% of being linear in λ. Thus, when a rubber band characterized by (3.3) is stretched sufficiently between the fingers of both hands and plucked, the pitch will vary only slightly while approaching a limiting value ν_∞ as the stretch is increased indefinitely. Therefore, our theoretical result for a neo-Hookean material supports and explains the observations of Baker [9]. Notice that the frequency ratio ν/ν_∞ in (3.4) is a universal function valid for every neo-Hookean material. Of course, other constitutive models may yield similar behavior.

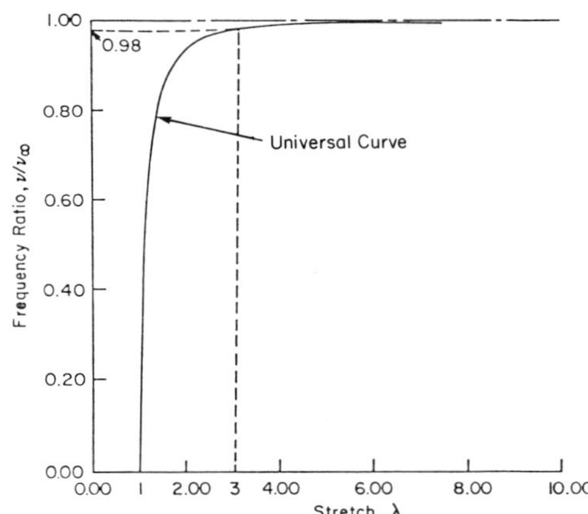

Fig. 16: The frequency ratio ν/ν_∞ is a universal function of λ; it is the same for every neo-Hookean material. See [12].

3.3B. The James-Guth Model

The James-Guth theory of isotropic rubber elasticity was derived from a non-Gaussian statistical analysis of a finitely extensible, long chain molecular network model of rubber. The relation between the tension and the stretch for this theory was first obtained by James and Guth in 1943 [14]. They found the complex expression

$$T(\lambda) = \frac{CA_o}{3} \sqrt{n}\ \left[L^{-1}\left(\frac{\lambda}{\sqrt{n}}\right) - \lambda^{-3/2} L^{-1}\left(\frac{1}{\sqrt{\lambda n}}\right) \right] \quad \text{with } \sqrt{n} > \lambda \geq 1, \quad (3.5)$$

where n is the average number of links in a molecular chain, C is a material constant, and $L^{-1}(y)$ is the inverse Langevin function defined by

$$L^{-1}(y) = x \quad \text{with } y = L(x) \equiv \coth x - \frac{1}{x}. \qquad (3.6)$$

Notice that (3.5) shows that for this model the maximum network extensibility is less than $\sqrt{n}$. The constant C is related to Young's modulus in a complicated way which is of no concern here.

It seems intuitively clear that the ultimate extensibility of a chain, i.e. the ratio of the fully extended length between chain ends to the distance between chain ends in the natural state, increases with the number of its links. Therefore, the number n, as emphasized in (3.5), is a material parameter that characterizes the degree of elastic extensibility of the rubber material. As n grows indefinitely large, the formula (3.5) reduces to the neo-Hookean case.

Use of (3.5) in (3.2) yields the transverse vibrational frequency of a rubber string characterized by the James-Guth model, namely,

$$\nu(\lambda) = \nu_\infty \left[\frac{\sqrt{n}}{3\lambda} \left(L^{-1}\left(\frac{\lambda}{\sqrt{n}}\right) - \lambda^{-3/2} L^{-1}\left(\frac{1}{\sqrt{\lambda n}}\right) \right) \right]^{1/2} \qquad (3.7)$$

for $1 \leq \lambda < \sqrt{n}$. Here, by definition,

$$\nu_\infty = \frac{1}{2} \left[\frac{CA_o}{M\ell_o} \right]^{1/2}. \qquad (3.8)$$

In view of the convexity of the inverse Langevin function, it can be shown that the transverse vibrational frequency for the James-Guth model is a monotonically increasing function of λ. The theoretical result (3.7) is shown in Fig. 17 for some typical values of n.

Unlike the neo-Hookean case, as λ increases but remains bounded above by $\sqrt{n}$, the frequency does not approach a constant limit; rather, the frequency grows indefinitely as $\lambda \to \sqrt{n}$. Nonetheless, there exists a fairly straight intermediate range R in the neighborhood of the intercept $\nu(\lambda_o) = \nu_\infty$ at $\lambda = \lambda_o$ such that the frequency increases but little as λ increases through R. The extent of R, as seen from Fig. 17, depends on n. In this fairly flat range the frequency response is consistent with the characteristics of the Baker–von Lang phenomenon as described previously in the experiments by Beatty and Chow [12].

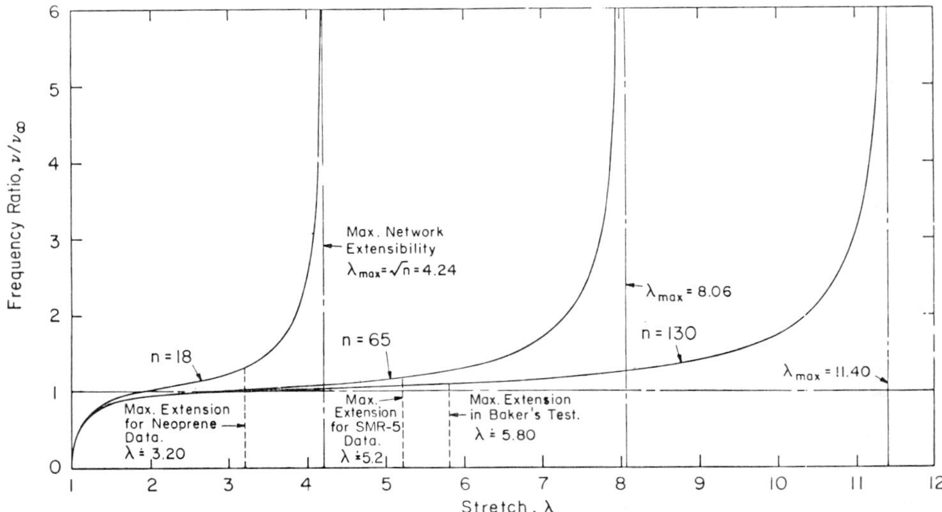

Fig. 17: Frequency ratio vs. stretch for a James–Guth
string plotted for some typical values of the
average number of molecular chain links n.

The vibrating string data mentioned before were studied on the basis of this molecular model. A computer program was devised to determine simultaneously from the data the constant C and the apparent number of links n in the molecular chain structure of each of several varieties of rubber strings. The final result is summarized in Fig. 18. Notice that the gum and buna-N rubbers have almost the same apparent number of links n but different C values, and these curves coincide as predicted by (3.7), within expected experimental differences.

The experiments show that the nearly constant pitch phenomenon is a molecular network finite extensibility effect. It was observed for all the materials studied that after the stretch had about doubled, the frequency increased only very slightly as the tension was increased, and this slight rate of growth of the pitch is smaller for

those rubbers for which the apparent number of chain links in the molecular network is larger. We shall see further on an important connection between these data and the results of the following problem of
longitudinal oscillations [15-16].

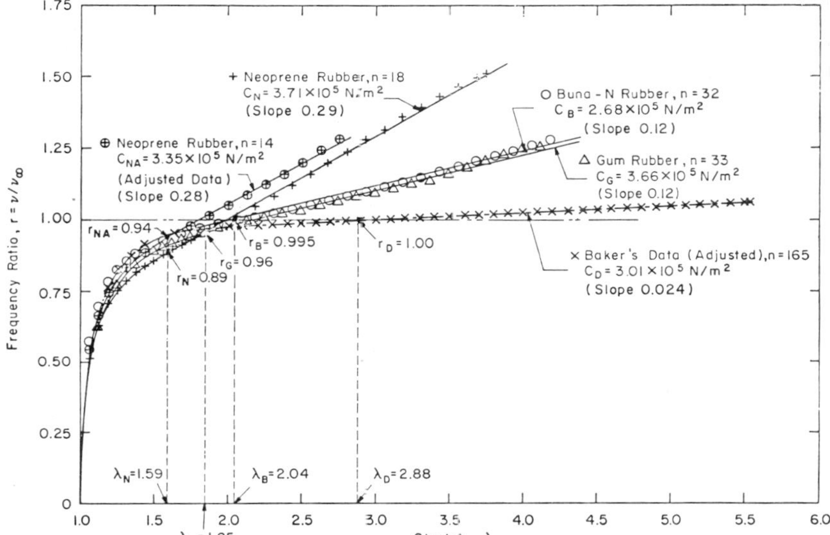

Fig. 18: Normalized frequency ratio vs. stretch for
        data averaged from four tests on each of
        three rubber materials and for estimates
        obtained from Baker's data. See [12].

4. LONGITUDINAL VIBRATIONS OF A MASS ON A RUBBER SPRING.

Various kinds of rubber spring mountings are used as suspension
or compression supports for machines where only small amplitude oscillations commonly are encountered. Consequently, it is no surprise that
in vibration analyses the affect of the nonlinearly elastic behavior
of the rubber material has been ignored, even though the initial
static deformation of the springy body may be substantial. It is easy
to show that the undamped, small amplitude, free vibration of a mass
superimposed on a finite static extension or compression of an ideal
rubber spring leads to a simple harmonic solution for the attached
load, but the fundamental frequency of the vibrations will depend upon
the initial deformation. Hence, even in this case considerable error
may be introduced into dynamical measurements of mechanical properties

when the initial deformation is ignored, and the opportunity to dis-
cover other interesting results may be lost.

The exact solution for the undamped, finite amplitude, free
oscillations of a heavy mass supported vertically by a rubber suspension spring made of an ideal neo-Hookean material will be described
next. We wish to determine the frequency (or the period) and to characterize the amplitude of the vibration.

4.1. Finite Oscillations of a Mass on an Ideal Rubber Spring

Everyone probably has seen a child's paddle ball toy which is
sketched in Fig. 19. There is no doubt that the motion of the ball,
which is attached to the end of a highly elastic rubber string, exe-
cutes large amplitude oscillations. Of course, the problem is compli-

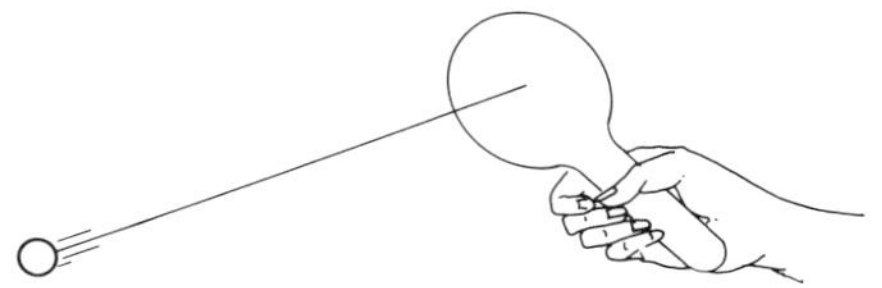

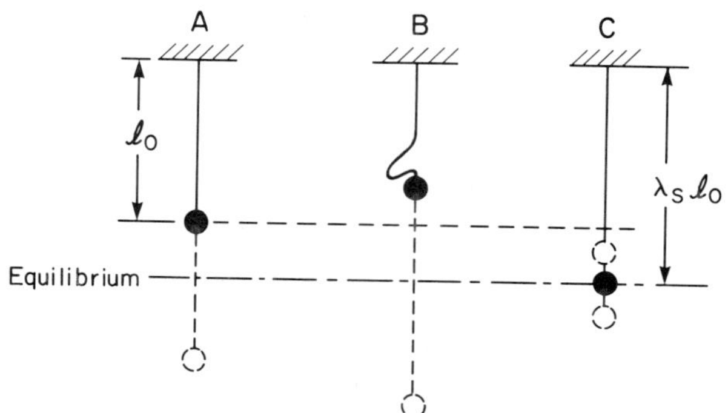

Fig. 19: Finite amplitude oscillations related to a paddle
ball toy. Diagram A shows a finite amplitude, free
vibration from the unstretched state; B illustrates
the finite amplitude, free vibration in which free
flight occurs as the string collapses; and C shows
the small amplitude, free vibration from a static
equilibrium state with finite stretch λ_S.

cated by the motion and impact induced by the paddle, and, to my knowledge, this nonlinear oscillation problem has never been fully investigated.   Beatty and   Chow [15] have provided a partial solution for
the ideal case when the paddle is held fixed and the ball misses the
board   altogether in the ideal manner suggested in Fig. 19B.   We shall
not consider this problem here, but the general tools needed to effect
the idealized solution are similar to those that enter the problem of
the   finite   amplitude,   free vibration of a mass attached to a rubber
string in pure stretch. (See Fig. 19A.)   This case is   outlined   next,
and some general properties of the oscillation are described.

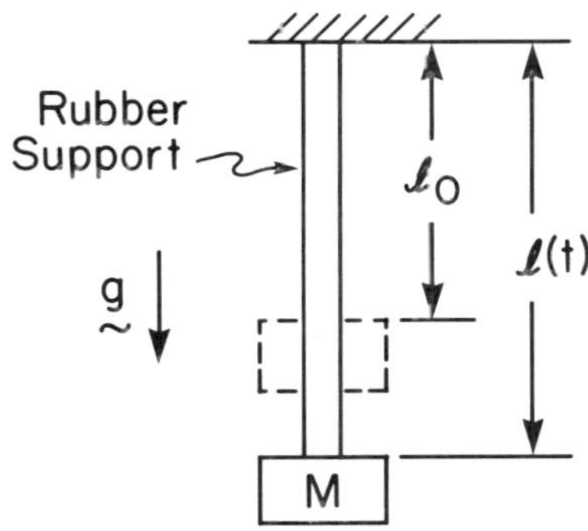

Fig. 20: Finite oscillations of a
mass on a rubber spring.

A neo-Hookean rubber spring of undeformed length ℓ_o,   cross   sectional   area A_o,   and   fixed at one end is subjected to a homogeneous
uniaxial extension by a mass M attached vertically to its   other   end,
as shown in Fig. 20. Let $\ell(t)$ denote the deformed length of the spring
at time t. Then $\lambda(t) \equiv \ell(t)/\ell_o$ is the amount of stretch at time t. The
mass of the spring is assumed negligible compared with M, as usual, so
that its motion and the effects of waves propagating in the spring itself   may be ignored.   The influence of these continuum effects on the
finite motion of M is an unsolved problem. Of course, if   the   spring
is   a   rubber   band or strip, compression states are not allowed; this
gives rise to the special paddle ball problem in Fig. 19B.
    Since the only forces that act on M are gravity and the hyper-
elastic spring force, the system is conservative with constant total
energy E given by

$$E = \frac{1}{2} M \ell_o^2 \dot{\lambda}^2 + \ell_o \int_{\lambda_o}^{\lambda} T(\lambda) \, d\lambda - Mg\ell_o(\lambda - \lambda_o). \qquad (4.1)$$

Here g denotes the acceleration of gravity, a dot denotes differentiation with respect to time, and $\lambda_o = \lambda(0)$ is the initial stretch at which the initial "speed" is $v_o = \dot{\lambda}(0)$. Use of (3.3) in (4.1) gives the speed $\dot{\lambda}(t)$ of the mass M supported by a neo-Hookean spring as

$$\dot{\lambda}(t) = v(t) = \pm \left\{ - \frac{2g\kappa}{\ell_o \lambda} \left[\lambda^3 - \frac{\lambda^2}{\kappa} - \frac{V\lambda}{\kappa} + 2 \right] \right\}^{1/2} \qquad (4.2)$$

for $\lambda \in [\alpha, \beta]$, where α and β are zeros of the right-hand side of (4.2). The energy constant V in (4.2) is defined by

$$V = V(\lambda_o, v_o) \equiv \frac{\ell_o v_o^2}{2g} + \kappa(\lambda_o^2 + 2\lambda_o^{-1}) - \lambda_o, \qquad (4.3)$$

and $\kappa \equiv A_o E/6Mg$. Hence, the time required for M to move from λ_o to $\lambda(t)$ follows from an integration of (4.2) in which the appropriate sign is to be used:

$$t = \pm \int_{\lambda_o}^{\lambda} \frac{d\lambda}{v(\lambda)}. \qquad (4.4)$$

Thus, in the motion between the two admissible extreme stretch states $\lambda(t_1) = \alpha$ and $\lambda(t_2) = \beta > \alpha$ defined by the physical condition that the speed $v(\lambda)$ must vanish at these extremes, the period $\tau \equiv 2(t_2 - t_1)$ of the finite amplitude, free vibration of M is given by

$$\tau = 2 \int_{\alpha}^{\beta} \frac{d\lambda}{v(\lambda)}. \qquad (4.5)$$

It is seen from (4.2) that (4.4) is a general elliptic integral. Hence, the motion is indeed periodic. The standard form for (4.4) may be found by use of the transformation [15-16]

$$\phi(\lambda) \equiv \sin^{-1}\left[\frac{\beta(\lambda - \alpha)}{\lambda(\beta - \alpha)}\right]^{1/2} \quad \text{so that} \quad \lambda = \alpha[1 + n \sin^2\phi]^{-1}, \quad (4.6)$$

in which $0 \leq \phi \leq \pi/2$ and the integral parameters are defined by

$$n \equiv \frac{\alpha - \beta}{\beta}, \quad k^2 \equiv \frac{\gamma(\beta - \alpha)}{\beta(\alpha + \gamma)}, \quad 0 < k < \sqrt{-n} < 1, \quad \alpha\beta\gamma = 2. \quad (4.7)$$

When this is done, it turns out that the travel time (4.4) is given by

$$t = \pm \frac{\tau^*}{2}[\Lambda(\phi\ n,k) - \Lambda(\phi_0;n,k)] \qquad (4.8)$$

in which $\Lambda(\phi;n,k)$ is the Heuman Λ-function [15, 17], $\phi_0 \equiv \phi(\lambda_0)$ is given by (4.6), $\tau^* \equiv 2\pi(\ell_0/g\kappa)^{1/2}$, and the sign is chosen appropriately. It is seen from (4.6) that for $\lambda_0 = \alpha$, $\phi_0 = \phi(\alpha) = 0$; and clearly $\phi(\beta) = \pi/2$. Since the Λ-function has the properties

$$\Lambda(0;n,k) = 0, \quad \Lambda(\tfrac{\pi}{2};n,k) = \Lambda_0(\xi;k) \qquad (4.9)$$

in which $\xi \equiv \sin^{-1}[(1 + k^2/n) \div (1 - k^2)]^{1/2}$. It follows that the period has the simple, elegant form

$$\tau = \tau^* \Lambda_0(\xi;k) \qquad (4.10)$$

expressed in terms of Λ_0, the tabulated, complete Λ-function [17].

It follows from these relations that *the finite amplitude, free vibrational period (frequency) of a neo-Hookean oscillator is always less (greater) than the period (frequency) of a linear spring oscillator having the same constant stiffness* $k_0 = AE/3\ell_0$; *that is,* $\tau < \tau^*$. More precisely, for $\beta \in (0,\pi/2)$ and for all $k \in (0,1)$, the period τ, hence the frequency p, is bounded as follows:

$$2\beta/\pi < \tau/\tau^* = p^*/p < \sin\beta. \qquad (4.11)$$

This is a consequence of the properties of the complete Λ-function.

It can be shown that *the mass oscillates asymmetrically about its equilibrium state.* We recall that a universal property of every linear oscillator is that the dynamical deflection always is exactly twice the static deflection. It turns out that, *unlike the linear oscillator in a vertical vibration starting from its undeformed rest state, the magnitude of the dynamic deflection for a rubber spring suspension support always is larger than twice the static deflection.* These features are evident in Fig. 21.

These results provide a better view of the paddle ball problem, which remains unsolved. We leave this for another place, and turn next to a much simpler problem that has a useful application in life science. The frequency relation for small amplitude motions superimposed on a finitely deformed static state, as illustrated in Fig. 19C, will be described next.

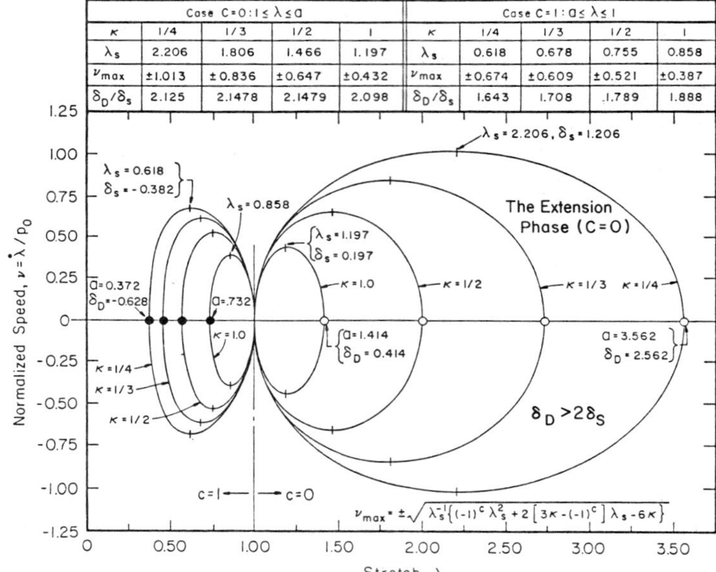

Case C=0: $1 \leq \lambda \leq a$					Case C=1: $a \leq \lambda \leq 1$				
κ	1/4	1/3	1/2	1	κ	1/4	1/3	1/2	1
λ_s	2.206	1.806	1.466	1.197	λ_s	0.618	0.678	0.755	0.858
ν_{max}	±1.013	±0.836	±0.647	±0.432	ν_{max}	±0.674	±0.609	±0.521	±0.387
δ_D/δ_s	2.125	2.1478	2.1479	2.098	δ_D/δ_s	1.643	1.708	.1.789	1.888

Fig. 21: Phase plane graph of the finite amplitude periodic motion of a mass attached to a neo-Hookean spring for various values of κ and initial data $\lambda_o = 1$, $v_o = 0$. See [12]. Two principal characteristics of the motion are evident: 1) the mass oscillates asymmetrically about its equilibrium state λ_S; and 2) the dynamic deflection of the mass from the undeformed rest state is always greater than twice the static deflection, whereas for the same data a linear oscillator always satisfies $\delta_D = 2\delta_S$.

4.2. Small Oscillations Superimposed on an Equilibrium State

Let us suppose that a small displacement $\delta(t)$ is superimposed on an assigned finite static stretch λ_S of a neo-Hookean spring so that $\lambda(t) = \lambda_S + \delta(t)$. Substituting this relation into (4.2), retaining only terms to the second order in δ, and then differentiating the result with respect to time, we obtain the differential equation for the small amplitude oscillations of the mass superimposed on an initial finite static stretch of the spring, namely, $\ddot{\delta} + p^2(\lambda_S)\delta = 0$, wherein the small amplitude circular frequency is identified by

$$p(\lambda_S) = p_o \left[\frac{\lambda_S^3 + 2}{\lambda_S(\lambda_S^3 - 1)} \right]^{1/2} \tag{4.12}$$

where $p_o \equiv \sqrt{g/\ell_o}$. The static equilibrium stretch λ_S is determined by

$$\lambda_S^3 + \frac{1}{2\,\kappa}\,\lambda_S^2 - 1 = 0. \tag{4.13}$$

Of course, (4.13) has at most one positive solution. An important result follows [15]: *Equation (4.12) is a universal relation valid for every neo-Hookean material independently of its elastic modulus; hence, no small amplitude longitudinal oscillation test can distinguish one neo-Hookean material from any other. On the other hand, if (4.12) is not satisfied in every superimposed small amplitude, uniaxial motion of an isotropic, incompressible material, that hyperelastic material can not be modeled as a neo-Hookean material.*

The graph of the universal function (4.12) for the extension case is shown in Fig. 22. If we imagine in Fig. 19C that the ball is loaded statically by a weight W and given a small initial longitudinal disturbance, the frequency of the small amplitude oscillation may be readily counted and timed with the aid of a stopwatch. Suppose next that the weight is doubled and then tripled, and so forth, and the frequency is measured as before. It will be observed that the small amplitude frequency will decrease nonlinearly as the load is increased, that is, as the static stretch is increased. It is seen in Fig. 22 that the universal function (4.12) predicts this nonlinear frequency effect very nicely. Of course, materials that are not neo-Hookean in their response necessarily will behave somewhat differently. This will be seen in the next section where some experimental results having application in biology are reviewed.

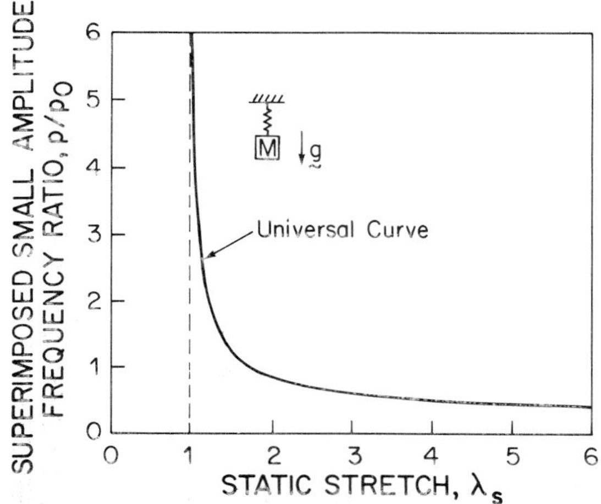

Fig. 22: Universal frequency ratio vs. equilibrium stretch for small amplitude oscillations superimposed on a finite static stretch of any neo-Hookean spring-mass support system. The frequency ratio decreases with increasing static stretch. See [15].

4.3. Experimental Results: An Application to Life Science

Lawton and King [18] have shown that the universal relation (4.12) describes nicely the small amplitude, free vibrational frequency response curves obtained from test data for certain unfilled gum rubber strips when the static stretch is not greater than approximately 2 to 2.5. Their data are shown in Fig. 23. All the experimental curves lie above the universal graph (4.12), and for small deformations λ_S near unity, the data show good agreement with the theory. But for larger deformations, there is considerable discrepancy from the monotonic decreasing behavior characterized by the universal curve. For the gum rubber specimens A, B, and D, the departure from the ideal case widens as the material stiffness is increased by additional sulfur and filler content. The data exhibit distinct minima, the frequency decreasing with increasing static stretch until a certain critical stretch is attained, after which the frequency increases with the stretch. The cause of this behavior is unknown, but it seems clear from the data that the minimum in the frequency response curve is affected by the stiffness. As the material stiffness increases, the corresponding critical stretch decreases and the minimum in the frequency response curve becomes sharper and moves upward and closer to the frequency axis.

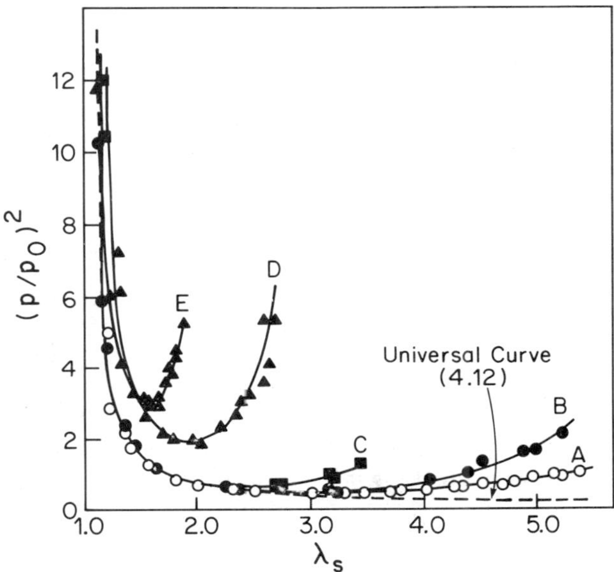

Fig. 23: Experimental frequency response curves of Lawton and King [18]. (A) Gum rubber stock with medium sulfur content, (B) Gum stock with high sulfur content, (C) Strip from a household rubber band, (D) Gum stock A with carbon black filler, and (E) Circumferential strip of human thoracic aorta (21 years). The universal frequency response curve of Fig. 22 is plotted as the dotted line.

Data for an aortic strip cut from the thoracic region of a 21-year old male post mortem shows greater stiffness than any of the rubber materials studied. The affect of aging on males may be seen in Fig. 24 in which the response from an aortic strip taken from the cadaver of a 70-year old male is compared with the aortic strip from the 21-year old male. It is seen that the old aorta is stiffer than the young one, which may be expected intuitively. I would suggest that it is not much different for women either. This simple test, therefore, supports the conclusion [18] that *aging arterial walls may lose their*

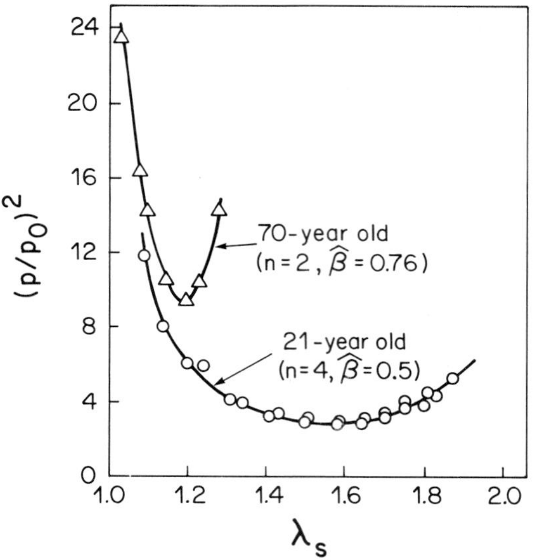

Fig. 24: Small amplitude frequency response vs. static
stretch showing the affect of aging on the
frequency response of male aortic tissue [18].

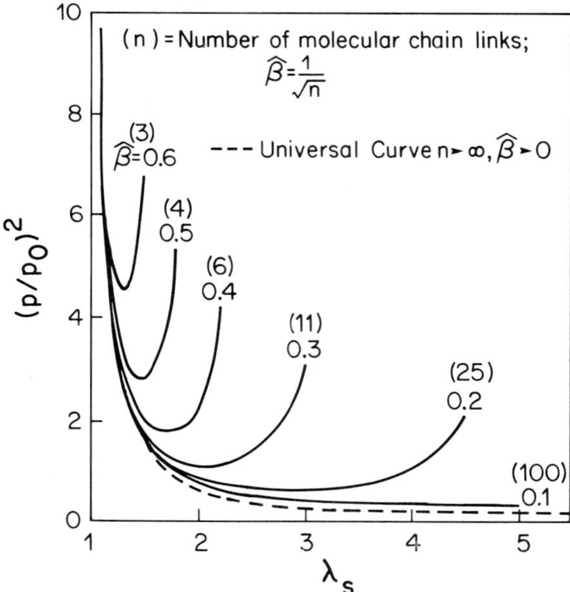

Fig. 25: Theoretical frequency response curves of the
James-Guth model for values of the molecular
chain link parameter n from 3 to ∞, i.e. for
$\widehat{\beta}$ from 0 to 0.60. See reference [18].

extensibility due to increased interchain bonding and deposition of filler-like particles. I think this effect is known more commonly as *hardening of the arteries.*

The discrepancy between the test data and the universal neo-Hookean prediction may be resolved to a considerable, though not completely satisfactory degree by use of the more sophisticated James-Guth model. It can be shown that the frequency of the superimposed small vibrations for the James-Guth continuum (3.5) is determined by the rule

$$p(\lambda_S) = p_o \left\{ \frac{\hat{\beta}[2\lambda_S^3 D(\eta) + D(\mu)] + 3\lambda_S^{1/2} L^{-1}(\eta)}{2[\lambda_S^3 L^{-1}(\eta) - \lambda_S^{3/2} L^{-1}(\mu)]} \right\}^{1/2}, \qquad (4.14)$$

in which

$$\hat{\beta} \equiv \frac{1}{\sqrt{n}}, \qquad \eta \equiv \hat{\beta}\lambda_S, \qquad \mu \equiv \frac{\hat{\beta}}{\sqrt{\lambda_S}}, \qquad (4.15)$$

and

$$D(z) \equiv \frac{d}{dz} L^{-1}(z) = \frac{z^2 \sinh^2 z}{\sinh^2 z - z^2}. \qquad (4.16)$$

We recall that n is the hypothetical average number of links in the molecular chain structure, a measure of the finite extensibility behavior of rubberlike materials. It happens that equation (4.14) reduces to (4.12) when $n \to \infty$.

The theoretical relation (4.14) as a function of λ_S is shown in Fig. 25 for several values of n. It is seen that the James-Guth model is able to account for the observed minimum in the superimposed, small amplitude frequency response curve. In this case, the frequency relation for the superimposed small oscillations is not universal, rather the frequency varies with the elastomeric parameter n, the apparent number of links in the molecular chain structure. In fact, in the curve fitting scheme used by Lawton and King [18], the value of n is selected so that the theoretical curve based on (4.14) has a minimum value of $(p/p_o)^2$ equal to its minimum found in the test. The experimental data shown in Fig. 26 for the gum rubber stock compare favorably with the theoretical result; but for the stiffer, reinforced gum rubber and biological materials, it turned out that the agreement was poor, the data curves being similar but shifted substantially along

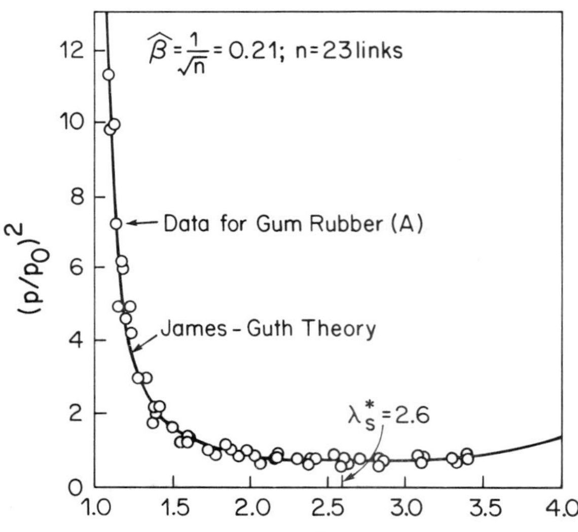

Fig. 26: Experimental frequency response data for a gum rubber of low sulfur content compared with (4.14) for the James-Guth model [18].

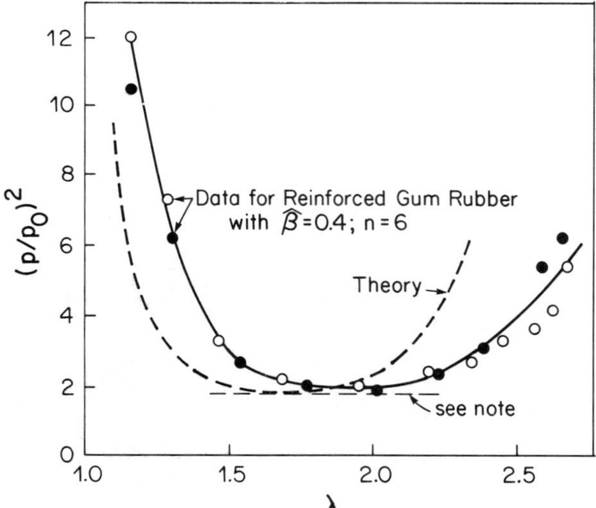

Fig. 27: Frequency response curve for two specimens of the same filled, gum rubber stock. The result shows that the discrepancy grows with increased material stiffness. Note: n was chosen to fit the theoretical minimum value of $(p/p_o)^2$ in (4.14) to its measured minimum value. See [18].

the stretch axis. The discrepancy for the reinforced gum rubber is shown in Fig. 27. Comparison with the data for the two aortic tissues was not provided in [18]. It should be mentioned that the values of n determined in the manner described above are significantly smaller than those found by other static methods; hence, the curve fitting scheme used in [18] may be in serious question. In some rough and unchecked preliminary tests on other elastomers, I have found that for data corresponding to the values of n found in the transverse vibration experiments described earlier, the frequency curves move upward as n decreases, as predicted. However, the curves for the gum and buna-N, which have virtually the same n values, did not lie as close together as I expected they should, based on the transverse vibration tests. Instead, there was a considerable horizontal shift in the buna-N data. It is clear that more work, both analytical and experimental, needs to be done toward understanding these basic dynamical phenomena.

5. CONCLUDING REMARKS.

Although the physical principles underlying the operation of the ink jet printer were understood in the 19th century, it was virtually 100 years before this mathematical physics was transformed into useful technology. Moreover, it is evident that the development of the ink jet printer for word processing required the solution of many interdisciplinary dynamical problems in the continuum mechanics of fluids and solids, aerodynamics, acoustics, electrostatics, chemistry, and materials science. The foundations of rubber elasticity, on the other hand, are relatively new, and there is need for their further development. While some of the experiments described herein were reported in 1951 and studied on the basis of a theory developed as recently as 1943, the analysis of the vibration problems summarized above were published only recently in 1983 and 1984. It was shown that these results are able to explain an unusual acoustical phenomenon, and they have implications for the material science of rubberlike materials, including the behavior of biological tissues. It is clear from these few examples that investigation of these problems also demands an interdisciplinary approach. Additional applications are being explored. Indeed, their connection with a vehicle stability problem, which I learned of in conversation with industrial colleagues following my seminar presentation at the research laboratories of a major tire and rubber manufacturer, currently is under investigation. I had a similar experience concerning the ink jet printer.

I first learned of the ink jet printer invention at a Sunday, summer afternoon piano recital held at the home of an IBM engineer. Our casual shoptalk led to a discussion of a new IBM product, and I was presented with a copy of the July 1976 issue of the Office Products Division Development Newsletter [6] in which this innovative technology was formally introduced. I am easily excited by good ideas; and when suddenly I realized myself the simplicity and basic principle of its operation, I confess that the ink jet printer captured my enthusiasm. A short time thereafter, during the Fall Term 1976, in a manner similar to that sketched above, I presented these elementary ideas for the first time to my undergraduate class in dynamics [19].

I relate these experiences because they exemplify how informal discussion of "hot topics" may be effective in creating interest and enthusiasm for further investigation and search for understanding of new or unexplored areas in applied mathematics, engineering science, and technology. Since my arrival at the University last September, I have witnessed in myself and in others at the Institute for Mathematics and its Applications a parallel experience generated through participation in seven IMA Workshops held throughout the year. By bringing together engineers, physicists, chemists, mathematicians and others, both analysts and experimenters, and by providing a forum for the mutual exchange of ideas and the open discussion among university and industrial researchers of important technical problems, it is conceivable that the use of costly cut and try methods in timely industrial applications may be reduced and the time gap between the discovery and understanding of physical phenomena and the conversion of this knowledge to useful applications and to technology may be narrowed. To accomplish this goal, mutual interaction between industrial and university personnel must be stimulated and vigorously promoted by both groups. I am pleased to have had the opportunity to help promote this kind of collaboration through presentation of this lecture to both IMA Workshop participants and to our industrial colleagues through the medium of television.

I wish to end with a story told, I believe, in a political speech by John Kennedy. The circumstances are unimportant. The message is clear.

The French General Lautey one day remarked to his gardener that he wished a special variety of tree to be planted, so he directed: "Gardener, plant this tree tomorrow." And the gardener replied; "But,

Sir, that variety won't bear fruit for possibly a hundred years." "In that case," replied the General, "plant it today."

<u>Acknowledgement</u>. This work was completed in the course of a research study supported by the National Science Foundation, whose support I gratefully acknowledge. I wish to thank Professor Jerry Bona for his helpful editorial suggestions. I am grateful to the Institute for Mathematics and its Applications for support of my fellowship at the University of Minesota during 1984–1985; and I especially thank Professors Jerry L. Ericksen and David Kinderlehrer for inviting my participation.

I appreciate permission granted me to use my adaptations of illustrations published in other works. Figures 5, 6, and 7 appeared originally in Mr. Buehner's IBM Office Products Division Technical Report TR 08.092, Copyright 1977; figure 8B was adapted from his article in the Office Products Division Development Newsletter, Copyright 1976; and figures 8A, 10, and 12 were adapted from the IBM Journal of Research and Development, Copyright 1977. All copyrights are by International Business Machines Corporation; reprinted with permission, for which I express my gratitude. I thank the IBM Thomas J. Watson Research Center for use of figure 11, Copyright 1979 by Scientific American. Figures 23 through 27 are adapted from the Journal of Applied Physics, Copyright 1951 by the American Institute of Physics. I thank Dr. Allen L. King for permission to use the figures from his work with Dr. Richard W. Lawton, whom, regretfully, I was unable to locate.

REFERENCES

[1] J.W.S. Rayleigh, On the instability of jets. *Proc. London Math. Soc.* **10** (1878), 4–13.
[2] J.W.S. Rayleigh, On the capillary phenomena of jets. *Proc. Roy. Soc.* **29** (1879), 71–79.
[3] W. L. Buehner, The application of ink jet technology to an office product. *IBM Office Prod. Tech Rpt.* TR 08.092, June 1977.
[4] W.L. Buehner, J.D. Hill, T.H. Williams and J.W. Woods, Application of ink jet technology to a word processing output printer. *IBM J. Res. Develop.* **21** (1977), 2–9. This entire issue is devoted to various aspects of ink jet printing technology.
[5] R.G. Sweet, High frequency recording with electrostatically deflected ink. *Rev. Scient. Instru.*, Series 2, **36**, (1965), 131–136.
[6] L. Kuhn and R.A. Myers, Ink-jet printing, *Scientific Amer.* **240** (1979), 162–178.

[7] W.L. Buehner, Ink jet printing for the IBM 46/40. *Office Products Division Development Newsletter*, July 1976.

[8] W.T. Pimbley and H.C. Lee, Satellite drop formation in a liquid jet. *IBM J. Res. Develop.* 21 (1977), 21-30.

[9] T.J. Baker, The frequency of transverse vibrations of a stretched India rubber cord. *Phil. Mag.* 49 (1900), 347-351.

[10] V. von Lang, Ueber Transversale Töne von Kaulschukfäden. *An. Phys. u. Chem.* 68 (1899), 335-342.

[11] J.W.S. Rayleigh, *The Theory of Sound*, The Macmillan Co., London, 1944.

[12] M.F. Beatty and A.C. Chow, On the transverse vibration of a rubber string. *J. Elasticity* 13 (1983), 317-344.

[13] L.R.G. Treloar, *The Physics of Rubber Elasticity.* Clarendon Press, Oxford, 1958, 3nd Ed.

[14] H.M. James and E. Guth, Theory of the elastic properties of rubber. *J. Chem. Phys.* 11 (1943), 455-481.

[15] M.F. Beatty, Finite amplitude oscillations of a simple rubber support system. *Arch. Rational Mech. Anal.* 83 (1983), 195-219.

[16] M.F. Beatty and A.C. Chow, Free vibrations of a loaded rubber string. *Inter. J. Nonlinear Mech.* 19 (1984), 69-81.

[17] C. Heuman, Tables of complete elliptic integrals. *J. Math. & Phys.* 20 (1941), 127-206. Errata. *Ibid.* 336

[18] R.W. Lawton and A.L. King, Free longitudinal vibrations of rubber and tissue strips. *J. Appl. Phys.* 22 (1951), 1340-1343.

[19] M.F. Beatty, *Principles of Engineering Mechanics. Volume 2: Dynamics - The Analysis of Motion.* Plenum Press (To appear)

EXISTENCE AND ASYMPTOTIC BEHAVIOR FOR STRONG SOLUTIONS
OF THE NAVIER-STOKES EQUATIONS IN THE WHOLE SPACE

H. Beirao da Veiga

Departimento di Matematica
Università di Trento
38050 Povo, Trento, Italy

We shall consider the initial-value problem for the non-stationary
Navier-Stokes equations in all of space, namely

$$
\begin{aligned}
v' - \mu \Delta v + (v \cdot D)v = f - \nabla p &\qquad \text{in }]0,T[\times R^n, \\
\nabla \cdot v = 0 &\qquad \text{in }]0,T[\times R^n, \\
v(0,x) = a(x) &\qquad \text{in } R^n, \\
\lim_{|x| \to +\infty} v(t,x) = 0 &\qquad \text{for } t \in \,]0,T[,
\end{aligned}
\tag{1}
$$

where $T \in \,]0,+\infty]$, μ is a positive constant, $v' = \partial v/\partial t$, and

$$
(v \cdot \nabla)v = \sum_{i=1}^{n} v_i \frac{\partial v}{\partial x_i}.
$$

The initial velocity $a(x)$ and the external forces $f(t,x)$ are given. The
pressure is determined by the condition $\lim p(t,x) = 0$, as $|x| \to +\infty$. Moreover,
it is assumed that $\nabla \cdot f = 0$ and $\nabla \cdot a = 0$.

By a solution of problem (1), we mean a divergence-free vector $v(t,x)$ such
that

$$
\int_0^T \int [v \cdot \phi' + \mu v \cdot \nabla \phi + (v \cdot \nabla)\phi \cdot v + f \cdot \phi]dx \, dt =
$$

$$
= -\int a \, \phi|_{t=0} \, dx,
$$

for every regular divergence free vector field $\phi(t,x)$ having compact support with
respect to the spatial variables, and such that $\phi(T,x) \equiv 0$.

Notation. $L^\alpha = L^\alpha(R^n)$, $|\ |_\alpha = \|\ \|_{L^\alpha(R^n)}$,

$$
N_\alpha(v) \equiv \int |\nabla v|^2 \, |v|^{\alpha-2} \, dx.
$$

Local existence. Let $\alpha > n$, $a \in L^{\alpha}$, $f \in L^1(0,T;L^{\alpha})$. Then, there exists a unique solution v of (1), $v \in C([0,T_{\alpha}];L^{\alpha})$, where T_{α} is the time of existence of the solution $y(t)$ of the o.d.e.

$$y'(t) = c_{\mu}^{-(\alpha+n/\alpha-n)} y(t)^{(3\alpha-n/\alpha-n)} + |f(t)|_{\alpha},$$
$$y(0) = |a|_{\alpha}.$$

Moreover,

$$|v(t)|_{\alpha} \leqslant y(t), \quad \forall \, t \in [0,T_{\alpha}[.$$

In particular, if $v \in L^{\beta}(0,T; L^{\alpha})$, where

$$\frac{2}{\beta} + \frac{\mu}{\alpha} = 1, \quad \alpha > n,$$

then $v \in C(0,T; L^{\alpha})$. Hence, if $f \in C^{\infty}(]0,T[\times R^n)$, then $v \in C^{\infty}(]0,T[\times R^n)$.
In particular, if $v \in L^{\beta}(0,T;L^{\alpha})$, where

$$\frac{2}{\beta} + \frac{n}{\alpha} = 1, \quad \alpha > n,$$

then $v \in C(0,T;L^{\alpha})$. Hence, if $f \in C^{\infty}(]0,T[\times R^n)$, then $v \in C^{\infty}(]0,T[\times R^n)$.

An existence result related to that stated above, was proved by Fabes, Jones and Riviere [2]. They showed that the solution v lies in $L^p(0,T^*;L^{\alpha})$, for some $T^* > 0$, and every finite p.

Our main concern will be the asymptotic behavior of the solutions. We prove the following results:

Theorem 1 Given $\alpha > n$, there exists two constants c_1 and c_2, depending only on α and n, such that the following statements hold.
Let $T \in \,]0,+\infty]$, and let $a \in L^{\alpha} \cap L^2$ and $f \in L^{\infty}(0,T;L^{\alpha}) \cap L^1(0,T;L^2)$. Assume that

$$[|a|_2 + \|f\|_{L^1(0,T;L^2)}]^{(2(\alpha-n)/\alpha(n-2))} \, |a|_{\alpha} \leqslant c_1 \, \mu^{(n(\alpha-2)/\alpha(n-2))}, \tag{2}$$

and that

$$[|a|_2 \|f\|_{L^1(0,T;L^2)}]^{(6\alpha-2n/\alpha(n-2))} \|f\|_{L^\infty(0,T;L^\alpha)} < c_2 \mu^{(2(\alpha n+\alpha-n)/\alpha(n-2))}. \tag{3}$$

Then, there exists a (unique) solution $v \in L^2(0,T;H^1) \cap C([0,T];L^\alpha \quad L^2)$ of the Navier Stokes equation (1).

Moreover,

$$\|v\|_{C([0,T];L^\alpha)} < c_1 \mu^{(n(\alpha-2)/\alpha(n-2))} [|a|_2 + \|f\|_{L^1(0,T;L^2)}]^{-(2(\alpha-n)/\alpha(n-2))}. \tag{4}$$

In the absence of external forces, we have the following decay estimate:

Theorem 2. Given $\alpha > n$, there exist positive constants c_3, c_4 and c_5, depending only on α and n, such that if $f \equiv 0$, $a \in L^\alpha \cap L^2$, $\nabla \cdot a = 0$, and

$$|a|_2^{(2(\alpha-n)/\alpha(n-2))} |a|_\alpha < c_3 \mu^{(n(\alpha-2)/\alpha(n-2))}, \tag{5}$$

then there exists a (unique) solution $v \in L^2(0,+\infty;H^1) \cap C(0,+\infty;L^\alpha \cap L^2)$ of problem (1). Moreover,

$$|v(t)|_\alpha < |a|_\alpha [1 + c_4 \beta \mu |a|^{-\beta} |a|^\beta t]^{-1/\beta}, \tag{6}$$

for every $t \in [0,+\infty[$, where $\beta = 4\alpha/(\alpha-2)n$. In particular,

$$|v(t)|_\alpha < c_5 |a|_2 \left(\frac{1}{\mu t} \right)^{((\alpha-2)n/4\alpha)}, \quad \forall\, t > 0. \tag{7}$$

We also consider the limiting case $\alpha = n$, for which one has the following result.

Theorem 3. Let $\alpha \in L^n \cap L^2$ and $f \in L^1(0,T;L^2) \cap L^\infty(0,T;L^n)$. Then there exist positive constants c_6 and c_7, depending only on n, such that if

$$|a|_n < c_6 \mu, \tag{8}$$

and

$$[|a|_2 + \|f\|_{L^1(0,T:L^2)}]^{4/(n-2)} \|f\|_{L^\infty(0,T:L^n)} < c_7\, \mu^{2n/(n-2)}, \tag{9}$$

then

$$|v(t)|_n < c_6 \mu, \quad \forall\, t \in [0,T], \tag{10}$$

where $v(t)$ is the solution of (1).

Moreover, _if_ $f \equiv 0$, _and if_ (8) _holds, then_

$$|v(t)|_n \leq |a|_n [1 + c_\mu |a|_2^{-(4/(n-2))} |a|_n^{4/(n-2)} t]^{-(n-2)/4}, \tag{11}$$

for every $t \in [0,+\infty[$. _In particular,_

$$|v(t)|_n \leq c|a|_2 (\frac{1}{\mu t})^{(n-2)/4}, \quad \forall\, t > 0. \tag{12}$$

Some results, related to those presented here, can be found in Fabes, Jones and Riviere [2], and in Kato [5]. In this last paper some asymptotic estimates are given in the case $a \in L^n \cap L^p$, $p < n$, and $f \equiv 0$. It is interesting to note that by setting $p = 2$ and $q = n$ in estimate (1.5) of reference [5] one has

$$|v(t)|_n \leq 0(1/t^{(n-2)/4}),$$

as $t \to +\infty$, which is just the asymptotic behavior implied by our estimate (12). However, in [5] the result is proved under the assumption that $(n-2)/4 < 1$.

For other results, more or less related to ours, see e.g. [3], [4], [6], [7].

Remarks (i) It is well known that if $v(x,t)$, $p(x,t)$, is a solution of (1), then $v_\lambda(x,t) = \lambda v(\lambda x, \lambda^2 t)$, $p_\lambda(x,t) = \lambda^2 p(\lambda x, \lambda^2 t)$, is also a solution of (1), with data $a_\lambda(x)$ $\lambda a(\lambda x)$, $f_\lambda(x,t) = \lambda^3 f(\lambda x, \lambda^2 t)$. Note that estimates (2), (3), (5), (8), and (9), are invariant under that change of scale in space-time.

(ii) By setting $\alpha = n$ in the statements of theorems 1 and 2, we obtain the statement of theorem 3. However, the proofs of theorems 1 and 2 are not valid for $\alpha = n$.

(iii) For bounded domains the asymptotic estimates are stronger. In fact, in that case, the solutions have an exponential decay as $t \to +\infty$. (We will return to this point in the sequel.)

The results stated above were proved in [1]. In what follows, the main ideas utilized in the proofs of the _a priori_ estimates are presented. For simplicity, it is assumed here that $f \equiv 0$.

Multiply both sides of equation $(1)_1$ by $|v|^{\alpha-2}v$ and integrate over R^n. After suitable integrations by parts there appears the identity

$$\frac{1}{\alpha}\frac{d}{dt}\,|v|_{\alpha}^{2} + \mu N_{\alpha}(v) + 4\mu\frac{\alpha-2}{\alpha^{2}}\int|\nabla|v|^{\alpha/2}|^{2}dx = -\int\nabla p\,\cdot\,v|v|^{\alpha-2}dx. \tag{13}$$

In particular,

$$\frac{1}{\alpha}\frac{d}{dt}\,|v|_{\alpha}^{\alpha} + \mu N_{\alpha}(v) \leqslant (\alpha - 2)\int|p|\,|v|\,|v|^{\alpha-2}dx$$

$$\leqslant \frac{(\alpha-2)^{2}}{2\mu}\int p^{2}\,|v|^{\alpha-2}dx + \frac{\mu}{2}\,N_{a}(v).$$

Hence, by Hölder's inequality,

$$\frac{1}{\alpha}\frac{d}{dt}\,|v|_{\alpha}^{\alpha} + \frac{\mu}{2}\,N_{\alpha}(v) \leqslant \frac{(\alpha-2)^{2}}{2\mu}\,|p|_{\frac{\alpha+2}{2}}^{2}\,|v|_{\alpha+2}^{\alpha-2}$$

Now, by applying the divergence operator to both sides of equation $(1)_{1}$, one gets (since $\nabla\cdot v = 0$)

$$-\Delta p = \sum_{i,j=1}^{n}\frac{\partial^{2}}{\partial x_{i}\,\partial x_{j}}\,(v_{i}v_{j}). \tag{14}$$

By using a Calderon-Zygmund inequality one obtains

$$|p|_{\frac{\alpha+2}{2}} \leqslant c|v|_{\alpha+2}^{2}. \tag{15}$$

Consequently,

$$\frac{1}{\alpha}\frac{d}{dt}\,|v|_{\alpha}^{\alpha} + \frac{\mu}{2}\,N_{\alpha}(v) \leqslant \frac{c}{\mu}\,|v|_{\alpha+2}^{\alpha+2}. \tag{16}$$

On the other hand, one can prove the following inequality:

$$|v|_{\alpha+2}^{\alpha+2} \leqslant c|v|_{\alpha}^{\alpha-n+2}\,[N_{\alpha}(v)]^{n/\alpha}. \tag{17}$$

Hence, by Young's inequality,

$$\frac{c}{\mu}\,|v|_{\alpha+2}^{\alpha+2} \leqslant c\mu^{-((n+\alpha)/(\alpha-n))}|v|_{\alpha}^{(\alpha/(\alpha-n))(\alpha-N+2)} + \frac{\mu}{4}\,N_{\alpha}(v).$$

Consequently,

$$\frac{1}{\alpha}\frac{d}{dt}\,|v|_{\alpha}^{\alpha} + \frac{\mu}{4}\,N_{a}(v) \leqslant c\mu^{-(n+\alpha)/(\alpha-n)}|v|_{\alpha}^{(\alpha(\alpha-n+2))/(\alpha-n)} + |f|_{\alpha}|v|_{\alpha}^{\alpha-1}, \tag{18}$$

where (for completeness) we inserted the term corresponding to the external forces f.

In particular, by dividing by $|v|_\alpha^{\alpha-1}$, one has

$$\frac{d}{dt}\,|v(t)|_\alpha \leq c\mu^{-(n+\alpha)/(\alpha-n)}|v(t)|^{(3\alpha-n)/(\alpha-n)} + |f(t)|_\alpha, \qquad |v(0)|_\alpha = |a|_\alpha. \qquad (19)$$

From (19) and standard comparison theorems for ordinary differential equations there follows the local existence result and the estimate for $|v(t)|_\alpha$.

Let $\dfrac{2}{\beta} + \dfrac{n}{\alpha} = 1.$ Since $\dfrac{3\alpha-n}{\alpha-n} = 1+\beta \left(^{(1)}\right),$

equation (1.9) yields $v \in C(0,T;L^\alpha)$, whenever $v \in L^\beta (0,T; L^\alpha)$.

<u>Global Existence</u>

<u>Remark</u>.

For a bounded domain Ω one easily shows that

$$N_\alpha(v) \geq c(\alpha,n,\Omega)|v|_\alpha^\alpha, \quad \text{if} \quad v = 0 \quad \text{on} \quad \partial\Omega, \qquad (20)$$

which is a Poincare-type inequality. Formulae (18) and (20) together imply that (assume for convenience that $f \equiv 0$)

$$z' \leq -k_0(1 - k_1 z^{2/(\alpha-n)})z,$$

where $z(t) \equiv |v(t|^\alpha$, and k_0, k_1 are positive constants, which depend on α, n, Ω, μ. Hence, if $k_1 z(0)^{2/(\alpha-n)} \leq 1/2$, i.e. if

$$|a|_\alpha \leq (k_1/2)^{(\alpha-n)/2\alpha},$$

then $|v(t)|_\alpha^\alpha \equiv z(t) \leq |a|_\alpha^\alpha e^{-(k_0/2)t}$, $\forall\, t \in [0,+\infty[$. This proves the exponential decay of $|u(t)|_\alpha$, as $t \to +\infty$, provided the initial data $a(x)$ is small in the L^α norm. However, for a bounded domain, some devices must be introduced in order to obtain estimates, like (15), not obtainable from (14) alone.

Let us return to the case $\Omega = R^n$. One can show that the following Sobolev type inequality

---

$(^1)$ The author is grateful to P. Secchi, for calling his attention to this property.

$$N_\alpha(v) \geqslant c|v|_{\alpha n/(n-2)}^\alpha, \tag{21}$$

holds. On the other hand, by interpolation,

$$|v|_\alpha \leqslant |v|_2^{4/(4+(\alpha-2)n)} \; |v|_{\alpha n/(n-2)}^{((\alpha-2)n)/(4+(\alpha-2)n)},$$

since $2 < \alpha < \alpha n/(n-2)$. Hence,

$$N_\alpha(v) \geqslant c|v|_2^{-\beta}|v|_2^{\alpha+\beta}, \tag{22}$$

where $\beta = 4\alpha/(\alpha-2)n$. A well known erergy estimate for the solution of (1) says that

$$|v(t)|_2 \leqslant |a|_2, \quad \forall \, t > 0.$$

Consequently,

$$N_\alpha(v) \geqslant c|a|_2^{-\beta} \; |v|_\alpha^{\alpha+\beta}. \tag{23}$$

By replacing $N_\alpha(v)$ in (18) by the right-hand side of (23), and then dividing both sides of the inequality so obtained by $|v|_\alpha^{\alpha-1}$, one gets (in case $f \equiv 0$)

$$\frac{d}{dt} \, |v|_\alpha \leqslant -c_0[c_\mu|a|_2^{-\beta} - \mu^{-(n+2)/(\alpha-n)}|v|_\alpha^\gamma]|v|_\alpha^{1+\beta}, \tag{24}$$

where t_0 and c are suitable positive constants, depending at most on α and n, and

$$\gamma = \frac{2\alpha^2(n-2)}{n(\alpha-2)(\alpha-n)} \; .$$

If

$$\mu^{-(n+2)/(\alpha-n)}|a|_\alpha^\gamma \leqslant \frac{1}{2} \, c_\mu \, |a|_2^{-\beta} \; ,$$

(which is just condition (5)), then $d|v|_\alpha/dt < 0$, $|v(t)|_2$ decreases, and in particular

$$\mu^{-(n+\alpha)/(\alpha-n)}|v(t)|_\alpha^\gamma \leqslant \frac{1}{2} \, c_\mu|a|_2^{-\beta}. \tag{25}$$

Consequently, by using (24) and (25) one gets

$$\frac{d}{dt} |v|_\alpha \leq -\frac{c_0}{2} c\mu |a|_2^{-\beta} |v|_\alpha^{1+\beta}.$$

Formula (6) now follows from standard comparison theorems for ordinary differential equations.

The Case $\alpha = n$.

By setting $\alpha = n$ in equations (16), and (17), one obtains the relations

$$\frac{1}{n} \frac{d}{dt} |v|_n^n + \frac{\mu}{2} N_n(v) \leq \frac{c}{\mu} |v|_{n+2}^{n+2}, \tag{26}$$

and

$$|v|_{n+2}^{n+2} \leq c|v|_n^2 N_n(v), \tag{27}$$

respectively. Hence

$$\frac{1}{n} \frac{d}{dt} |v|_n^n - \frac{\mu}{2} N_n(v) \left[1 - \frac{c}{\mu^2} |v|_n^2\right]. \tag{28}$$

If

$$\frac{c}{\mu^2} |a|_n^2 \leq \frac{1}{2} \tag{29}$$

(which is just assumption (8)), then one easily shows (arguing as before) that

$$\frac{c}{\mu^2} |v(t)|_n^2 \leq \frac{1}{2}, \quad \forall\, t > 0.$$

Consequently, if (29) holds, then

$$\frac{1}{n} \frac{d}{dt} |v|_n^n \leq -\frac{\mu}{4} N_n(v).$$

Now, using the inequality (23) with $\alpha = n$, one obtains

$$\frac{1}{n} \frac{d}{dt} |v|_n^n \leq -c\mu|a|_2^{-4/n+2} |v|_n^{(n+2)/(n-2)},$$

and so

$$\frac{d}{dt} |v|_n \leq -c\mu|a|_2^{-4/(n+2)} |v|_n^{(n+2)/(n-2)}.$$

By the usual comparison theorems for ordinary differential equations, one deduces (11) and hence (12).

References

[1] H. Beirao da Veiga, "Existence and asymptotic behavior for strong solutions of the Navier-Stokes equations in the whole space", to appear.

[2] E.B. Fabes, B.F. Jones, N.M. Riviere, "The initial value problem for the Navier-Stokes equations with data in L^p", Arch. Rat. Mech. Anal., __45__ (1972), 222-240.

[3] Y. Giga, T. Miyakawa, "Solutions in L_r to the Navier-Stokes initial value problem", Arch. Rat. Mech. Anal., to appear.

[4] Y. Giga, "Weak and strong solutions of the Navier-Stokes initial value problem", Publ. RIMS, Kyoto Univ. __19__ (1983), 887-910.

[5] T. Kato, "Strong L^p-solutions of the Navier-Stokes equation in R^n, with applications to weak solutions", Math. Z., __187__ (1984), 471-480.

[6] K. Masuda, "Weak solutions of Navier-Stokes equations", Tohoku Math. J., __36__ 9184), 623-646.

[7] F. Weissler, "The Navier-Stokes initial value problem in L^p", Arch. Rat. Mech. Anal., __74__ (1980), 219-230.

A CONFLUENCE OF EXPERIMENT AND THEORY

FOR WAVES OF FINITE STRAIN IN THE SOLID CONTINUUM

James F. Bell

The Johns Hopkins University
Baltimore, Maryland

TABLE OF CONTENTS

1. Introduction

The widespread erroneous conviction that nearly all plastic deformation in
crystalline solids is viscoplastic, and that Prandtl-Reuss type analyses are to be
preferred, is the consequence of a century of emphasis in one domain where those
assumptions have some support in observation, namely, when the plastic strain components are sufficiently small to be of the same order of magnitude as the initial
linear elastic strain components. The two are non-homogeneously distributed among
the variously oriented crystallites and thus their interaction becomes a source of
complexity.

At finite strain, neither the assumption of viscoplasticity nor the introduction of Prandtl-Reuss constitutive statements squares with careful experiment. In
the present discussion we are concerned only with finite strain: measured strain
components from 1/4% to 60%, measured rigid body rotations from 0 to 45 degrees,
measured rates of strain from 10^{-6} sec^{-1} to 10^{+4} sec^{-1}, and measured constitutive
statements at ambient temperatures from near absolute zero to 95% of the melting

point of the solid.

The context is the experimentally observed unity in the finite strain response of fifty solids. These include seven different crystal structures with a sufficient variety of initial states, alloying mixtures, purities, and thermal, chemical, and mechanical prior histories to define an entire class of ordered solids. Derived from this unity is a compatible, internally consistent continuum theory. Dynamical problems in continuum physics for the large or very large deformation of ordered solids, hence, can be considered from the unified perspective of both experiment and theory.

Herein is a systematic presentation of fourteen specific problems. The confluence of continuum experiment and continuum theory they provide, sums what I believe can be said with confidence on the continuum physics of dynamic plasticity at finite strain when observation and compatible explanation are paired.

Before examining these dynamical problems, a brief reference to the underlying theory obviously is in order.

2. Continuum Experiment vs. Continuum Theory

With T_R, $R^T = R^{-1}$, $V = V^T$, as first Piola-Kirchhoff stress tensor, rigid body rotation, and pure homogeneous deformation respectively, experiment on finite strain plasticity discloses that great order and simplification prevail. One perceives this order not by attempting to relate a Cauchy stress tensor with Cauchy-Green strain tensors, which leads to great complication in the interpretation and unification of data, but by introducing and relating a stress tensor $\sigma = RT_R^{\ T}$ and strain tensor $E = V - 1$ for isotropic solids in which the strain energy depends only on the invariants of the finite strain E. Large plastic deformation in ordered solids is subject to an internal constraint, trace $E = 0$, that characterizes every aspect of experiment. We note that $\sigma = \sigma^T$ and $E = E^T$.

From this internal constraint, trace $E = 0$, we obtain a stress which does no work. The total stress is the sum of this stress and σ. The form of the stress which does no work is obtained from experiment at all strain levels. It is compatible with the internal constraint, trace $E = 0$, and thus is admissible (BELL,1983b). Letting S be this total stress, we have

$$S = \sigma - 1/3(\text{trace } \sigma)1 \tag{1}$$

where, of course, the stress which does no work, $-1/3(\text{tr } \sigma)1$, is not the usual hydrostatic stress. We note that $S = S^T$. The internal constraint, trace $E = 0$, implies a small change of volume, a matter established in exhaustive experiment over two decades ago, (see Bell, 1985, Section A).

At finite strain, constitutive equations for loading and unloading are radically different. This difference is the source of the characteristic permanent deformation. To describe the interaction of loading and unloading waves, we shall have to consider both sets of constitutive equations. First, however, we shall assume that the second invariant of the total stress S is either equal to or greater than zero, i.e., $dT \geqslant 0$, where $T^2 = 2II_S$.

Briefly recapitulating the statements leading to the present theory for finite strain plasticity, (BELL 1979, 1983a, 1983b, 1985), we may write for the strain energy W,

$$dW = \text{trace } T_R^T \, dF. \tag{2}$$

From the polar decomposition theorem $F = VR$, where the deformation gradient F has components $\partial x^k / \partial X_M$, with x and X referring to the deformed and undeformed reference configurations respectively, we may rewrite Eq. (2),

$$dW = \text{trace } RT_R^T \, dV + \text{trace } T_R^T \, VdR. \tag{3}$$

For _isotropic_ solids in which the strain energy W depends only upon the invariants of the strain E, the second term, trace $T_R^T \, V \, dR$, vanishes. (See BELL, 1983b, Section 5.)[*] That ordered solids are isotropic for the domain of _finite_ plastic strain is amply demonstrated by direct experiment. (See, for example, BELL, 1985, Section B. It is irrelevant that for some of these solids which are isotropic at finite plastic strain, elastic constants at infinitesimal elastic strain are anisotropic.) It was during the 1960's, that experiment in my laboratory established that the strain energy W depended only on the invariants

---

[*] Again I should like to acknowledge my indebtedness to Jerald Ericksen for clarifying these matters.

of E. The observed parabolicity relating functions of stress and strain invariants, which establishes this dependence, was a driving force in the search for a plausible and comprehensive continuum theory to describe finite strain plasticity in ordered solids. (See BELL, 1985, Sect. C.)

With the second term in Eq. (3) equal to zero, with our definition of $\sigma = RT_R^T$ given above, and with $dV = dE$, we may write Eq. (3) as Eq. (4)

$$dW = \text{trace } \sigma \, dE = \text{trace } S \, dE, \tag{4}$$

or from the internal constraint trace $E = 0$, as Eq. (5)

$$dW = \sigma_{ij} dE_{ij} = S_{ij} dE_{ij} = Td\Gamma, \tag{5}$$

where $T^2 = 2II_S$ and $(d\Gamma)^2 = 2II_{dE}$.

For $dT \geqslant 0$, and for loading paths of arbitrary composition and direction, diverse experiment on 50 ordered solids, high strength metal alloys and pure metals, have provided the ratios of Eq. (6) for the components of S and dE,

$$dE_x/S_x = dE_y/S_y = dE_z/S_z = dE_{xy}/S_{xy} = dE_{yz}/S_{yz} = dE_{zx}/S_{zx} = d\Gamma/T = 2dT/\beta_a^2 \tag{6}$$

where β_a is a measured material constant designating a series of relative reference configurations found from experiment. Specified in such terms are the stable and unstable aspects of a quantum structure implicit in the finite plastic strain of all 50 of the ordered solids thus far considered in the laboratory. At infinitesimal plastic strain, the ratios of Eq. (6) in the present theory of finite strain include classical theory as an approximation.

From the normality condition implied in Eq. (6) one obtains from experiment not only the parabolic relation between the second stress and second strain invariants, Eq. (7), but also, for $dT \geqslant 0$, the incremental constitutive equations, Eqs. (8). Equations (7) and (8) unite an internally consistent continuum theory with global experimental results for finite strain in the class of ordered solids.

$$d\Gamma = 2TdT/\beta_a^2. \tag{7}$$

$$dE = 2S \, dT/\beta_a^2. \tag{8}$$

In terms of the first Piola-Kirchhoff stress tensor, the equations of motion are given in Eqs. (9), where b are the body forces,

$$\partial T_R / \partial X + \rho_R b = \rho_R \ddot{x}. \tag{9}$$

Since $\sigma = RT_R^T$ with $\sigma = \sigma^T$, we obtain for the first Piola-Kirchhoff stress tensor, $T_R = \sigma R$. Equations (9) thus become

$$\partial \sigma R / \partial X + \rho_R b = \rho_R \partial v / \partial t, \tag{10}$$

where v are the particle velocities, and ρ_R is the mass density in the undeformed reference configuration.

3. The Quantized Structure of Relative Reference Configurations

The undeformed reference states for the large deformation of ordered solids are specific. They are a quantized series of related, relative reference configurations. The metallurgical recipes which produce high strength metal alloys are in fact recipes for stabilizing the solid in one of these quantized states.

The importance of this discovery for the finite deformation of ordered solids, was recognized at its inception in the early 1960's; at that time the phenomenon appeared to be only a second order transition structure relating the large deformation of the fully annealed polycrystal and the corresponding single crystal. Inasmuch as over twenty papers, a _Springer Tract in Natural Philosophy_ (BELL, 1968a), and a part of Chapter 4 of my _Handbuch der Physik_ treatise (BELL, 1973) were devoted to the analysis of hundreds of measurements of this particular phenomenon, I shall utilize here only a few of the results.

The measured material constant, β_a, may be written as Eq. (11) to reflect these observations:

$$\beta_a = \lambda_N^{3/2} \beta, \tag{11}$$

where $\lambda_N = 1 + \varepsilon_N$ is a parameter designating the series of relative reference configurations and β is the measured material constant for the fully annealed parent polycrystalline solid and/or corresponding single crystal in a "genesis" reference configuration for which $\lambda_N = 1.000$; other states in the quantum structure are referred to this genesis reference configuration. The value of Γ_N from which, since trace $E = 0$, $\varepsilon_N = [2/3]^{1/2}\Gamma_N$ is derived from the measured values of the second strain invariant,

$$(II_E)_N = 1/4(2/3)^N, \quad \text{or} \quad \Gamma_N = (2)^{-1/2}(2/3)^{N/2}, \tag{12}$$

with N found in experiment to have the values $N = 0,2,4,6,8$, for the structural metal alloys of steel, aluminum, and copper. For annealed solids, where $N = 0,2,4,6,8,10,13$, and 18, the material constant is given by

$$\beta = (2/3)^{r/2}\mu(0)B_0(1 - \theta/\theta_M). \tag{13}$$

The deformation mode index r defines the sequence $r = 1,2,3,4\ldots$ for the quantized slopes which characterize the response function on either side of the second order transitions in the fully annealed solid; $\mu(0)$ is the shear modulus at $0°K$; $B_0 = 0.0207$ is a dimensionless universal constant for 50 ordered solids; and θ and θ_M are the ambient temperature and melting point, respectively.

For the purpose of discussion, let the pure homogeneous strain for the genesis reference configuration where $\lambda_N = 1.000$ be designated as $\overline{V}$. The study of the response to proportional and non-proportional loading paths in twenty structural metal alloys has provided the relative reference sequence of

$$V = \lambda_N^{-1}\overline{V}. \tag{14}$$

Correspondingly, we have $dE = \lambda_N^{-1}d\overline{E}$ and hence $d\Gamma = \lambda_N^{-1}d\overline{\Gamma}$. From $F = VR$ we obtain $R = V^{-1}F$. Substituting in the definition of $\sigma = RT_R^T$, we obtain $\sigma = \lambda_N\overline{\sigma}$, and consequently $S = \lambda_N\overline{S}$ and $T = \lambda_N\overline{T}$. Writing Eqs. (7) and (8) for the genesis reference configuration, we have

$$d\overline{\Gamma} = 2\overline{T}d\overline{T}/\beta^2 \tag{15}$$

and

$$d\overline{E} = 2\overline{S}\,d\overline{T}/\beta^2. \tag{16}$$

Substituting dE, $d\Gamma$, σ, S and T fcr a relative reference configuration in terms of the "genesis" configuration defined above, we obtain Eq. (7) relating the second invariants of the strain and the total stress, and the incremental constitutive equations (8) defining the entire class of ordered solids. In Eqs. (7) and (8) β_a is given by Eq. (11).

Relative reference configurations for stable states thus far have been identified definitively for five of the eight known critical strains, namely, for $N = 0,2,4,6,$ and 8. These five experimental values of λ_N, obtained from the statistical analysis of a very large amount of data, are: 1.577, 1.385, 1.260, 1.170, and 1.114, respectively.

These parameters for metal alloys would be only a convenient empirical arrangement were it not for the following experimental fact: the series of five values of λ_N in ordered solids are row found, twenty years after their discovery as unstable strains in annealed solids, to be such that a unit cube in the "genesis" reference configuration is 1/4, 3/8, 1/2, 5/8, and 3/4 of the volume, $[\lambda_N]^{-1/3}$, of the respective reference cube in the quantized relative reference configurations. (See BELL, 1979, 1983b.)

One other aspect of the response of high strength metal alloys should be mentioned, since it also appears in dynamical problems in such solids. Between the inner yield surface, T_Y, and an upper bound, T_c, the initial plastic deformation for an "intermediate plastic region" is given by the present theory for large deformation, but the origin for the parabolicity in equations (7) and (8) is at the elastic limit T_Y and Γ_Y. At the bound T_c there begins a "totally plastic region", also described by the present theory with the same value of the material constant β_a found in the intermediate region. In this domain the origin of parabolicity in Eqs. (7) and (8) abruptly shifts to the strain abscissa at a point located a distance Γ_N from the generalized strain Γ_c at the bound T_c. This totally plastic region continues to failure. (We again note that $\Gamma_N = \overline{\Gamma}_N/\lambda_N$.)

The generalized theory applies to both proportional and non-proportional loading paths, as has been amply demonstrated in experiment. For our analysis here of problems in dynamics of the continuum, we may confine the discussion to the proportional loading situation of Eq. (17) for the intermediate region, and Eq. (18) for the totally plastic region.

$$T = T_Y + \beta_a (\Gamma - \Gamma_Y)^{1/2} \qquad T_Y \leqslant T \leqslant T_c, \qquad (17)$$

$$T = \beta_a (\Gamma - \Gamma_c + \overline{\Gamma}_N / \lambda_N)^{1/2} \qquad T \geqslant T_c. \qquad (18)$$

For $\Gamma = \Gamma_c$

$$T_c = \beta_a (\overline{\Gamma}_N / \lambda_N)^{1/2}. \qquad (19)$$

From II_E we find that $\overline{\Gamma}_N = (3/2)^{1/2} \varepsilon_N$. Since $\lambda_N = 1 + \varepsilon_N$, and, since T_c and β_a are readily measured in a T^2 vs. Γ plot, λ_N is specified in any measurement of large deformation in a metal alloy or hardened metal.

Obviously, from the above, we see that $dW = \overline{T} d\overline{\Gamma} = T d\Gamma$. For proportional loading this becomes $W = \int \overline{T} d\overline{\Gamma} = 2/3\ \overline{T}\ \overline{\Gamma}$. Substituting β_a from Eq. (11) into Eq. (19) we find that $T_c = \lambda_N \overline{T}_N$. By definition $\Gamma_N = \overline{\Gamma}_N / \lambda_N$, hence we obtain the important energy equivalence of Eq. (20).

From all this, what we learn is that the metallurgical recipes that produce a stable structural metal alloy may be conceived as equivalent to loading the "genesis" parent metal to, and stabilizing it at, the transition strain $\overline{\Gamma}_N$ and transition stress $\overline{T}_N$. From the quantized structure for proportional loading in the genesis reference configuration where $W_N = (2/3)\beta\ (\overline{\Gamma}_N)^{3/2}$, the strain energy W_N for these stable states in the finite strain of the high strength structural metal alloy is given by

$$W_N = (2/3)\ \overline{T}_N \overline{\Gamma}_N = (2/3)\ T_c \Gamma_N. \qquad (20)$$

From Eq. (12) $\overline{\Gamma}_{N+2} = (2/3)\ \overline{\Gamma}_N$, we note further that

$$W_{N+2} = (2/3)^{3/2} W_N. \qquad (21)$$

Equation (21) represents a series of energy levels for $N = 0,2,4,6$, and 8, that characterize the known stable states in a quantum structure for ordered solids in the continuum.

4. On The Second Yield Surface in Finite Strain Plasticity

The existence of two yield surfaces was unknown in classical plasticity. Denoting the classical yield surface, or bound on linear elasticity, as an "inner" yield surface, the "outer" yield surface is one established by the last maximum of the second total stress invariant II_S. It has the property of retaining a memory of II_S for any arbitrary unloading-reloading history.

At the inner yield surface, plastic strain components are small, of the order of magnitude of linear elastic strain components. Hence the inner yield surface, despite its importance in infinitesimal strain plasticity, is of little interest in the present study of the finite strain domain. On the other hand, when the loading path re-enters and extends beyond the outer yield surface, $T = (S_{ij}S_{ij})^{1/2}$, which contains the previous maximum of the second strain invariant, $T = (2II_S)^{1/2}$, the finite deformation again continues. The outer yield surface thus is of prime importance for the study of finite strain plasticity. (MOON, 1973, 1975, 1976; PHILLIPS AND MOON, 1977; PHILLIPS AND LEE, 1979; BELL, 1983a; PHILLIPS, 1984.)

Whatever are the unloading and reloading histories inside the outer yield surface, or wherever occurs the re-entry point, or whatever is the direction of re-entry, the present theory, Eqs. (7) and (8), applies in close detail after the stress path has crossed and continued beyond the outer yield surface.

5. Dynamical Problems in Continuum Physics for Finite Strain Plasticity

In Appendix I are specific references containing experimental data and analysis for each of the following problems, from I through XIV.

I. One Dimensional Waves of Finite Strain: Uniaxial Stress (Compression)

For proportional loading in uniaxial stress, σ_1, with $R = 1$, and $b = 0$, Eqs. (10), become

$$\rho_R \, \partial v / \partial t = \partial \sigma_1 / \partial X = \{d\sigma_1/dE_1\}\{\partial E_1/\partial X\}. \tag{22}$$

Referring to the stress and strain invariants II_S and II_E for uniaxial stress, either Eq. (7) or (8) becomes

$$\sigma_1 = (3/2)^{3/4} \beta_a E_1^{1/2}. \tag{23}$$

Combining with Eq. (22), where v is the particle velocity in the X direction, we obtain

$$\rho_R \, \partial v / \partial t = (3/2)^{3/4} \beta_a (2E_1)^{-1} \partial E_1 / \partial X. \tag{24}$$

In the waste of war in the early 1940's, five different theorists in three different nations independently provided solutions for Eq. (22).

Wave speeds $C(E_1)$ and particle velocities $v(E_1)$ are specified as known functions either of strain or of stress. In terms of strain for the present theory, Eqs. (7) and (8), we have

$$C(E_1) = \{(3/2)^{3/4} \beta_a / 2\rho_R\}^{1/2} \{E_1\}^{-1/4}, \tag{25}$$

and

$$v(E_1) = \{(8/9)(3/2)^{3/4} \beta_a / \rho_R\}^{1/2} \{E_1\}^{3/4}. \tag{26}$$

By 1948 the secrecy surrounding such solutions finally had terminated, and the challenge to obtain direct experimental evidence could begin. Neither the post-deformation measurements of Pol Duwez, D.W. Wood, and D.S. Clark between 1942 and 1947, nor my incremental wave measurements between 1949 and 1951, provided such direct evidence. With the development of an entirely new experiment in 1956, using high density diffraction gratings to measure the profiles of finite strain waves (BELL, 1956), there was strong laboratory evidence for the solutions of Eq. (22). By 1960, for a free-flight, symmetrical impact experiment, a density in excess of 30,000 lines per inch in a ruled diffraction grating, made laboratory evidence definitive: a one dimensional theory for which stress is a single valued function of strain does indeed apply to the finite deformation of crystalline solids. (BELL, 1960b.)

In this experiment, two identical cylinders of a known solid are in axial collision. All measurements are made during free flight. They include not only

all of the detail of the wave front but also maximum values. Analysis of the
latter added emphasis to the accord and established bounds for the theory.

Finite strain and surface angle are determined by optically observing the
variation of the density of 30,000 lines per inch ruled diffraction gratings
caused by the wave; particle velocities, by measuring the induced voltages in fine
wires traversing a calibrated magnetic field; displacements, by optical sensing
of the motion of the boundary between reflecting and non-reflecting surfaces; and
maximum stresses, either by observing very thin piezo-crystals responding to
pressure on the impact face of the struck specimen or by observing a conven-
tionally instrumented hard load bar in axial impact with an annealed specimen of
the same solid. By the use of quartz windows in specially constructed furnaces,
finite strain is measured by diffraction gratings at ambient temperatures to
within 95% of the melting point θ_M in Eq. (13) (BELL, 1962a). Without introducing viscoplasticity, constitutive statements for finite strain measured at
strain rates of 10^{-6}sec^{-1} still apply when finite strain is measured at strain
rates of 10^{+4}sec^{-1}; i.e., the constitutive statements apply over a range of
strain rates of 10^{10}.

In the 1960's, such wave propagation measurements on the annealed metals --
aluminum, copper, zinc, alpha brass, magnesium, iron, and lead -- demonstrated the
universality of the agreement between experiment and the non-linear, one dimensional wave theory. Whatever is the diameter of the specimen, by a distance the
equivalent of one diameter length from its impact face the collapse of the initial
uniaxial plane strain front produces an uniaxial plane stress wave (Section X
below). At that distance from the impact face, the resultant shape of the strain
wave profile, as in Fig. 1, becomes the initial condition for the subsequent
uniaxial stress wave. The dispersion of this uniaxial stress wave in the next few
diametral lengths along the bar has been the focus of experimental study.

Indicative of the universality of Eqs. (13), (25), and (26), the solid lines
in Fig. 1 are wave profiles from diffraction grating measurements on annealed
zinc; the crosses are the calculation from Eq. (13) from tests in annealed aluminum at the same locations. Knowing θ_M, ρ_R, r, and $\mu(0)$ for both aluminum and
zinc permits the comparison shown in Fig. 1.

Fig. 2 is a single load bar measurement on annealed copper; it illustrates that the measured maximum stress is in close correspondence with Eqs. (13), (23), and (26). In Fig. 2 the mode index $r = 4$ for annealed copper at strain rates of 10^{+3}sec^{-1} provides, from Eq. (13), the identical parabola coefficient, $\beta = 37$ kg/mm^2, which characterized all of the many measurements in this solid at strain rates of 10^{-4}sec^{-1}, described in (BELL, 1983b), and in (BELL and KHAN, 1980).

II. One Dimensional Waves of Finite Strain: Uniaxial Stress (Tension)

For one dimensional uniaxial stress, Eqs. (22) through (26) make no distinction, other than the positive or negative sign, between uniaxial compression and uniaxial tension. Equation (23) is an odd function. In <u>quasi-static</u> measurement, with σ_1 and E_1 determined for uniaxial stress from the invariants of II_S and II_E in the present theory, experiment has established that there is no difference other than sign. Tension and compression response functions are identical as components of the stress tensor σ and the strain tensor E. (BELL, 1973, section 4.19.)

Between 1970 and 1972, Akhtar S. Khan undertook the difficult experimental problem of extending this quasi-static study to the finite strain domain at the high strain rates of dynamic plasticity. Using the same high density diffraction grating apparatus described in Section I above, he succeeded in obtaining precise optical data on tensile wave profiles. He simultaneously induced finite strain waves of tension in the tubular section of a specimen, and compression waves at finite strain in a solid section of the same bar, (KHAN, 1973, 1974). This study reveals that wave speeds and maximum strains are in close accord with the above theory for tension and for compression waves propagating in opposite directions in the same specimen. Eqs. (22) through (26) of the one dimensional theory describe the observed response for both tension and compression. The material constant β_a and the response function $\sigma(E)$ are the same for both positive and negative stress, a fact having important significance for any theory of finite strain, particularly in view of the long history of contrary conjecture.

III. On the Interaction of Waves of Loading and Unloading

Beyond the region of the uniaxial stress wave is a region where the linear elastic precursor, reflected from the free end as an unloading wave, interacts with the dispersive non-linear loading wave. E.H. Lee (LEE, 1953), using the method of characteristics, examined this problem for an arbitrary one dimensional response function, $\sigma_1(E_1)$. Lee assumed linear elastic constitutive equations for unloading. He found that for a stress-strain function concave to the strain axis, the reflected unloading wave would be absorbed as indicated in

$$\sigma_b - \sigma_a = \sigma_Y - (1/2)(\rho_R C_o v_b - \sigma_b) \tag{27}$$

where σ_b and v_b refer to the values of the stress and particle velocity before wave interaction, and σ_a refers to the stress after interaction; σ_Y is the elastic limit of the solid (inner yield surface); ρ_R is the mass density in the undeformed reference configuration; and $C_o = (E/\rho_R)^{1/2}$ is the bar wave speed, with E the elastic bar modulus of the solid. As v_b increases, the right hand side of Eq. (27) vanishes, and $\sigma_a - \sigma_b = 0$. Thus the unloading wave precursor seems to be absorbed. In fact, of course, it is reflected back to the free end by the opposing dispersive finite amplitude strain wave front.

In the present theory $v_b(\sigma_b)$ is exactly defined. Combining equations (23) and (26), we have

$$v_b = \{(8/9)/\rho_R\}\sigma_b^{3/2}\{(3/2)^{3/4}\beta_a\}^{-1}. \tag{28}$$

Using Eq. (28), one can determine the position along the impacted bars where this total absorption occurs. For annealed 1100-0 aluminum at high strain rates, $r = 2$ in Eq. (13), (BELL, 1985 Figs. 11 and 12b). With other parameters known, $\beta = 30.5$ kg/mm^2. The measured dynamic elastic limit is $\sigma_Y = 0.7$ kg/mm^2. Introducing these quantities into Eq. (27), the location for complete absorption is a distance of 0.43 of the length, L, of the rod. One of those rare moments of exultation in the laboratory occurred in 1960 when a measurement of strain histories at 0.6L, 0.5L, and at 0.4L revealed that the reflected wave in this solid, as shown in Fig. 3 from those data, did indeed disappear precisely where predicted, namely, between 0.4L and 0.5L from the impact face. (BELL, 1961a)

With respect to these experimental data, some years later Nicolae Cristescu and I reported upon a computer calculation of the details of the rest of this loading-unloading wave interaction until the specimens separated. (CRISTESCU and BELL, 1970) The entire process of interaction between loading and unloading waves is compatible with expectation from the present incremental theory of finite strain.

IV. Wave Propagation at Finite Strain in a Relative Reference Configuration

For the experiments described in Sections I and III above, consider a rod of a very high strength steel alloy with an unusually high elastic limit, $\sigma_Y = 151$ kg/mm^2 (215,000 psi). For the experiments described in I and II above, the observed strain rates, $\dot{E}_1$, ranged from 10 sec^{-1} to 10^{4}sec^{-1}. The data for an instrumented, high strength steel alloy rod were determined from directly measured wave profiles with a maximum measured strain rate of $\dot{E}_1 = 6 \times 10^4$sec^{-1}. This rod collided axially with a steel block at an impact velocity of 1 km/sec (2,237 miles per hour). The data are those of G.E. Hauver (HAUVER, 1978) who measured wave profiles by telemeter at three distances from the impact face from which he determined wave speeds $C(E_1)$. Assuming the one dimensional theory, Eq. (22), for an unknown response function, Hauver substituted these measured wave speeds into the expression relating wave speed, stress, and strain obtained from the integration of $C(E_1) = \{(d\sigma_1/dE_1)/\rho_R\}^{1/2}$,

$$\sigma_1 = \int_{E_Y}^{E_1} \rho_R\, C(E_1)^2 dE_1 + \sigma_Y. \tag{29}$$

Similarly, by the integration of Eq. (30), he determined particle velocities $v(E_1)$ from measured wave speeds $C(E_1)$ when $\sigma_1(E_1)$ is assumed to be a single valued unspecified function of strain:

$$v(E_1) = \int_{E_Y}^{E_1} C(E_1) d\, E_1 + v_Y, \tag{30}$$

where $v_Y = \sigma_Y/\rho_R C_o$. The curves Hauver obtained by integrating Eqs. (29) and (30) in 1978, are compared with the above theory. (See Figs. 6(a) and 6(b) in BELL,

1985.) At room temperature, this steel (BELL, 1985, Section VI) deforms in a relative reference configuration for which $N = 0$, and $r = 1$. With $\mu(0) = 8580$ kg/mm^2 for the "genesis" parent metal iron, and a melting point of $\theta_M = 1809°$K, at room temperature Eq. (13) provides a parabola coefficient of $\beta = 121$ kg/mm^2. The relative reference configuration determined from low strain rate quasi-static tests, for $N = 0$, is $\lambda_N = 1.577$. Substituting into Eq. (11), we obtain $\beta_a = 240$ kg/mm^2. With this value of β and from the second invariants for stress and strain $T = (2II_S)^{1/2}$ and $\Gamma = (2II_E)^{1/2}$, i.e., $\sigma_1 = (3/2)^{1/2}T$, and $E_1 = (2/3)^{1/2}\Gamma$, for metal alloys we may write Eq. (17) in terms of the uniaxial stress and strain σ_1 and E_1, for proportional loading in the intermediate region:

$$\sigma_1 = 151 + 326(E_1 - 0.007)^{1/2} \text{ kg/mm}^2 \tag{31}$$

where $E_Y = 0.007$ is the measured axial strain at the elastic limit or inner yield surface $\sigma_Y = 151$ kg/mm^2. Similarly, with β_a known, one can determine $v(E_1)$ from Eq. (26).

To measured strains of 16% at strain rates of 6×10^4, the stress vs. strain response, $\sigma_1(E_1)$, and the particle velocity vs. strain response, $v_1(E_1)$, are in precise accord with the present theory for the intermediate region of Eq. (17), in the form of the response function of Eq. (31) and the velocity statement of Eq. (26). The comparison of theory and measurement emphasizes that independent of strain rate, whether for a wave propagation or a quasi-static situation, one needs only to determine the relative reference configuration, λ_N, from a simple quasi-static test (BELL, 1979, 1985).

V. Finite Strain Waves at a Linear Elastic-Plastic Boundary

The convenient mathematical concept of a fixed boundary is an inconvenient fiction in the laboratory. This is particularly evident when finite amplitude waves are involved; experiment reveals that their behavior is very sensitive to local conditions. To study approximate "fixity", is to consider a situation in which one side of a boundary remains in a linear elastic state in infinitesimal strain, while the other sustains waves of large finite strain in parabolic plasticity. For linear elasticity, the uniaxial response function is $\sigma = E\epsilon$, with E the bar modulus and ϵ the infinitesimal strain. For a fully annealed bar of the same solid in the genesis reference configuration, the uniaxial response function for parabolic plasticity is given by Eq. (23). That is, $\sigma_1 = (3/2)^{3/4} \beta E_1^{1/2}$, where $\lambda_N = 1.000$.

The momentum statement for the maximum stress and particle velocity, v, for a rectangular wave front in a linear elastic bar is

$$\sigma = \rho_R C_o v. \tag{32}$$

Combining Eqs. (25) and (26) with $\lambda_N = 1.000$, the momentum statement for a dispersive wave front in a parabolic plastic bar at finite strain, is

$$\sigma_1 = (3/2)\rho_R C(E_1) v. \tag{33}$$

Let a one dimensional, approximately rectangular wave pulse, with an amplitude σ_I, be induced in the linear elastic bar. The direction of propagation is toward the elastic-plastic boundary. The parabolic plastic bar is on the same axial line and is in contact with the linear elastic bar. From the present theory, if one dimensional waves of finite strain prevail, we should have the following boundary conditions, where σ_R is the amplitude of the reflected wave in the elastic bar and σ_T is the amplitude of the transmitted wave in the plastic bar, with the corresponding designations for maximum particle velocities, v_R and v_T:

$$\sigma_I - \sigma_R = \sigma_T \tag{34}$$

and

$$v_I + v_R = v_T. \tag{35}$$

Substituting Eqs. (32) and (33) in Eqs. (34) and (35) and eliminating σ_R, we obtain Eq. (36),

$$\sigma_I = \{1/2 + C_o/3C(E_1)\}\sigma_T. \tag{36}$$

With σ_I assigned, C_o known, and $C(E_1)$ as a function of σ_1 also known from

Eqs. (23), (25) and (26), σ_T can be determined as a function of σ_I from Eq. (36).

In Fig. 4 is a Lagrangian diagram of the entire <u>measured</u> history of the response, including loading and unloading in each bar. The initial wave approaching the boundary in the linear elastic bar is the rectangular pulse shown. (These details are given in BELL, 1968b) (Also see DILLON, 1967.)

VI. Non-Symmetrical Impact: the Analysis of an Historical Problem

For over a century, measurements on the impact of solids have been beset with unsubstantiated assumptions. Using the diffraction grating method of measuring strain, I have explored non-symmetrical impact phenomena whose history of accompanying <u>ad hoc</u> assumption goes back to the middle of the 19th century. A cylindrical bar of one solid, labeled the projectile or impact bar, strikes a stationary or struck bar of the second solid. The measurements are made on the struck bar. Assuming one dimensionality, the boundary conditions become

$$\sigma_1 = \sigma_2 \quad \text{and} \quad v_H - v_1 = v_2, \tag{37}$$

where v_H is the projectile velocity prior to impact. (BELL, 1969).

In the terms described above, situations for which continuum theory and experiment are in accord are:

1) Same solid, but projectile (#1) at low (room) temperature; struck specimen (#2) at high temperature (say 1000°F).

2) Projectile, a linear elastic solid (#1); struck specimen (#2), a parabolic plastic solid of the same or different composition. (see Fig. 4 above.)

3) Projectile, (#1), a parabolic plastic solid; struck specimen (#2), a linear elastic solid of the same or different composition. (see Fig. 2 above.)

4) Both specimens parabolic plastic but different solids with different values of $\mu(0)$, θ_M and r in Eq. (13), and ρ_R in Eqs. (25) and (26).

5) Projectile, a single crystal of known orientation (#1); struck specimen (#2) a linear elastic polycrystal.

6) Projectile, a linear elastic polycrystal (#1); struck specimen (#2), a single crystal of known orientation.

7) Projectile, a single crystal of one orientation (#1); struck specimen (#2), a single crystal of the same solid with a different orientation.

The initial response can be markedly different in the two specimens. In the first 5 or 10 microseconds there may be differences in the maxima of the initial dilatational waves and the timing of their collapse. In addition, there may be differences in phase when the one dimensional uniaxial stress waves develop from sidewall reflections in the first diametral length from the impact face. Either of these possibilities can provide an unpredicted division of energies between the two specimens before one-dimensional uniaxial stress waves are formed. It is not sufficient to attempt to determine the division only from a knowledge of the velocity of the projectile prior to impact.

Laboratory efforts have addressed every case from (1) to (7) above. A partial summation of these data is given in (BELL, 1969). <u>With</u> <u>one</u> <u>exception,</u> after the first five or ten microseconds, experimental observations for <u>all</u> of these non-symmetrical impact situations are in detailed accord with the incremental theory of finite strain presented above.

The exception in the above group is for situation (4), but only when there is a large difference in the mass densities of the two plastically deforming bars. Fully annealed copper specimens and fully annealed aluminum specimens alternated as projectile and struck specimen, is the case most studied (BELL, 1969). A similar situation has been found for other highly different mass densities, such as magnesium vs. aluminum or zinc vs. aluminum in non-symmetrical impacts.

VII. The Incremental Wave in Pre-Stressed Solids: Metals

The first experiment to measure incremental waves in a pre-stressed metal bar or string was performed in 1949 on a long rod of steel. (BELL, 1951). Incremental waves in pre-stressed rubber had been studied in the 1870's. (See BELL, 1973, Section 3.33.) The non-linear theory, Eq. (22), predicted that the wave speeds of a dispersive plastic wave would be given by the slopes of some unspecified uniaxial response function for which stress is a single valued function of strain.

My idea in 1948 was to study an incremental wave on a bar under a pre-stress. A six foot long mild steel bar was loaded quasi-statically to failure in a specially designed testing machine. (See BELL, 1973, Section 4.27) At intervals during the loading, small amplitude incremental loading waves were sent down the rod. These incremental waves always passed through a "plastic sea" in which an "elastic island" was instrumented to record arrival times and thus provide a measure of wave speeds. At every level of pre-stress, I found that the small incremental waves do <u>not</u> have wave speeds given by the tangent moduli of any non-linear response function for large plastic deformation, as would be anticipated by the non-linear theory. At every level of pre-stress the incremental wave propagated with a velocity approximating the linear elastic bar velocity, C_o, where $C_o = (E/\rho_R)^{1/2}$.

During the next few years, similar results were found in copper, lead, and aluminum. By 1962, with diffraction gratings used to determine strain, it became possible to study the problem for a dynamic pre-stress. Incremental waves were superimposed upon an initial large amplitude wave which served as the pre-stress. The incremental wave was introduced behind the front of the main wave so that the passage of both waves could be observed. By changing the location of the measurement, the proximity of the two wave fronts could be altered until they met.

In the early 1970's, I developed an experiment that increased the magnitude of the incremental wave; I found that only the initial leading tit, or precursor of the wave, proagates at the velocity of the linear elastic bar. (BELL, 1973, Section 4.34) The rest of the wave propagates at the far slower plastic wave speeds. One unanticipated observation was that the incremental wave does not pass through the main wave front as some theorists had conjectured it should, but, rather, on arrival, it becomes absorbed as part of that front. Such a larger amplitude incremental wave on top of a dynamic pre-stress from this later series is shown in Fig. 5.

The "staircase" phenomenon is seen only in dead weight loading; its vertical steps have been known for a century and a half. Savart, in 1837, suggested the importance of the phenomenon. In my opinion, the steps, discussed in Section XIII below, are responsible for the high velocity of the precursor of the incremental wave. Dead weight loading dominated 19th century testing and is rarely used in the 20th century. The popular simplified relaxation oriented interpretations of three decades ago as to the source of the high velocity for incremental waves, have not withstood the test of time and further experiment. No large elastic front generating plastic deformation behind it as a relaxation phenomenon has been directly observed by anyone.

VIII. The Incremental Wave in Pre-Stressed Solids: Rubber

An important improvement on a late 19th classic, Franz Exner's incremental wave experiment on a rubber string in 1874, was described by H. Kolsky in a brief note in the journal, Nature. (KOLSKY, 1969) In such experiments on rubber, a long (say 10 feet) rectangular cross-sectioned rubber string is initially stretched to one, two, three, or even four times its undeformed length. A small section at one end is subjected to a prescribed additional tension. The fusion of a metal wire introduces an incremental tension pulse in the string. Imbedded small wires attached to the rubber string at prescribed locations, move through a calibrated magnetic field and provide a detailed velocity profile of the non-linear wave front at each of four positions.

Kolsky's observation qualitatively exhibited the growth of a shock front in a progressive wave in a solid. (KOLSKY, 1969) The wave is developing into a shock as it propagates along the string. No analytical discussion accompanied Kolsky's brief presentation of those data in 1969 or Exner's in 1874. In 1978 R. Khanwalkar corrected the omission with an analysis of both experiment and theory (KHANWALKAR, 1978).

Since the mid-19th century the quasi-static response of rubber has been known to have an inflection point separating an initial portion concave toward the strain axis from a second portion concave toward the stress axis. Khanwalker, in an elegant study of the corresponding dynamical problem, discovered a similar stress-strain function for the dynamic response of rubber string. However, the location of the inflection point differs from that in quasi-static measurement. He found that the measured waves speeds were constant for a given particle velocity, and hence the resulting dispersion demonstrated the applicability of the one dimensional, non-linear wave equation, Eq. (22), on both sides of the inflection point.

The governing stress-strain function is such that stress is either a "hard" or "soft" function of strain. For a pre-stress below the inflection point, the wave front did not steepen; it spread out. For pre-stresses above the inflection point the wave front steepened toward a shock front in the manner indicated by Kolsky in 1969 and Exner in 1874. In both instances, the wave speeds for a given value of particle velocity were constant. Hence, from measuring the wave speeds, a dynamic stress strain curve was determined on either side of the inflection point.

In Fig. 6 are shown Khanwalkar's wave speeds for an incremental tensile wave of 6%, traveling between 123 cm. and 222 cm. from its source. The pre-stretch is λ = 2.81. The change of slope in this C vs. v response is characteristic of an inflection point. For larger pre-stretches from λ = 3.34 to λ = 4.27, the constant wave speeds provide only a developing shock front.

Khanwalkar used thermocouples to measure the temperature profiles of the incremental wave in the rubber string. He observed a curious large delay for the thermal wave front behind the corresponding particle velocity wave front. This, of course, is of importance for further research on the thermodynamics of the very rapid loading of rubber.

IX. The Surface Angle Measurement of Large Radial Motion for Waves in Rods

The internal constraint, trace E = 0, tells us that the plane axial wave front in a rod is accompanied by a radial component with a strain amplitude at least half as large. In diffraction grating measurement, the finite strain wave passes through a high density grating and produces a change in the spacing of the grating lines and, correspondingly, a change in the diffraction angles. I had shown (BELL, 1956, 1959) that the determination of the response of two first order diffraction images provides an optically accurate measure of wave profiles of strain and simultaneously of surface angle while the wave propagates along a rod.

Denoting E_R, u_R and R_o as radial strain, radial displacement, and initial radius, respectively, for a cylindrical bar, the surface angle α at a strain amplitude E_1 is given for a compression wave by

$$\alpha = (1 - E_1)^{-1} \partial\, u_R / \partial X. \tag{38}$$

Assume the existence of a plane wave front of uniaxial stress. From trace $\mathbf{E} = 0$, radial symmetry provides $E_R = -E_1/2$. Since $E_R = \partial u_R / \partial R$, we may integrate and obtain

$$\alpha = \{R_o\, \partial E_1 / \partial X\} \{2(1 - E_1)^{-1}\} \tag{39}$$

Since $v(E_1)$ is a single valued function and since compatibility requires that $\partial v / \partial X = \partial E_1 / \partial t$, we obtain

$$\partial E_1 / \partial X = \dot{E}_1 / C(E_1). \tag{40}$$

Substituting in Eq. (39), we have

$$\alpha = \{R_o\, \dot{E}_1 / C(E_1)\} \{2(1 - E_1)\}^{-1}. \tag{41}$$

With R_o given, $C(E_1)$ known from Eq. (25) of Section I, and $\dot{E}_1$ measured at any measured strain amplitude E_1 from a single diffraction grating wave profile, every quantity on the right hand side of Eq. (41) is given by experiment. Since the surface angle α on the left side of Eq. (41) is measurable by the same diffraction grating observation during wave propagation, every quantity in Eq. (41) is determined. The assumption of a one-dimensional, plane wave front thus is amenable to demonstration in experiment. The examples shown in Fig. 7 typify the close correlation between theory and experiment. (BELL, 1968b).

If we had assumed incompressiblity instead of the internal constraint trace $\mathbf{E} = 0$, then, as I showed in Section 4.28 of a treatise on the experimental foundations of solid mechanics (BELL, 1973), the quantity $(1 - E_1)$ in Eq. (41) is replaced by $(1 - E_1)^{5/2}$, a significant difference for strains above 10%. The accurate determination of the profiles of very small surface angle waves introduces a new, measurable variable to consider in general wave theories for solids.

X. Wave Propagation in a Three Dimensional Axially-Symmetric Field

An initial uniaxial strain wave front propagates away from the impact face during the first one or two microseconds. The measurement of stress is made across the impact face itself by a 0.005 inch thick, calibrated quartz piezo crystal cemented to the impact face of the struck specimen. For an _uniaxial strain_ wave, since $E_2 = E_3 = 0$, the internal constraint trace $E = 0$ implies that the wave front is not one in finite strain plasticity but, rather, is one in elasticity. Compatible with the internal constraint trace $E = 0$, but for measured strain wave profiles along a generator on the surface of the cylindrical specimen approximately a centimeter away from the impact face, one may make an _empirical_ correlation (suggested by Jerald Ericksen in 1961) between observation (BELL, 1962b) and certain of C. Truesdell's results in his general theory of waves in non-linear elasticity. (TRUESDELL, 1961, 1962).

The Truesdell theory includes statements for wave speeds for small amplitude waves propagating in a non-linear elastic solid under a state of stress. If this state of stress were purely hydrostatic, the relation between longitudinal wave speeds $U_\parallel$ and transverse wave speeds $U_\perp$ given by Truesdell would be,

$$U_\parallel^2 = (4/3)U_\perp^2 + dp/d\rho. \tag{42}$$

For more general stress states, Truesdell found the transverse wave speed U_{12} to be given by

$$(\rho/v_1^2)U_{12}^2 = (\sigma_1 - \sigma_2)/2(v_1^2 - v_2^2), \tag{43}$$

where $v_i = 1 + \varepsilon_i$ are the stretches in principal directions. The corresponding principal stresses are σ_1 and σ_2.

With $\sigma_2 = 0$, the empirical parallel arises from the fact that when the wave speeds determined for each level of strain from traverse times between 0.63 cm and 1.27 cm are introduced first into Eq. (42) for the shock front and then into Eq. (43) for the rest of the wave front, one obtains the remarkable empirical correlation (shown in Fig. 8) of stress vs. time histories as measured at the impact face (triangles and squares) and as calculated at a distance of 0.95 cm away from the impact face (circles).

To minimize the effects of dispersion, I used 0.025 mm long, 30,720 lines per inch diffraction gratings to obtain strain wave profiles at five positions, from 0.05 cm to 1.27 cm distance from the impact face. The measurements were along a principal axis (BELL, 1962b). P.W. Bridgman's data in aluminum provided $dp/d\rho = 4.25 \times 10^{10}(in/sec)^2$. The measured dilatational wave speed and the measured transverse wave speed are $U_{\parallel} = 3.2 \times 10^5 in/sec$ and $U_{\perp} = 2.12 \times 10^5 in/sec$. Both of them are much higher than those given for this solid at zero pre-stress! At zero pre-stress, $U_{\parallel} = 2.5 \times 10^5 in/sec$ and $U_{\perp} = 1.25 \times 10^5 in/sec$ for aluminum. The ratios of measurements at zero pre-stress to measurements in wave propagation under a pre-stress are 0.78 and 0.59 respectively. In (BELL, 1962b) are the details of the calculation. The circles from the calculated results of Eqs. (42) and (43) indicate later time than the squares and triangles of the piezo-crystal measurement, because a few microseconds are required to traverse the intervening distance.

With every strain-time profile obtained by high density diffraction grating measurement, there is a surface-angle profile as shown in Fig. 7. Whatever may be the distance from the impact face, whatever may be the maximum strain, the crystalline solid, or the magnitude of the surface angle, the surface-angle maximum invariably occurs at an intermediate strain, $E^* = 0.63\ E_{1max}$. (BELL, 1963) Since E^* is independent of the location and maximum strain, it must have its origin in the formation of the initial wave.

For $\lambda_N = 1.000$, the strain energy, Eq. (20) is

$$W = (2/3)\beta(\overline{\Gamma})^{3/2}. \tag{44}$$

Referring to the second stress and strain invariants, II_S and II_E, this becomes for uniaxial stress σ_1 and E_1

$$W = (2/3)(3/2)^{3/4}\beta(E_1)^{3/2}. \tag{45}$$

For the strain at the surface angle maximum we have

$$W^* = (2/3)(3/2)^{3/4}\beta(0.63)^{3/2}(E_{1max})^{3/2}, \tag{46}$$

or, since $(0.63)^{3/2} = 1/2$, there is an <u>equipartition of energy</u>. When the initial uniaxial strain wave collapses to the uniaxial stress wave, sidewall reflections produce, as part of the process, an equipartition of energy between dilatation and shear:

$$W^* = 1/2 \ W_{max}. \tag{47}$$

We shall summarize the findings reported upon in several papers in the early 1960's. In terms of Huygens principle, rays from the impact face arrive at an arbitrary point on the surface. Fcr linear elasticity, when rays from all angles between 0 and $\pi/2$ degrees impinge at a point, the energies of the reflected dilatation and shear waves at a free boundary are equipartioned. (BELL, 1960a, 1961b, 1963). For $E_Y \leqslant E_1 \leqslant E^*$, the shear wave forms the initial portion of the wave profile. The remainder of the front to E_{1max} is delayed until the successive sidewall reflections of the dilatation complete the formation of the uniaxial stress wave. This process has been accomplished at a distance from the impact face equivalent to a one diameter length. For details on these matters, see (BELL, 1960a, 1960c, 1961b, 1962b, 1963, 1973 Section 3.38, pp. 347-351; BELL and SUCKLING, 1962).

XI. The Elastic Limit and Outer Yield Surface in Dynamic Plasticity

On rare occasions one has the interesting laboratory experience that a simple experiment conceived over a weekend and requiring less than a month to bring to fruition can approach in importance, complex experiments which consumed the better part of a decade to perfect. My prime example of one such simple experiment arose from an idea generated in 1960 by my interest in the interaction of loading and unloading waves (Section III above).

A thin line of light of uniform intensity lies in a diameter plane of two cylindrical specimens before, during, and after an axial impact. As the projectile specimen approaches the stationary specimen, the total amount of light decreases. The slope of the intensity vs. time curve provides a direct measure of the relative velocity before impact. In a second interval, namely that between the first contact of the specimens and their first separation, we measure the time of contact, t_c. After the specimens have separated, the increasing intensity of the light is a measure of the relative velocity after impact. One thus obtains a direct measure of the time of contact, t_c, and the coefficient of restitution, e.

Section 3.35 of Handbuch der Physik volume VIa/1 (BELL, 1973), and a subsequent paper (BELL, 1982), call attention to the 19th century observation that in linear elasticity the symmetrical axial impact of two identical rods, t_c decreases with increasing impact velocity.

Those same sources refer to the discovery, also from experiment, (BELL, 1961a, 1961b), that in the finite strain domain, t_c increases with increasing impact velocity. To determine the dynamic elastic limit one need only impact pairs of identical specimens from very low velocities in the linear elastic domain to very high velocities in the parabolic plastic domain. The measured minimum for t_c occurs at a particle velocity, v_Y, at the dynamic elastic limit; the stress at the elastic limit or inner yield surface, σ_Y, immediately follows. (See BELL, 1982.)

That both the inner and outer yield surfaces described in Section 4 above for low strain rates appear in the response at the high strain rates of dynamic plasticity, is also established in this study. (BELL, 1982)

XII. On the Role of Dimensionless Universal Parameters at Finite Strain

Equation (12) and particularly Eq. (13) imply an order of more than casual importance for the finite deformation among crystalline solids. The Springer Tract in Natural Philosophy entitled The Physics of Large Deformation of Crystalline Solids, (BELL, 1968a), was written to describe this discovery. Because of the focus of the present paper, I have chosen an illustrative dynamical problem from that volume.

Repeating Eq. (13), we have for $\lambda_N = 1.000$,

$$\beta = (2/3)^{r/2} \mu(0) B_0 (1 - \theta/\theta_M), \tag{48}$$

for the parabola coefficient in the constitutive equations for the present theory,

Eqs. (7) and (8). Shown in Fig. 1 were the wave profiles of the hexagonal metal, annealed zinc, at three locations from the impact face (solid lines) for the symmetrical, free-flight impact experiment. Similar impact tests, not shown, were performed on the face-centered cubic metal, annealed aluminum, at the same locations. In Eq. (48) for zinc we have a melting point of θ_M = 694°K, and the shear modulus of $\mu(0)$ = 4700 kg/mm^2 at absolute zero, while for aluminum the melting point is θ_M = 932°K and the shear modulus at 0°K is $\mu(0)$ = 3110 kg/mm^2.

The difference in mass densities also is of importance in relating the wave speeds. For aluminum we have ρ_R = 0.000253 lb.sec^2/in.^{4}; for zinc we have the much higher value of ρ_R = 0.000665 lb.sec^2/in.^{4}. The dimensionless universal constant B_o = 0.0207 is the same for both metals, as is the mode index r = 2. Perhaps one picture is indeed worth a thousand words. In Fig. 1, measured diffraction grating profiles of strain in annealed zinc (solid lines) were compared with prediction (crosses) from the present theory (Section I), using Eq. (48) as the common link to determine the response in zinc from the response measured in aluminum.

At 0°K, aluminum and silver have the same shear modulus $\mu(0)$ and the same deformation mode, r = 2. From Eq. (48), i.e. Eq. (13) describing the universality of the present theory, the finite strain response functions of these two metals, which differ markedly at room temperature, should coincide for tests near 0°K. That this is indeed exactly the situation, may be seen in the data from the tests of Fig. 9 at 20°K! The solid line is the predicted common response.

XIII. McReynolds-Dillon Slow Waves During Finite Strain in Ordered Solids

Well before 1850, A.P. Masson, quoted a prophetic pronouncement of his mentor, Félix-Savart. According to Masson, Savart believed that the subject of finite strain plasticity never would be completely understood until one could explain the "staircase phenomenon" observed in dead weight loading of metals. After over a century of developing machines to suppress this phenomenon, or of postulating its existence as due to an unfortunate attribute of testing machines in general, there now is a sizeable, increasing number of individuals who view the

"Savart-Masson" effect as having the importance foreseen in 1837.

Twentieth century studies of this phenomenon in "dead weight loading" are those of A. McReynolds in 1949, O. Dillon in 1962, W. Sharpe in 1966, and my own experiments between 1962 and 1973. (McREYNOLDS, 1949; DILLON, 1962; SHARPE, 1966; BELL AND STEIN, 1962; BELL, 1968a, 1973).

To illustrate the nature of the Savart-Masson "staircase phenomenon" in the dead weight loading of annealed aluminum, Fig. 10 shows a test of mine in 1968. The high velocity of incremental waves in metals (Section VII above), has its origin in the vertical section of the steps (BELL AND STEIN, 1962). An illustration from Dillon's detailed and definitive study of the "slow wave" phenomenon is in Fig. 11. These waves travel at a few centimeters a second. They may propagate in opposite directions from a common source, as shown in Fig. 11, or they may propagate from one end of the specimen to the other, as one may observe in Dillon's data. (DILLON, 1966).

Both Dillon and I attribute the McReynolds-Dillon slow waves in particular, the "Savart-Masson effect" in general, and my incremental waves of 1951 as a special case, to nonhomogeneity in the finite strain of crystalline solids. Our independent research in this vein is tangential to this discussion of purely dynamical problems in continuum physics.

XIV. On the Generation of Radial Shear Waves at a Plastic-Elastic Boundary

In Section V above we considered the response at a boundary of two solids, one of which permitted large finite strain, the other being restricted to infinitesimal linear elastic strain. Since the wave was initiated in the linear elastic bar, the reflection from the boundary occurred in that bar and hence superposition simplified the problem.

A far more complex situation is found when the wave is initiated in the parabolic plastic bar. The reflected loading at finite strain interacts with the remainder of the highly dispersive initial wave. Both are governed by a nonlinear wave equation, and hence superposition no longer applies.

In this plastic-elastic boundary experiment two annealed cylindrical bars are in axial collision, like those in Sections I and III above. Now, however, the struck bar at its far end is in contact with a long, hard, linear elastic bar that remains at infinitesimal strain. During the non-linear wave interaction, measurements along the struck bar are made for finite strain, surface angle, particle velocity, and displacement. The stress at the interface is measured in the hard bar. Post-deformation measurements are made of final diameters along both the struck specimen and the projectile. High density diffraction gratings are used to measure the finite strain of incident and reflected waves in the struck bar, while electric resistance gages are used to measure the infinitesimal strain of the transmitted wave in the hard bar. The length of the hard bar is varied to produce unloading at any desired time or level of strain at the interface of the hard and soft bars. The conditions at the interface are altered in several ways, including changes in the relative initial diameters, and changes of surface conditions from glued to variously lubricated.

Having been informed on more than one occasion of the formidable amount of computer time required to attempt to describe the complex detail of this non-linear finite strain wave interaction, I have resorted to the amassing of experimental detail to better serve the same objective. In every year since the first measurements in 1961, I have added the results of diverse experiments to further understand this fascinating problem. They are contained in the manuscript of a forthcoming publication summarizing nearly 25 years of observation on this problem.

One of these observations of particular interest here, is introduced as a conjecture based upon experiment. I hypothesize that there can be a large amplitude radial shear wave in a cylindrical rod. To generate it, we use the plastic-elastic boundary experiment with a high pressure silicone lubricant at the interface. A dispersive incident wave in the struck bar whose response is described by Eqs. (22) through (26) impinges on the plastic-elastic interface. The wave response is very sensitive to the conditions at the interface itself. We compare two extremes, a glued interface and one having a silicone lubricant. For an incident wave having a maximum strain of 0.022, a maximum stress of 8300 psi, and a maximum particle velocity of 800 in/sec (to use the dimensions of the original data), diffraction grating measurements of strain and surface angle are made at many positions along the struck bar including 1/4 inch from the interface. Resistance gage measurements of the transmitted wave in the hard bar also are measured at several positions.

For the input wave just described for the glued interface, the measured maximum stress due to the reflection at the interface is 12,000 psi. For the same incident wave, for the lubricated interface the maximum stress of the reflected wave front is 8,770 psi, approximately the same as that for the incident wave maximum of 8,300 psi. For the reflected wave in the plastic bar, the maximum strain near the boundary for a glued interface is 0.031 and 0.039 a diameter length away; it has the much higher value of 0.052 for a lubricated interface and 0.052 for a diameter length away. When post deformation measurements of the diameters of the projectile and struck specimens are made along their entire lengths, we find for a glued boundary that the diameters of the struck specimen at all positions are smaller than those of the projectile. Of much more importance, for the lubricated interface the diameters of the struck specimen are much larger than those of the projectile. Finally, the measured surface angle for the wave front reflected from the lubricated interface undergoes a dramatic reversal in slope.

In sum: at an appropriately lubricated plastic-elastic interface, the axial component of the incident wave is constrained and the reflection should include an increase in stress as well as in strain. The radial component under minimal constraint reflects as if it were at a free boundary, i.e. the stress decreases while the strain increases far beyond the maximum value found for a glued boundary. The net stress for the reflection is nearly zero: 8,770 psi compared to 8,300 psi for the incident wave. The increase of strain from 0.031 for glued and unlubricated boundaries to 0.052 with lubrication is compatible with my conjecture, as is the large change in the response of the surface angle. Finally, the post-deformation difference in diameters for projectile and struck specimen is consistent with the concept of a radial shear wave that would be unable to pass from struck specimen to projectile when the reflected wave front has reached the impact face.

Assuming a radial shear wave is kinematically acceptable, its incorporation in the present or any other non-linear wave theory has yet to be accomplished.

REFERENCES

1949   McReynolds, Andrew W. Plastic Deformation Waves in Aluminum. Transactions of the American Institute of Mining and Metallurgical Engineers. vol. 185, 32-45.

1951   Bell, James F. Propagation of Plastic Waves in Pre-Stressed Bars. Technical Report No. 5, U.S. Naval Contract. The Johns Hopkins University.

1953   Lee, E.H. A Boundary Value Problem in the Theory of Plastic Wave Propagation. Quarterly of Appl. Math., vol. 10 (4), 335-346.

1956   Bell, James F. Determination of Dynamic Plastic Strain through the Use of Diffraction Gratings. Jnl. Appl. Physics, vol. 27 (190), 1109-1113.

1959   Bell, James F. Diffraction Grating Strain Gauge. Proceedings, Spring Meeting of the Soc. for Experimental Stress Analysis, Washington, D.C. vol. 17 (2), 51-64.

1960a   Bell, James F. Study of Initial Conditions in Constant Velocity Impact. Jnl. of Appl. Physics, vol. 31 (12), 2188-2195.

1960b   Bell, James F. Propagation of Large Amplitude Waves in Annealed Aluminum. Jnl. of Appl. Physics, vol 31 (2), 277-282.

1960c   Bell, James F. The Initial Development of an Elastic Strain Pulse Propagating in a Semi-Infinite Bar. Technical Report, No. 6, U.S. Army, Ballistics Research Laboratories, Aberdeen Proving Ground. The Johns Hopkins University.

1961a   Bell, James F. An Experimental Study of the Unloading Phenomenon in Constant Velocity Impact. Jnl. Mech. & Physics of Solids, vol. 9, 1-15.

1961b   Bell, James F. Experimental Study of the Interrelation between the Theory of Dislocation in Polycrystalline Media and Finite Amplitude Wave Propagation in Solids. Jnl. of Appl. Physics, vol. 32 (10), 1982-1993.

1961   Truesdell, Clifford A. General and Exact Theory of Waves in Finite Elastic Strain. Archive for Rational Mech. & Analysis, vol. 8(3) 263-352. [Corrected reprint in Continuum Mechanics, Part IV, Problems of Non-linear Elasticity; Internat'l Science Review Series. NY-London-Paris, Gordon & Breach, 1965].

1962a   Bell, James F. Experimental Study of Dynamic Plasticity at Elevated Temperatures. Experimental Mechanics, vol. 2 (1) 1-6.

1962b   Bell, James F. Experiments on Large Amplitude Waves in Finite Elastic Strain. Symposium on Second-Order Effects in Elasticity, Plasticity, and Fluid Dynamics. Internat'l Union of Theoretical and Applied Mechanics, Haifa, Israel. Proceedings (Pergamon Press, Oxford-Paris-NY; 1964; 173-186.)

1962   Bell, James F. and Stein, Albert. The Incremental Loading Wave in the
        Pre-Stressed Plastic Field. Jnl. de Mecanique, vol. 1 (4), 395-412.

1962   Bell, James, F. and Suckling, J.H. The Dynamic Overstress and the
        Hydrodynamic Transition Velocity in the Symmetrical Free Flight Plastic
        Impact of Annealed Aluminum. Proceedings, Fourth U.S. Nat'l Congress of
        Appl. Mechancis, 877-883.

1962   Dillon, Oscar W., Jr. An Experimental Study of the Heat Generated during
        Torsional Oscillations. Jnl. Mech. & Physics of Solids, vol. 10,
        235-244.

1962   Truesdell, Clifford A. Second-Order Theory of Wave Propagation in
        Isotropic Elastic Materials. Symposium on Second-Order Effects in
        Elasticity, Plasticity, and Fluid Dynamics. Internat'l Union of
        Theoretical and Applied Mechanics, Haifa, Israel. Proceedings (Pergamon
        Press, Oxford-Paris-NY; 1964; 187-199.)

1963   Bell, James F. The Initiation of Finite Amplitude Waves in Annealed
        Metals. Symposium on Stress Waves in Anelastic Solids, Internat'l Union
        of Theoretical and Applied Mechanics, Brown University, Providence, Rhode
        Island; Proceedings (Springer-Verlag: Berlin-Göttingen-Heidelberg);
        1964; 166-182.

1966   Dillon, Oscar W. Jr. Waves in Bars of Mechanically Unstable Materials.
        Jnl. Appl. Mechanics, vol. 33, 267-274.

1966   Sharpe, William N. The Portevin-le Chatelier Effect in Aluminum Single
        Crystals and Polycrystals. Jnl. Mech. & Physics of Solids, vol. 14,
        187-202.

1967   Dillon, Oscar W. Jr. The Dynamic Elastic-Plastic Interface and Related
        Topics. Jnl. Mech. and Physics of Solids, vol. 15, 341-358.

1968a   Bell, James F. The Physics of Large Deformation of Crystalline Solids.
        Springer Tracts in Natural Philosophy, vol. 14. Springer-Verlag:
        Berlin-Heidelberg-NY.

1968b   Bell, James F. Large Deformation Dynamic Plasticity at an
        Elastic-Plastic Interface. Jnl. Mech. & Physics of Solids, vol. 16,
        293-313.

1969   Bell, James F. The Dynamic Plasticity of Non-Symmetrical Free-Flight
        Collision Impact. Internat'l Jn. Mech. Science, vol. 11, 633-657.

1969   Kolsky, H. Production of Tensile Shock Waves in Stretched Natural
        Rubber. Nature, vol. 224 (5226), p. 1301.

1970   Cristescu, N. and Bell, James F. On Unloading in the Symmetrical Impact
        of Two Aluminum Bars. Battelle Memorial Institute, Symposium, Columbus,
        Ohio; Proceedings, 397-421.

1973   Bell, James F. The Experimental Foundations of Solid Mechanics.
        Handbuch der Physik, vol. VIa/1. Springer-Verlag: Berlin-Heidelberg-NY.
        [Reprinted 1984: entitled Mechanics of Solids, Vol. 1: The Experimental
        Foundations of Solid Mechanics.]

1973   Khan, A.S. Behavior of Aluminum during the Passage of Large-Amplitude
        Plastic Waves. Internat'l Jnl. Mech. Science, vol. 15, 503-516.

1973   Moon, Hahngue. An Experimental Study of the Outer Yield Surface and of the Incremental Response Function in the Totally Plastic Region for Annealed Polycrystalline Aluminum. Ph.D. dissertation, The Johns Hopkins University, Baltimore, Maryland.

1974   Khan, A.S. A study of the One Dimensionality and Isochoric Deformation during the Passage of Tensile Waves. Experimental Mechanics, vol. 14 (2), 57.

1975   Moon, Hahngue. An Experimental Study of the Outer Yield Surface for Annealed Polycrystalline Aluminum. Acta Mecanica, vol. 24 191-208.

1977   Phillips, Aris and Moon, Hahngue. An Experimental Investigation Concerning Yield Surfaces and Loading Surfaces. Acta Mecanica, vol. 27, 91-102.

1978   Hauver, George. Penetration with Instrumented Rods. Internat'l Jnl. Engineering Science, vol. 16, 871.

1978   Khanwalkar, R.K.T. Finite Amplitude Tensile Wave Propagation in Stretched Vulcanized Natural Rubber. Ph.D. dissertation, The Johns Hopkins University, Baltimore, Maryland.

1979   Bell, James F. A Physical Basis for Continuum Theories of Finite Strain Plasticity: Part I. Archive for Rational Mech. & Analysis, vol. 70, 319-338.

1979   Phillips, Aris and Lee, C.W. Yield Surfaces and Loading Surfaces; Experiments and Recommendations. Internat'l Jnl. of Solids & Structures, vol. 15, 715.

1980   Bell, James F. and Khan, A.S. Finite Plastic Strain in Annealed Copper during Non-Proportional Loading. Internat'l Jnl. of Solids and Structures, vol. 16 (8), 683-693.

1982   Bell, James F. On the Dynamic Elastic Limit. Experimental Mechanics, vol. 22 (7), 270-276.

1983a   Bell, James F. On the Evolution, from Experiment, of General Constitutive Equations in a Continuum Theory for Finite Plastic Strain. Plasticity Today Symposium, Udine, Italy. Proceedings (In press).

1983b   Bell, James F. Continuum Plasticity at Finite Strain for Stress Paths of Arbitrary Composition and Direction. Archive for Rational Mech. & Analysis, vol 84 (2), 139-170.

1984   Phillips, Aris. On the Experimental Foundation of the Two Surface Model of Plasticity and Viscoplasticity. Plasticity Today Symposium, Udine, Italy. Proceedings (In press).

1985   Bell, James F. Contemporary Perspectives in Finite Strain Plasticity. Internat'l Jnl. of Plasticity, vol. 1, 3-27.

APPENDIX I

Specific References for Problems I to XIV

I. One Dimensional Waves of Finite Strain: Uniaxial Stress (Compression)
     1956---Bell, James F.---J. Appl. Phys. v. 27, No. 10, pp. 1109-1113.
     1960---Bell, James F.---J. Appl. Phys. v. 31, No. 2, pp. 277- 282.
     1962---Bell, James F.---Experimental Mech. v. 2, No. 1, pp. 1-6.
     1962---Bell, James F. and Werner, W. Meade---J. Appl. Phys. v. 33,
               No. 8, pp. 2416-2425.
     1962---Sperrazza, Joseph---Proceedings 4th U.S. Congress of Applied
               Mechanics v. 2, pp. 1123-1129.
     1967---Bell, James F.---Experimental Mech. v. 7, No. 1, pp. 1-8.
     1968---Bell, James F.---The Physics of Large Deformation of Crystalline
               Solids, v. 14,Springer Tracts in Nat. Phil. Chapt.2

II. One Dimensional Waves of Finite Strain: Uniaxial Stress (Tension)
     1973---Khan, A.S.---Int. J. Mech. Sci. v. 15, pp. 503-516.
     1974---Khan, A.S.---Experimental Mechanics v. 14, No. 2, pp. 57.

III. On the Interaction of Waves of Loading and Unloading
     1953---Lee, E.H.---Quart. Appl. Math. v. 10, No. 4, pp. 335-346.
     1961---Bell, James F.---J. Mech.&Physics of Solids v. 9, pp. 1-15,
               pp. 261-278.
     1970---Cristescu, N. and Bell, J.F.---Battelle Memorial Institute
               Proceedings, McGraw Hill Book Co. pp. 397-421.

IV. Wave Propagation at Finite Strain in a Relative Reference Configuration
     1978---Hauver, George---Int. J. Engineering Sci., v. 16, pp. 871.
     1979---Bell, James F.---Arch. Rational Mech.&Analysis, v. 70, pp. 319-338.
     1985---Bell, James F.---Int. J. of Plasticity, v. 1, No. 1, pp. 3-27.

V. Finite Strain Waves at a linear Elastic Boundary
     1968---Bell, James F.---J. Mech. & Physics of Solids, v. 16, pp. 295-313.
     1980---Khan, A.S.---Journal of Strain Analysis, v. 15, No. 1, pp. 15-20.

VI. Non-Symmetrical Impact: The Analysis of an Historical Problem
     1969---Bell, James F.---Inter.J. Mech. Sci. v. 11, pp. 633-657.

VII. The Incremental Wave in Pre-Stressed Solids: Metals
     1951---Bell, James F.---Tech. Rpt. U.S.Navy Cont.,The Johns Hopkins Univ.
     1953---Sternglass,E.J. and Stuart,D.A.---J. Appl. Mech. v. 20,
               pp. 427-434.
     1956---Alter, B.E.K. and Curtis, C.W.---J.Appl.Phys.v. 27,No. 9,
               pp. 1079-1085.
     1962---Bell, J.F. and Stein, A.---Jnl. de Mecanique v. 1,No. 4,
               pp. 395-412.
     1963---Bianchi, G. IUTAM Symposium, Brown Univ. pp. 101-107, Springer.

VIII. The Incremental Wave in Pre-Stressed Solids: Rubber
     1969---Kolsky, H.---Nature v. 224, No. 5226, pp. 1301.
     1978---Khanwalkar, R.K.T.---Ph.D. Dissertation, The Johns Hopkins Univ.

IX. The Surface Angle Measurement of Large Radial Motion for Waves in Rods
     1968---Bell, James F.---J. Mech. & Physics of Solids, v. 16, pp. 295-313.
     1969---Bell, James F.---Int. J. Mech. Sci. v. 11, pp. 633-657.
     1973---Bell, James F.---Handbuch der Physik v. VIa/1, pp. 634-637.
     1974---Khan, A.S.---Experimental Mechanics, v. 14, No. 2, pp. 57.

X. Wave Propagation in a Three Dimensional Axially-Symmetric Field
        1961---Truesdell, C.A.---Arch. Rational Mech. Anal. v. 8, No. 3,
                    pp. 263-352.
        1962---Bell, James F.---Proceedings IUTAM Symp. Haifa, pp. 173-186,
                    Pergamon.

XI. The Elastic Limit and Outer Yield Surface in Dynamic Plasticity
        1982---Bell, James F.---Experimental Mech. v. 22, No. 7, pp. 270-276.

XII. On the Role of Dimensionless Parameters at Finite Strain
        1963---Bell, James F.---J. Appl. Physic, v.34, No. 1, pp. 134-141.
        1964---Bell, James F.---Phil. Mag. v. 10, No. 103, pp. 107-126.
        1965---Bell, James F.---Phil. Mag. v. 11, No. 114, pp. 1135-1156.
        1967---Bell, James F.---Proceedings Symposium on Mech. Behavior of
                        Solids Under Dynamic Loads, San Antonio Texas,
                        Springer.
        1968---Bell, James F.---Springer Tracts in Nat. Phil. v. 14, Springer.
        1970---Bell, James F.---Rendiconti Universita di Torino, v. 30, pp. **49-61**.

XIII.   1949---McReynolds, A.W.---Trans. Am. Inst. Mining & Met. Eng. v. 185,
                        pp. 32-45.
        1962---Dillon, Oscar W. Jr.---Jnl. Mech. & Physics of Solids v. 10,
                        pp. 235-244.
        1962---Bell, J.F. & Stein, A.---Jnl. de Mechanique, v. 1, No. 4,
                        pp. 395-412.
        1966---Dillon, Oscar W., Jr.---J. Appl. Mech. v. 33, pp. 267-274.
        1966---Sharpe, William N.---Jnl. Mech. & Physics of Solids, v. 14,
                        pp. 187-202.
        1968---Bell, James F.---Springer Tracts in Nat. Phil. v. 14, Chapt. VII.
XIV.   1985---Bell, James F.---Mansucript nearing completion,

NOTE: All of the above are described in condensed presentations in various sections of Springer Verlag's Handbuch der Physik VIa/1 of 1973, or its paperback
edition entitled Mechanics of Solids Vol. I in 1984, or in the Russian translation
by Nauk, also in 1984.

124

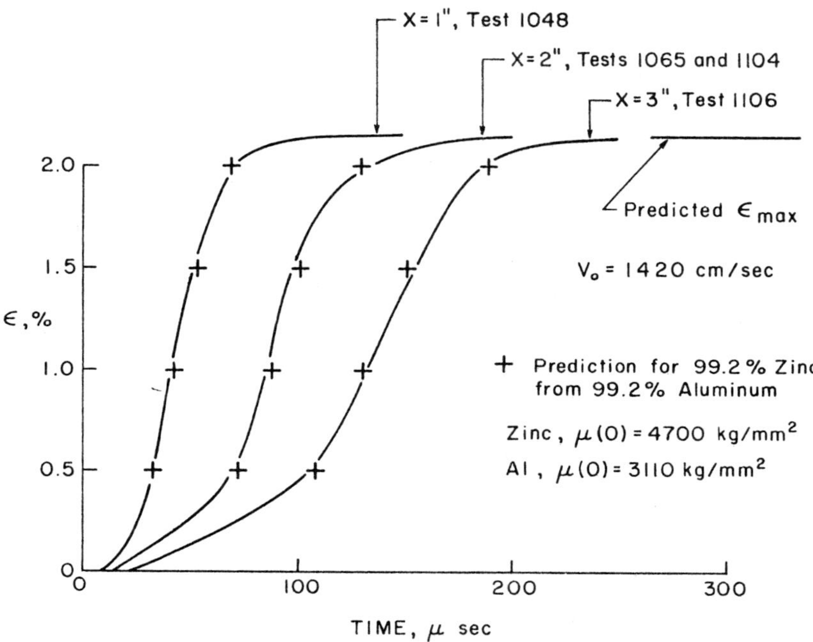

Fig. 1   Strain wave profiles from 30, 720 lines per inch diffraction grating measurements (BELL, 1968a) for the hexagonal metal annealed zinc (solid lines), compared with prediction from similar measurements at the same locations in terms of Eqs. (13), (25), and (26) for the face centered cubic metal, annealed aluminum.

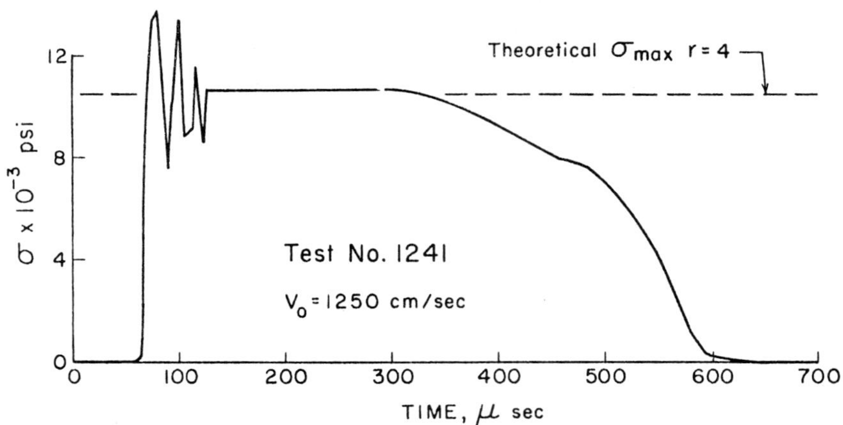

Fig. 2   A load bar measurement (BELL, 1968a) of the maximum stress for a wave propagating in a bar of annealed copper, compared with prediction for Eqs. (13), (23) and (26) of the incremental theory of finite strain.

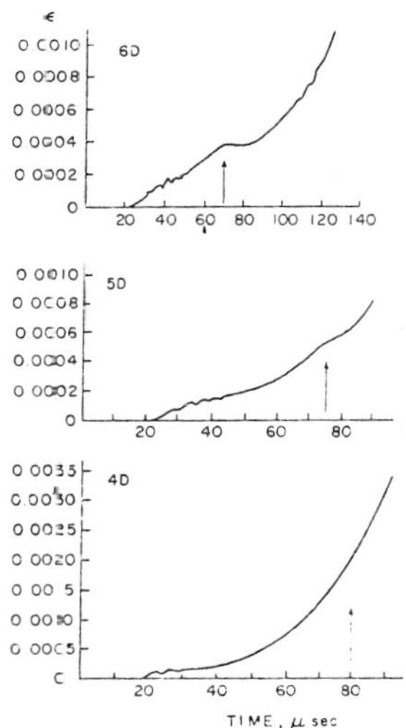

Fig. 3   The unloading wave arriving at 0.6L(6D) at the predicted time of 70 microsec.; arriving at 0.5L(5D) at the predicted time of 75 microsec.; it has disappeared in measurement at 0.4L(4D), as predicted from Eq. (27) for annealed aluminum. (BELL, 1961a).

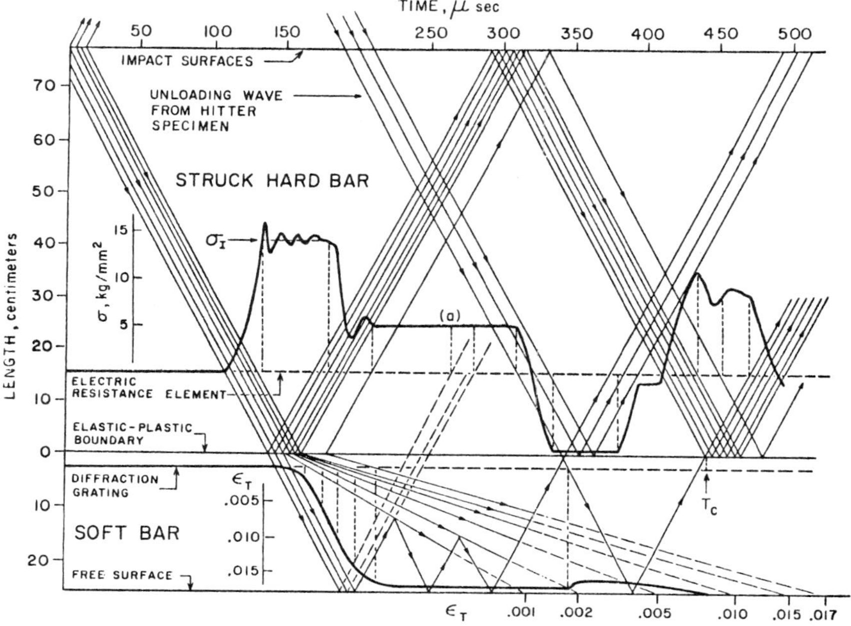

Fig. 4   The heavy solid lines are averaged experimental data for the incident and reflected waves in the hard bar, and for the transmitted wave in the soft bar (BELL, 1968b). The Lagrangian diagram plot reveals the very close correlation between experiment and the present theory for the timing sequence in such tests.

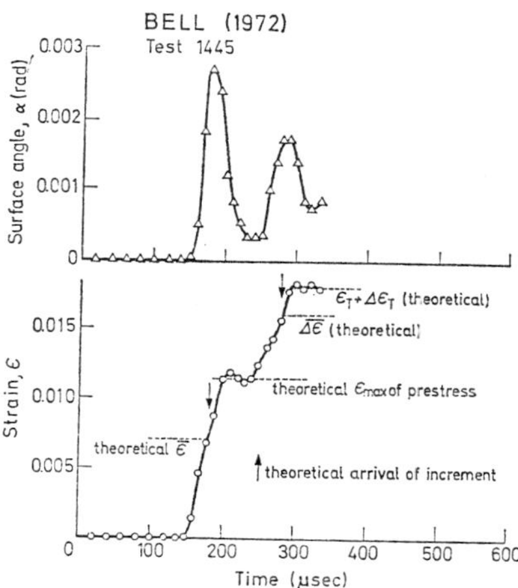

Fig. 5   An incremental wave of greater than infinitesimal amplitude from experiments in 1972 (BELL, 1973 page 700) in which the small precursor travels, as shown, at the linear elastic bar velocity, but the rest of the incremental front propagates at the slower plastic wave speeds.

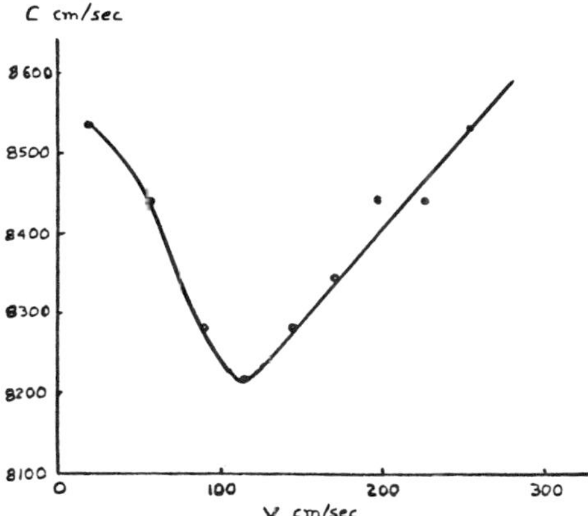

Fig. 6   Measured wave speeds as a function of particle velocity, from the data
of Khanwalkar (1978) for an incremental tensile wave in a rubber string
stretched 2.81 times its free length. The wave speeds are constant both
above and below the observed dynamic inflection point readily visible at
a particle velocity of 115 cm/sec.

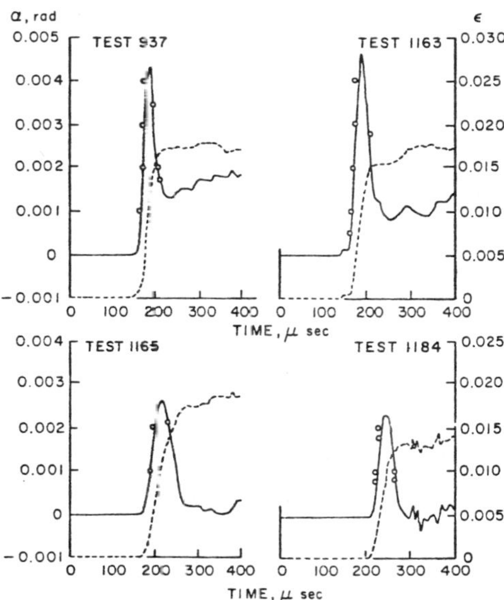

Fig. 7   Diffraction grating measurements (BELL, 1968b, 1973 pp. 635-637) of
surface-angle wave profiles in annealed aluminum (solid lines) compared
with prediction (circles) from Eq. (41). The measured strains are the
dashed lines. The test is of the type described in problem V above.
The maximum surface angle occurs at a strain of $0.63\ E_{1max}$, as described
in problem X below.

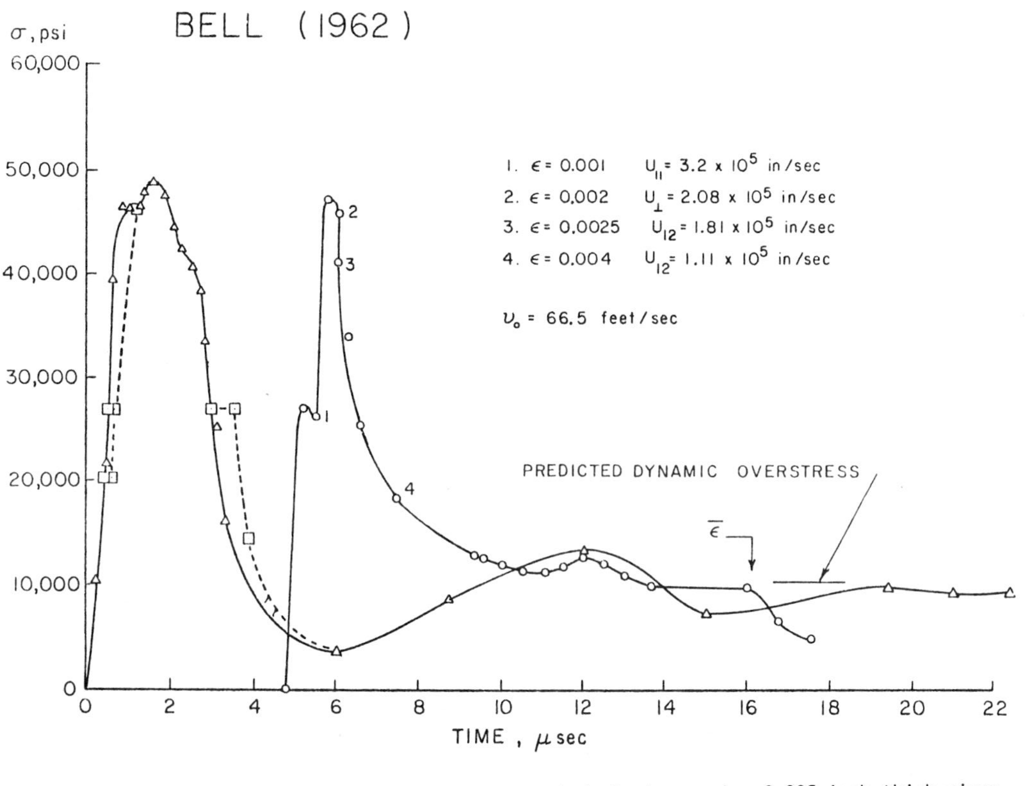

Fig. 8   Measurements of stress in annealed aluminum using 0.005 inch thick piezo crystals at the impact free (triangles and squares) (BELL, 1962b), compared with the calculation (circles) from TRUESDELL's Eqs. (42) and (43). The wave speeds are calculated from diffraction grating measured strain profiles between 1/4 and 1/2 diametral lengths from the impact face.

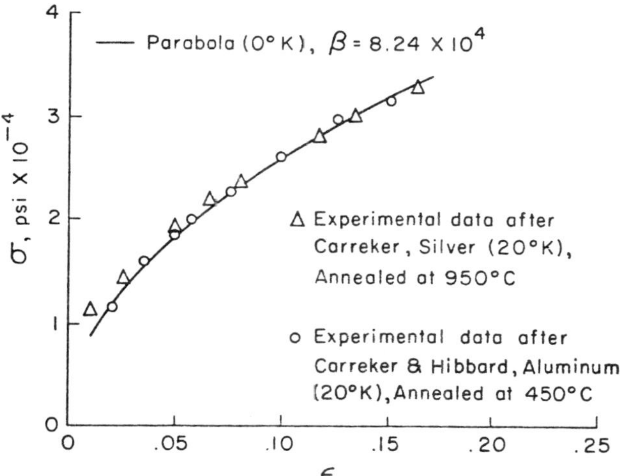

Fig. 9   CARREKER's response function for annealed silver for an uniaxial test at
20°K, compared with an uniaxial response function at 20°K for an annealed
aluminum test of CARREKER and HIBBARD. (See BELL, 1973, pp. 556-557.)
The predicted common response function at 0°K for solids with equivalent shear moduli at that temperature, (0), is shown as the solid line.

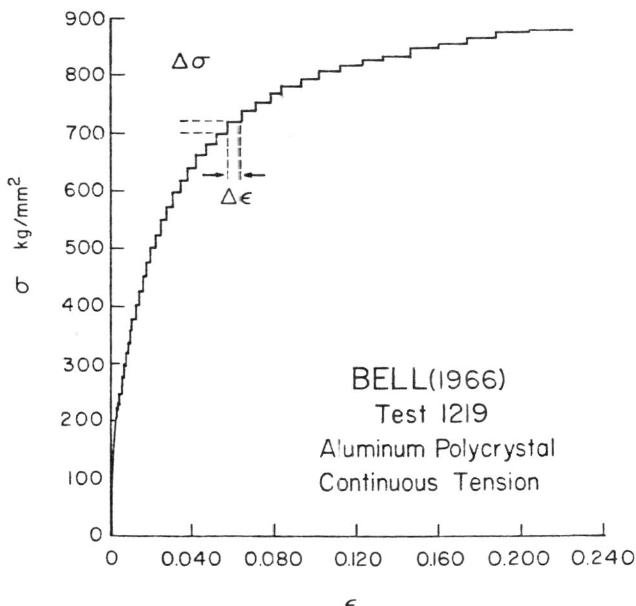

Fig. 10   A dead-weight loading uniaxial tension in annealed aluminum (BELL, 1968a,
Ch. VII) showing the "staircase" phenomenon of the Savart-Masson effect.
The vertical section departs from, and the horizontal section returns
to, the parabola of Eq. (23). For the incremental waves of problem VII
above, the high wave speeds are attributed to the steep vertical slope.

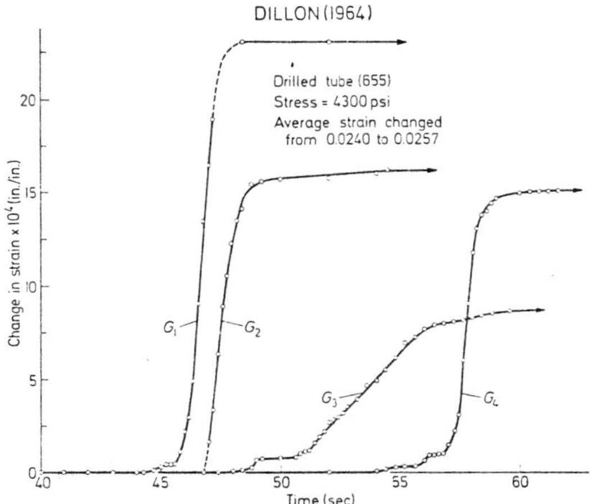

Fig. 11   "Slow" waves as measured in a test of DILLON's, for a torsion pre-stress
of 4300 psi. The successive locations, G_1 to G_4, are 3.81 cm. apart
on a tube of 1.27 cm diameter with 0.149 cm. wall thickness. The wave
speeds of the rapidly altering wave profiles are a few centimeters per
second.

INTERACTION OF SHALLOW-WATER WAVES AND BOTTOM TOPOGRAPHY

B. Boczar-Karakiewicz

INRS-Océanologie
Uriversité du Québec
Rimouski (Québec)
Canada G5L 3A1

J.L. Bona

Department of Mathematics
Pennsylvania State University
215 McAllister Building
University Park, PA 16802

D.L. Cohen

Massachusetts Institute of Technology
Lincoln laboratory
Lexington, Massachusetts 02173-0073

Abstract

A system of equations is proposed for the dynamical interaction of
progressive surface waves and bottom sediment in near-shore zones. This system is
shown to model the formation of stable, bar-like structures that are familiar
features of many coastal areas.

Introduction

In this paper, interest will be directed toward the description of surface
waves in comparatively shallow water and their mutual interaction with the topography over which they propagate. Because the water is shallow, the fluid motions
extend to the bottom and so the waves typically undergo a change of form as they
progress. If the fluid lies over a bed of loose sediment, such as sand, then the
passage of waves may in turn have an effect on the bottom. As the bed changes
shape, the surface waves will deform differently. Thus the entire system in view,
comprising both the fluid and the bed surfaces, admits the possibility of complex
self-interactions.

A primary motivation of this study is the desire to understand the formation of bars along the coasts of large bodies of water. We will accordingly concentrate on a two-dimensional situation wherein deep-water, plane, periodic wavetrains impinge on shallow-water zones. Field observations of situations that roughly correspond to this idealization have repeatedly identified quasi-static arrays of bars and troughs. In the more controlled environment of the laboratory wavetank, experiments have shown that periodic wavetrains incipient on an initially featureless bed of sand will eventually organize an equilibrium configuration of bars and troughs. An ocean coast may have incoming, deep-water waves with typical wavelengths of a hundred meters and sand bars spaced one-quarter to one-half a kilometer apart, whereas the scales that pertain in the laboratory may be a hundred times smaller. Despite the disparate scales, the laboratory and field phenomena of bar formation are strikingly similar, leading one to conjecture that such manifestations may depend upon some relatively simple and universal wave-bottom interaction. This optimistic view is quickly tempered by the variety that nature provides, and it must be candidly admitted that no single explanation is likely to cover all situations. Nevertheless, the ideas put forth below appear to have an interesting range of applicability.

The main accomplishment here is the derivation of a simple, but useful, model for two-dimensional, wave-bottom interaction. This model is analysed in various qualitative and quantitative aspects, and predictions stemming from the model are set against observations. The outcome of the comparisons justifies a somewhat sanguine appraisal of the prospects for such models, in spite of the very considerable restriction their use places on the wave and sediment regimes.

The rest of the paper is divided into four Sections. Section 1 is devoted to the derivation of a comparatively simple model that allows for time variations of both free surfaces. The equilibrium configurations of this system are studied in Section 2. A numerical scheme for the integration of the time-dependent equations is presented in Section 3 and used to show that solutions generally evolve into equilibrium formations. The forementioned comparisons between theory and observation are also made in Section 3. Some general commentary and suggestions for further development are given in the concluding Section.

1. The Mathematical Model

When attempting to model wave-bottom interaction it is very helpful to keep observational findings in mind. Especially useful for our purposes is a series of experiments performed in a rectangular channel with a paddle-type wavemaker mounted at one end and an energy absorber at the other. The bottom was laid with sand in various initial configurations before the channel was filled with water. The paddle was then driven at constant frequency and amplitude for many hours, and both the waveform and bedform were monitored over time. A detailed description of these tests is given by Boczar-Karakiewicz, Paplinska and Winiecki (1981), so we content ourselves with a summary of the salient points. The experimental set-up used by Boczar-Karakiewicz et al. (1981), along with a typical equilibrim bed configuration, is pictured in Figure 1.

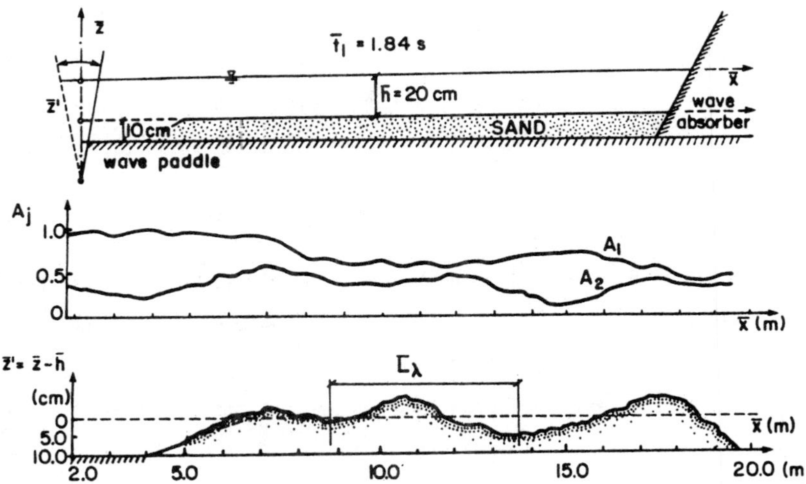

Figure 1. Typical equilibrium bottom profiles in a wave tank experiment:
a) initial bed configuration at $T = 0$,
b) equilibrium profile at $T = 80217\ t_1$; the A_j are the normalized harmonic amplitudes of the wavefield, $j = 1,2$, and t_1 is the wave period (after Boczar-Karakiewicz et al., 1981).

The principal point one observes is that in the laboratory scales appertaining to these tests (the flume was 0.5 meters wide, 21 meters long, the water depth was typically 0.3 meters at the wavemaker end, and the wavemaker period was usually about 1.8 seconds) the wave deforms in seconds, whereas the bed experiences significant alteration only over periods of hours. This is not surprising, but it does provide a firm basis for selecting two time scales for the description of the system, one for the evolution of individual waves, and another for the development of the bed. Because of the wide disparity between these two time scales, the bed appears fixed from the viewpoint of an individual wave. In consequence, the modelling may be approached in two stages. In Section 1.1 the wave field is modelled for time intervals that are short enough for the bottom topography to be considered fixed. In Section 1.2 a longer time variable is introduced and used to describe the waves' cumulative effect on the bottom.

1.1 Description of the Wave Field

In the first stage of the modelling, we seek a simple, but sensibly accurate description of a two-dimensional wavetrain as it propagates over a temporally fixed, but spatially variable depth. For the nonce, the spatial region R of interest is, therefore, considered fixed.

We begin by defining the spatial coordinates as follows. Let $\bar{z}$ denote the vertical coordinate, $\bar{x}$ the horizontal coordinate, and let the undisturbed free surface of the liquid be located at $\{(\bar{x},\bar{z}): \bar{z} = 0\}$. Here and in what follows, variables adorned with an overbar are dimensional and unscaled. The surface bounding the liquid from below is taken to be $\{(\bar{x},\bar{z}): = \bar{z} = -\bar{h}(\bar{x})\}$ (see Figure 2). Suppose also that the spatial region R of interest lies in the range where $0 < \bar{x} < \bar{M}$. The point $\bar{x} = 0$ will be identified with the physical point at which the incoming, deep-water wavetrain first comes into the purview of the model. The point $\bar{x} = \bar{M}$ is, in applications, thought of as the point closest to the shoreline where the model is to be applied. It must be emphasized that no attempt is made here to account for the zone very near the water line of a beach. Rather, the present development assumes the waves to be more or less completely dissipated in the very nearshore region. This point will be amplified later.

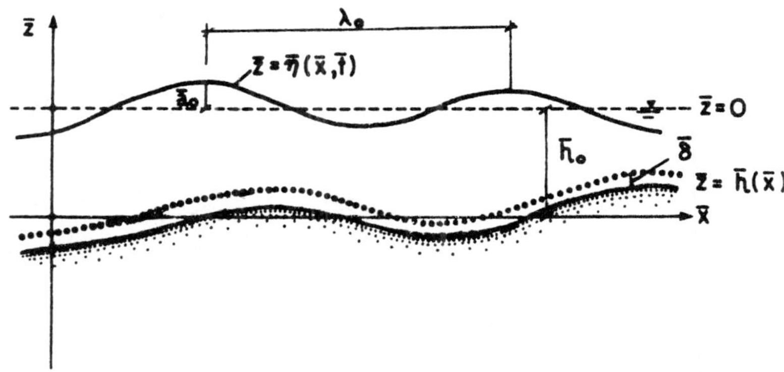

Figure 2. Definition sketch for the two dimensional model: here $\bar{a}_0$ is a typical
wave amplitude, λ_0 a typical wave length, $\bar{h}_0$ a typical water depth, $\bar{\delta}$
is the thickness of the boundary layer containing suspended sand,
$\bar{z} = -\bar{h}(\bar{x})$ the variable bottom, and $\bar{z} = 0$ denotes the rest position of
the free surface of the fluid.

An issue that immediately presents itself is wave breaking. Even on a gently
shelving beach, it is not uncommon for incoming progressive waves to break and
reform several times. The quantitative description of this process is beyond our
present capabilities. However, the experimental data suggests that breaking is
not a central ingredient for the sort of bottom structure whose genesis is considered here. For one thing, while the fluid under breakers tends to be more
sediment-laden, the general character of the bed deformation there does not appear
different than in non-breaking zones (cf. Boczar-Karakiewicz _et al._ 1981). The
sand movement seems to be closely related to the local harmonic content of the
flow, especially the first two harmonics. (This is not unexpected since the
higher harmonics are not in general energetic enough to move particles on the bottom unless the water is quite shallow; see Section 1.2 below). The evolution of
the harmonic content of the wave as it progresses, particularly as regards the
fluid motion near the bed, does not appear to be greatly affected by breaking (see

Figure 3). This may be because, if the area very near shore is excepted, breaking only affects the flow catastrophically near the surface; at depth the flow may be quite smooth (cf. Thornton and Schaeffer, 1978). Finally, it has been observed that breaking is not required in the formation of stable structures in a sandy bottom (cf. Benjamin, Boczar-Karakiewicz and Pritchard, to appear).

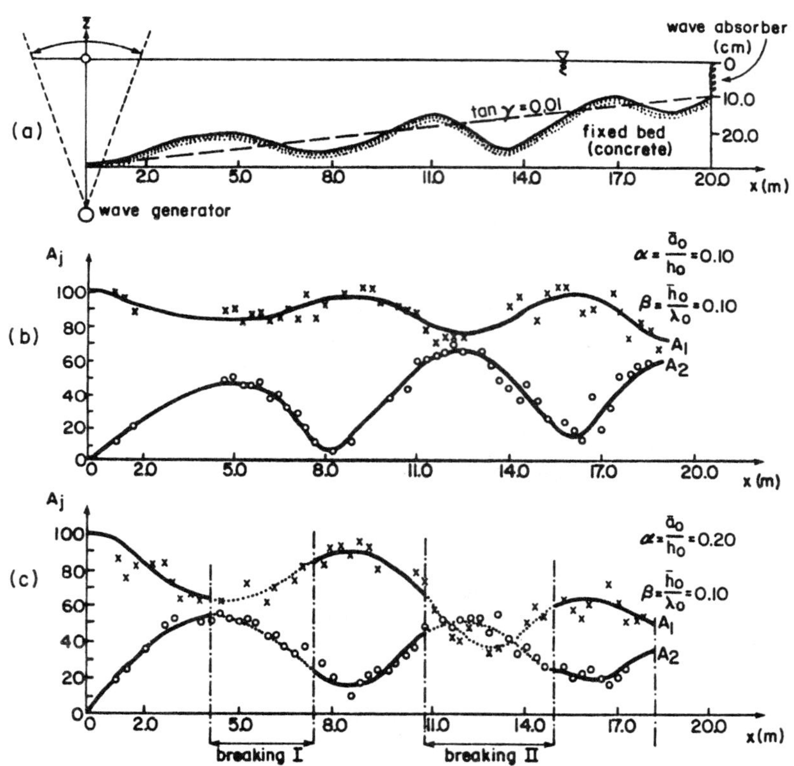

Figure 3. Wave-profile transformation over a fixed bed in a wavetank experiment: (a) experimental configuration, (b) spatial evolution of the two first harmonic amplitudes A_j (j=1,2) of a non-breaking wave, (c) spatial evolution of A_j for a breaking and reforming wave; the harmonic amplitudes A_j are normalized by $A_L^0 = A_j(0)$.

Analysis of the typical scales that obtain on a gradually-sloping beach indicates the shallow-water approximation to the Euler equations as a possible descriptive mode for the wave field (cf. Peregrine 1972, or Whitham 1974). Such a model is inexpedient because its solutions develop singularities near the shore where the actual wave breaks. Because of this feature, and because of the view enunciated above that breaking is incidental to the process that it is our ultimate interest to describe, we have elected to retain dispersion in the description of the wave motion. The resulting theory is due originally to Boussinesq (1871). It forms a basis for the accounting of the wave field that avoids the singularities inherent in the shallow-water equations. In following this line we are in good company (see e.g. Mei and Méhauté 1966, Madsen and Mei 1969, Lau and Barcilon 1972, Lau and Travis 1973, and Svendsen and Buhr-Hansen 1978).

The Boussinesq approximation to the Euler equations,

$$\bar{u}_{\bar{t}} + g\bar{\eta}_{\bar{x}} + \bar{u}\bar{u}_{\bar{x}} - \frac{1}{3}\bar{h}^2\bar{u}_{\bar{x}\bar{x}\bar{t}} = 0 \ , \tag{1.1}$$

$$\bar{\eta}_{\bar{t}} + [\bar{u}(\bar{h} + \bar{\eta})]_{\bar{x}} = 0,$$

is derived under the assumptions that nonlinear and dispersive effects are of the same, small order of magnitude and that the bottom varies gradually. Here g is the magnitude of the acceleration due to gravity, $\bar{t}$ is elapsed time, $\bar{\eta}(\bar{x},\bar{t})$ is the vertical deviation of the fluid surface from its undisturbed position at the point $\bar{x}$ at the time $\bar{t}$, and $\bar{u}(\bar{x},\bar{t})$ represents the depth-averaged horizontal velocity of the fluid above the point $\bar{x}$ at the time $\bar{t}$. This system of equations can be put into a variety of other forms by taking different dependent variables and by using the zeroth-order approximation to alter the first-order nonlinear or dispersive corrections; see Bona and Smith (1976), and the references contained therein. Here, it is convenient to work with the following non-dimensional variables;

$$x = \bar{x}/\lambda_0 \ , \quad t = \bar{t}\sqrt{gh_0}/\lambda_0, \quad u = \bar{u}\sqrt{h_0}/a_0\sqrt{g}$$

$$\zeta = \bar{\eta}/a_0, \qquad\qquad h = \bar{h}/h_0, \tag{1.2}$$

where a_0 and λ_0 denote a typical amplitude and wavelength, respectively, in the wavetrain, and h_0 is a representative depth in the region R of interest. In these variables the Boussinesq equations become

$$u_t + \zeta_x + \alpha u u_x - \frac{1}{3} h^2 \beta^2 u_{xxt} = 0,$$

$$\zeta_t + [u(a\zeta + h)]_x = 0, \tag{1.3}$$

where

$$\alpha = \frac{a_0}{h_0}, \qquad \text{and} \qquad \beta = \frac{h_0}{\lambda_0}. \tag{1.4}$$

Define also the parameters

$$\varepsilon = \frac{1}{\alpha} \max_{0 \leqslant x \leqslant \overline{M}} [|h'(x)|, |h''(x)|], \qquad \text{and} \qquad S_0 = \frac{\alpha}{\beta^2}, \tag{1.5}$$

where $\overline{M} = M/\lambda_0$. In the variables displayed in (1.2), u, ζ, and their partial derivatives with respect to x and t are assumed to be of order one. The formal validity of (1.3) as an approximation to the Euler equations subsists on the assumptions that the fluid is inviscid, irrotational, and incompressible, and the motion is planar, that α, β, and ε are small compared to 1, and that h and S_0 are order one. The Stokes number S_0 (Stokes 1847, Korteweg and de Vries 1895) is a measure of the relative importance of nonlinear and dispersive effects experienced by the wave train. Nonlinear and dispersive effects will contribute to order-one changes in the wave profile on temporal or spatial scales of order α^{-1} and β^{-2}, respectively. If the bulk of the wave motion is only in one direction, say in the direction of increasing x, then a simpler description is available, namely a variable-coefficient, Korteweg-de Vries-type equation (Johnson 1973, Svendsen and Buhr-Hansen 1978),

$$\zeta_t + c\zeta_x + \frac{3}{2} \frac{\alpha}{c} \zeta\zeta_x - \frac{1}{6} \beta^2 h^2 \zeta_{xxt} + \frac{1}{2} c_x \zeta = 0, \tag{1.6}$$

where $c(x) = [h(x)]^{1/2}$. Here, advantage has been taken of the leading-order relation $\partial_t = -c\,\partial_x$ to rewrite the dispersive term $\frac{1}{6} \beta^2 h^2 c \zeta_{xxx}$ appearing in the last-quoted references. As explained by Benjamin _et al._ (1972), this results in capturing more accurately the full, linearized dispersion relation for surface water waves. In the present study attention will be given exclusively to progressive waves, and so the just-mentioned simplification would be available. However, at a later stage we expect to allow for small reflection effects and so the level of generality inherent in the Boussinesq equations, the possibility for waves to propagate in two directions, has been retained.

The hypothesis that S_0 is of order one is troublesome. Even on moderately sloping beaches, say a mean slope of 0.5°, the local value of the Stokes number reaches 20 to 30 in regions of interest. As it happens, the Boussinesq-type approximation performs quite accurately in just this sort of regime. As a case in point, reproduced in Figure 4 there is a comparison made by Bona, Pritchard and Scott (1981) of a laboratory observation and an associated numerical solution of essentially equation (1.6) in a regime corresponding to the above-mentioned Stokes number. It should also be noted that this type of approximation is substantially correct even for considerably larger values of S_0 (cf. Zabusky and Galvin 1971, Hammack 1973, Hammack and Segur 1974, Svendsen and Buhr-Hansen 1978).

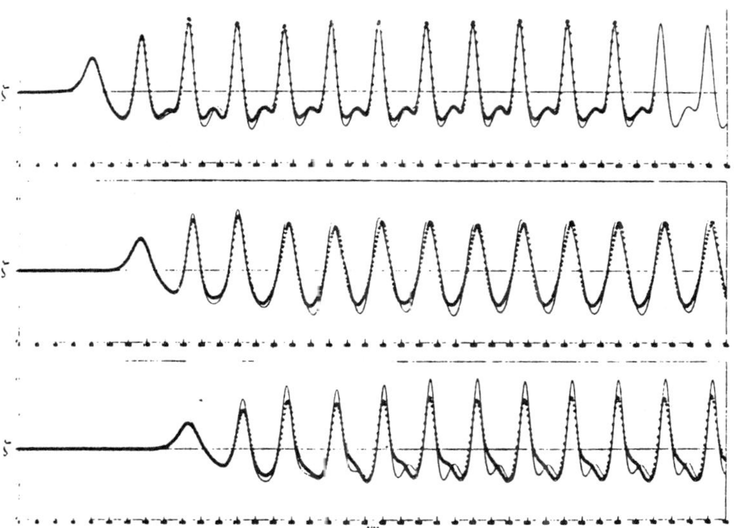

Figure 4. Wave measurements (the diamonds) of waves at Stokes number $S_0 = 26.3$ compared with computed values (the solid lines) of the free surface elevation using a KdV-type model, as a function of time at three different positions in the wavetank (after Bona <u>et al</u>.).

With approximate initial and boundary conditions, the system (1.3) constitutes a well-posed problem. A set of auxilliary specifications that suits the region R is the following:

$$\zeta(x,0) = g_1(x) \ , \ u(x,0) = g_2(x), \quad \text{for} \quad 0 < x < M,$$
$$\zeta(0,t) = \zeta_0(t) \quad \zeta(M,t) = \zeta_1(t), \quad \text{for} \quad t > 0, \tag{1.7}$$
$$u(0,t) = u_0(t) \quad u(M,t) = u_1(t),$$

where g_1, g_2, ζ_0, u_0 and u_1 are given functions. This corresponds to prescribing the surface displacement ζ and velocity distribution u at an initial instant of time, $t = 0$, and at the endpoints of the underlying horizontal domain $\{x: 0 < x < M\}$. For treatments of the well-posedness of (1.3) the reader may consult Schonbek (1978) and Amick (1984). For numerical schemes see Peregrine (1967), Madsen and Mei (1969) and Winther (1979). As regards similar issues for (1.6) wherein only ζ need be specified as in (1.7), see Svendsen and Burh-Hansen (1978), Bona and Dougalis (1980) and Bona, Pritchard and Scott (1981). For neither model is the theory in a completely satisfactory state. Even if g_1 and g_2 are set to zero, on the supposition that the initial wave configuration will have no sensible effect on the long-term status of the system, it is difficult to measure accurately the data exhibited in (1.7). In the field one certainly cannot expect this much information. A more likely provision would be an incoming wave spectrum at a point corresponding to $x = 0$.

For several reasons a further simplification of (1.3) or (1.6) will be carried out. The coarse-grained data that one can expect in a field situation means that the accuracy attained using (1.3) or (1.6) is somewhat illusory. Moreover, it must be remembered that the present model will be coupled to a model for the deformation of the bed in Section 1.2. In numerically approximating the equations for the liquid and sediment surfaces it will be required to predict the waveform many times during the slow evolution of the bedform. Consequently, while it is true that (1.3) can be efficiently integrated numerically, the computation becomes extensive when the erosion of the bed is considered. Finally, the forthcoming simplification has the salutory effect of allowing the use of analytic techniques to answer certain questions.

In the reduction to be effected now we explicitly follow the lead of Lau and Barcilon (1972) (see also Armstrong et al. 1962, and Mei and Ünlüata, (1972). The clue to this step is already evident in Figure 3, which emphasizes the relationship between the waves and their first few harmonic components. Assume now that the wave incident at the 'deep-water' end, $x = 0$, is a single-frequency, planar, steadily-propagating linear wavetrain. This presumption corresponds closely to the experimental conditions maintained by Boczar-Karakiewicz et al. (1981) cited earlier. It is also reasonably veritable on some real beaches. As an example there is shown in Figure 5 a representative energy density distribution measured off a Baltic Sea coast by Druet, Massel and Zeidler (1972). (Similar observations are reported by Guza and Thornton, 1980 and Elgar and Guza, 1984), using measurements obtained on a California beach.) Note especially the deep-water spectrum (solid line) that is sharply peaked about a single frequency ω_1 corresponding to a period of about 7 seconds. In this particular instance the

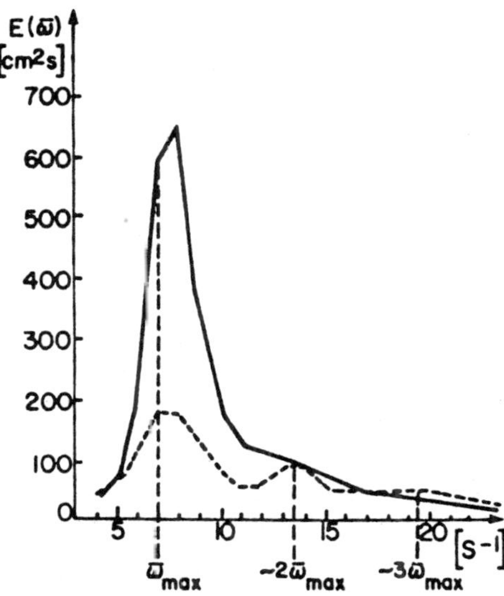

Figure 5. Energy density distribution off the coast of the Southern Baltic Sea: — offshore measurements ($\bar{h}_0 = 6.00$ m), --- nearshore measurements ($\bar{h}_0 = 2.00$ m).

waves were propagating normal to the beach and bathymetry showed that the assumption of the waves two-dimensionality was valid. As the waves propagated shoreward, there transpired a pronounced shift of energy to the second harmonic frequency, $\omega_2 = 2\omega_1$, (dashed line in Figure 5). This accords nicely with predictions we shall make later.

In agreement with the assumptions leading to (1.3), the bottom configuration h is taken to have the slowly-varying form

$$h(x) = 1 + \varepsilon f(x), \qquad (1.8)$$

for $0 \leqslant x \leqslant M$, where ε is the small parameter defined in (1.5). Since h has order one, we may take it that f has order one. However, (1.5) implies that $f'(x)$ has order α. Because of this, it is indicated to introduce a new horizontal variable X defined by

$$X = \alpha x.$$

Let $F(X) = f(X/\alpha)$. Then the foregoing assumptions concerning f and ε imply that h has the form

$$h(X) = 1 + \varepsilon F(X), \qquad (1.9)$$

where F and F' are both of order one, and ε is small compared to unity.

Following Lau and Barcilon (1972) a two-scale expansion will be introduced, treating x and X as independent variables. For this to be effective a decision regarding the relative sizes of the three small parameters α, β, and ε must be rendered. It will be presumed they all three have the same small size since this is in fact the case in many near-coastal areas.

This latter assumption has associated theoretical difficulties which are now explained. The supposition that ε/α is of order one presents no problem. But if α/β has order one, then the Stokes number $S_0 = \alpha/\beta^2$ has order $1/\beta$, and so is representative of a regime in which nonlinear effects predominate. Thus the specified scalings appear contradictory, and indeed a proper limiting procedure appropriate for (1.3) in which, say,

$$\varepsilon, \ \alpha, \ \beta \to 0, \ \text{with} \ S_0 \ \text{and} \ \frac{\varepsilon}{\alpha} \ \text{fixed},$$

would lead to a singularity in the forthcoming model. (This point is not emphasized in the earlier work of Lau and Barcilon 1972). The model will not be applied in such a limiting case, however, but rather to situations involving finite values of the parameters α, β, and ϵ. For the usual sort of values of β the difference between the variable X and the variable $Y = \beta^2 X$ more naturally associated with (1.3) is the moderate-sized multiplicative constant S_0. Looking ahead to the way in which X enters in the model, it becomes apparent that the choice in (1.8) is not crucial so long as fixed finite values of the parameters are in question. In fact, this choice of scales corresponds to assumptions, intermediate between those leading to the Boussinesq equations and those leading to nonlinear shallow-water theory, that emphasize nonlinearity, but do not ignore dispersion entirely. Presumably what is important is how accurately the overlying equation (1.3) or (1.6) predict wave phenomena in the regime that corresponds to a particular set of parameter values. The previously-quoted work of Svendsen and Buhr-Hansen (1978) and Bona, Pritchard, and Scott (1981) seems to assure that for the range of parameter values of utmost interest for the present study (α, β, ϵ about 0.10, S_0 in the range 10 to 30) the models predict rather well. (It should be acknowledged that the good agreement between numerical predictions and the outcome of laboratory measurements obtained in the just-referenced papers subsisted in part on appending to the model a dissipative mechanism. For laboratory-scale experiments there is little doubt that dissipation from the boundary layers and other sources is quite important. On field scales these effects are much less important, and so are ignored here. They will be incorporated in subsequent, more refined studies.)

Proceeding with the two-scale expansion, if r is a function of x and X, then by the chain rule,

$$\frac{d}{dx} r(x,X) = \frac{\partial r}{\partial x} + \alpha \frac{\partial r}{\partial X} . \qquad (1.10)$$

The relationship embodied in (1.10) is maintained even when x and X are treated as independent variables. Suppose the incoming wave-train is periodic with a frequency ω_1. The normal modes associated to the linearized Boussinesq equations ((1.3) with $\alpha = 0$) which propagate in the direction of increasing x **are**

$$e^{i(k_j x - \omega_j t)} ,$$

where $\omega_j = j\omega_1$ and $k_j > 0$ is determined by the linearized dispersion relation

$$\omega_j{}^2 = \frac{k_j{}^2}{1 + \frac{\beta^2}{3} k_j{}^2}. \tag{1.11}$$

In the absence of nonlinearity the general periodic solution of (1.3) propagating in the $+x$ direction would be represented as a linear combination of the normal modes. To take account of nonlinearity, the coefficients in the normal-mode decomposition are allowed to vary. Keeping in mind that nonlinear effects acumulate over time scales of order α^{-1}, it is natural to assume that the coefficients vary on the spatial scale embodied in the variable X. Hence, we write

$$\zeta(x,X,t) = \sum_j \zeta_j(X)e^{i(k_j x - \omega_j t)} + c.c.,$$

and

$$u(x,X,t) = \sum_j u_j(X)e^{i(k_j x - \omega_j t)} + c.c., \tag{1.12}$$

where we are following the usual method of representing a real function in terms of complex exponentials by letting c.c. stand for the complex conjugate of the sum it follows. As ζ and u are of order one, so are the coefficients ζ_j and u_j. On the basis of both field and laboratory measurements, attention is restricted to the first two harmonics, $j = 1,2$ (see again Figures 1,3 and 5). The resulting finite expansions are substituted into (1.3) and only terms containing modes of the frequency ω_1 or ω_2 are retained, the others being assumed negligible. Keeping in mind the relation (1.10), and writing

$$h(X) = 1 + \alpha G(X), \tag{1.13}$$

where $G(X) = \frac{\varepsilon}{\alpha} F(X)$, we obtain

$$u_t + \zeta_x - \frac{\beta^2}{3} u_{xxt} = -\alpha\zeta_X - \alpha u u_x + \frac{2\zeta\beta^2}{3} G u_{xxt}$$

$$+ \frac{2\beta^2\alpha}{3} u_{xXt} + order\ (\alpha^2), \tag{1.14}$$

$$\zeta_t + u_x = - \alpha u_x - \alpha(u\zeta)_x - \alpha G u_x + \text{order } (\alpha^2) \ .$$

From the second equation in (1.14) and assumptions (1.12) it is deduced that

$$u_j(X) = \frac{\omega_j}{k_j} \ \zeta_j(X) + \text{order } (\alpha), \qquad\qquad (1.15)$$

for $j = 1,2$. The variable ζ may be eliminated from the left-hand side of these equations, so coming to the single relation

$$u_{tt} - u_{xx} - \frac{\beta^2}{3} u_{xxtt} = \alpha\{- \zeta_{Xt} + u_{Xx} + G u_{xx} +$$

$$\frac{2\beta^2}{3} [G u_{xxtt} - u_{xXtt}] - \frac{1}{2} (u^2)_{xt} + (\zeta u)_{xx}\} + \text{order } (\alpha^2) \ . \qquad (1.16)$$

Note that whilst β is viewed as fixed and non-negligible relative to one, so dispersive effects enter, it is nevertheless small. Hence the dispersion relation (1.11) implies that $\omega_j \cong k_j$, $j = 1,2$, which in turn suggests that $k_2 - 2k_1 = \omega_2 - 2\omega_1 + \Delta k = \Delta k$ is small (it is of order β^2). Thus the variable $x \Delta k = (\Delta k/\alpha)X$ may be taken as comparable with X, and so independent of x. In equation (1.16) the terms of order α^2 are ignored, as are all temporal harmonics except ω_1 and ω_2. With these provisos, and because the first two harmonics are linearly independent, equation (1.16) means that the coefficients of the first two harmonics must each vanish. The result of this requirement is the Lau-Barcilon equations,

$$\zeta_1'(X) + iF_1\zeta_1(X) + iQ_1 e^{i \widetilde{\Delta k} X} \zeta_1{}^*(X) \zeta_2(X) = 0,$$

$$\qquad\qquad (1.17)$$

$$\zeta_2'(X) + iF_2\zeta_2(X) + iQ_2 e^{-\widetilde{\Delta k} X} \zeta_1{}^2(X) = 0,$$

where $*$ denotes complex conjugation,

$$F_j = \frac{k_j{}^3}{2\omega_j{}^2} (1 - \frac{2\beta^2}{3} \omega_j{}^2)G(X), \qquad j = 1,2,$$

$$Q_1 = (k_2 - k_1)(\frac{k_1}{k_2\omega_1})[\omega_1 - (k_2 - k_1)(\frac{k_2}{\omega_2} + \frac{k_1}{\omega_1})],$$

$$Q_2 = \frac{k_2{}^2}{8k_1} (1 - \frac{2k_1{}^2}{\omega_1{}^2}), \qquad\qquad (1.18)$$

and $\tilde{\Delta k} = \frac{\Delta k}{\alpha}$.

In deriving (1.17), repeated use of (1.15) has been made. Since (1.15) is applied to terms that are of order α, the error implicit in its use is of order α^2, and so is neglected in the approximation considered here. Note that discarding terms of order α^2 is consistent with the use of (1.3). Notice also that $\tilde{\Delta k}$ is of order $1/S_0$.

The representation (1.12) coupled with the Lau-Barcilon equations (1.17) and the relation (1.15) comprise our description of the wave field over a fixed, gradually-varying bottom. It is formally valid for times that are short compared to those over which the bed topography deforms.

The Lau-Barcilon equations must be supplemented with the appropriate analog of the auxilliary data (1.7). Because reflection has been ignored the boundary condition at $x = M$ is dispensible. The initial data has implicitly been set to zero and it is only necessary to account for the conditions at $x = 0$. Because of (1.15) it suffices to specify $\zeta(0,t)$, since $u(0,t)$ is then inferred. The assumed form (1.12) entails only the first two harmonics and consequently we need only specify the projection of $\zeta(0,t)$ onto these harmonics. That is, if

$$\zeta(0,t) = \sum_{j=1}^{\infty} a_j^0 e^{i\omega_j t} + c.c.,$$

then our presumption is that a_j is negligible for $j \geqslant 3$. Hence it is taken that

$$\zeta(0,t) = a_1^0 e^{i\omega_1 t} + a_2^0 e^{i\omega_2 t} + c.c.,$$

whereupon the equations in (1.17) are to be solved subject to

$$\zeta_1(0) = a_1^0 , \quad \zeta_2(0) = a_2^0. \tag{1.19}$$

A special case that corresponds to the experimenta' and field data described earlier obtains when $a^0 = 1/2$ and $a^0 = 0$, so that $\zeta(0,t) = \cos(\omega_1 t)$.

In using the model for the wave field described above, it is often convenient to return to the variable x throughout, the 'long' variable X having served its purpose. To this end define

$$a_j(x) = \zeta_j(X), \quad \text{for} \quad j = 1,2. \tag{1.20}$$

Then formulae (1.17 - (1.18) may be written in the useful form,

$$a_1'(x) + i\,\varepsilon f_1 f(x)a_1(x) + i\,\alpha Q_2 e^{i\,\Delta k x}a_1^*(x)a_2(x) = 0,$$

$$\tag{1.21}$$

$$a_2'(x) + i\,\varepsilon f_2 f(x)a_2(x) + i\,\alpha Q_2 e^{-i\,\Delta k x}a_1^2(x) = 0,$$

where, for $j = 1,2$,

$$f_j = \frac{k_j^3}{2\omega_j^2}\left(1 - \frac{2\beta^2}{3}\,\omega_j^2\right), \tag{1.22}$$

Q_1 and Q_2 are defined in (1.18), f is defined in terms of h in (1.8), and $\Delta k = k_2 - 2k_1$ as before. The boundary conditions (1.19) apply without change to the system (1.21).

1.2 Description of the Bottom Deformation

Still considering the bottom configuration as fixed and known, and assuming a_1 and a_2, and thereby ζ and u, to be determined, it is possible to infer the fluid velocity near the bed from inviscid theory. The horizontal velocity U at a depth z below the undisturbed free surface is, to the accuracy afforded by the approximations made in deriving the Boussinesq equations, given by the formula

$$U(x,t;z) = u(x,t) - \beta^2[\tfrac{1}{3}\,h^2(x) - zh(x) - \tfrac{1}{2}\,z^2)u_{xx}(x,t),$$

(cf. Peregrine, 1972). Define u_b to be the horizontal component of the fluid velocity at $z = -h(x)$. Then it follows that

$$u_b(x,t) = u(x,t) - \frac{\beta^2}{6}\,h^2(x)u_{xx}(x,t). \tag{1.23}$$

Using (1.12) and (1.15), u_b may be expressed in terms of a_1 and a_2 as

$$u_b(x,t) = \sum_{j=1}^{2} \frac{\omega_j}{k_j} \left(1 - \frac{\beta^2}{6} h^2(x) k_j^2\right) a_j(x) e^{i(k_j x - \omega_j t)} + c.c. \qquad (1.24)$$

The view is taken that the layer of fluid near the bottom is viscous and sediment-laden. Let $\bar{\delta}$ denote the thickness of this layer, and assume that $\bar{\delta} < h_0$. It is our purpose now to analyse the sediment movement in this boundary layer. The overall instantaneous deformation of the bed will then be identified with the differential movement of the sediment within the bottom layer of fluid. (This very simple model is ideal to illustrate principles. More realistic models for the sediment transport will be reported in subsequent studies of this general model).

To provide a basis for our discussion, it is worth reporting some detail of the laboratory observations of Boczar-Karakiewicz et al. (1981) in regard to the sediment movement. Very soon after subjecting an initially flat bed of sand to periodic wave motion, small scale ripples appear in the bed. The oscillatory motion of the fluid over such ripples generates small patches of vorticity that are very effective at lifting the sand from the bed (Sleath, 1978). These small-scale ripples persist, and from their inception onward there is a plentiful supply of sand in the lower reaches of the fluid. (An idealized analysis of an oscillating flow over a ripply bed has been given recently by Longuet-Higgens, 1981).

In the analysis to follow, no attempt will be made to describe the process whereby the sediment is suspended in the fluid. Rather, a distribution of sediment in the boundary layer will simply be postulated. It transpires that the resulting model is not highly sensitive to the assumed sediment distribution, and the boundary layer depth can be scaled out of the model equations.

As a first approximation the suspended particles are presumed to be transported with a velocity proportional to that of the ambient fluid, and their influence on the flow is neglected. Hence interest naturally turns to the mass-transport velocity of the fluid near the bed. The approach taken to the mass-transport velocity is based on the seminal work of Longuet-Higgens (1953). The general idea, recognized already by Stokes (1847), is that fluid-particle orbits generated by the propagation of finite-amplitude waves are not closed. Repeated passage of progressive waves therefore engenders a small drift velocity, called the mass-transport velocity.

As the suspended sediment is concentrated near the bed, and since under our assumptions a purely oscillatory motion would produce no mean movement of the sediment, it is the mass-transport velocity in the lower reaches of the fluid that is central. This is exactly what was studied by Longuet-Higgens in the setting of Stokes' waves over a laminar, viscous boundary layer. His methods carry over intact for the shallow-water waves of interest here. Of course, the motion near the bed is really turbulent, as the description of the experimental observations suggests. In such a situation the mean aspects of the motion could be described by way of a coefficient of eddy viscosity whose value varies with distance from the boundary. Using this sort of idea in the context of Stokes' waves, Johns (1970), showed that for both standing and progressive waves the predictions of the model only differed by a multiplicative factor from the description obtained via laminar boundary-layer theory. Such differences may be taken into account by applying the laminar theory with an eddy viscosity that is 10 to 100 times the ordinary kinematic viscosity.

We begin the analysis by concentrating attention around a particular point x on the horizontal scale. Longuet-Higgins' development simplifies considerably, because near x the bed may be treated as horizontal and without curvature on account of the assumption that the parameter ε defined in (1.5) is of order α. In addition to the horizontal coordinate x, the stretched variable

$$n = \frac{z + h(x)}{\delta} , \qquad (1.25)$$

where

$$\delta = \left[\frac{\overline{\nu}}{h_0(gh_0)^{1/2}} \right]^{1/2} \frac{\overline{\delta}}{h_0} , \qquad (1.25)$$

is used to describe the boundary layer. Here δ is a dimensionless boundary-layer thickness and ν refers to an appropriate, dimensional eddy viscosity. Observations in the laboratory indicate that typical values of δ are about 0.05.

Consideration is given to the mean horizontal velocity $U(x,n,t)$ in the boundary layer n units above the bed at x at time t. In the approximation afforded by the use of an eddy viscosity it is found (see Longuet-Higgins, 1953, formula (155)) that

$$[\beta \frac{\partial}{\partial t} - \frac{\partial^2}{\partial n^2}]u = \beta \frac{\partial u_b}{\partial t} , \qquad (1.26)$$

where, as in (1.24), u_b is the horizontal velocity at the bottom of the inviscid layer. Naturally, U satisfies the no-slip condition at $n = 0$, and must essentially agree with u_b for large values of n which correspond to the top of the viscous layer. The form of u_b in (1.24) compels the _ansatz_ that

$$U(x,n,t) = \sum_{j=1}^{2} P_j(x,n)e^{i(k_j x - \omega_j t)} + c.c. \qquad (1.27)$$

If the latter relation is substituted into (1.26), there results the equations

$$(- i\omega_j \beta - \frac{\partial^2}{\partial n^2})P_j = - i\beta \frac{\omega_j^2}{k_j} [1 - \frac{\beta^2}{6} h^2(x)k_j^2]a_j(x), \qquad (1.28)$$

for $j = 1,2$. The general solution of these ordinary differential equations is

$$P_j(x,n) = S_j(x)e^{\Lambda_j n} + T_j(x)e^{-\Lambda_j n} + \frac{\omega_j}{k_j} [1 - \frac{\beta^2}{6} h^2(x)k_j^2]a_j(x),$$

where

$$\Lambda_j = (1 - i)(\frac{\omega_j \beta}{2})^{1/2} ,$$

for $j = 1,2$, and the positive square root is meant. Since P_j must tend to the corresponding coefficient of u_b for large n and to 0 as n tends to 0, it is determined that

$$U(x,n,t) = \sum_{j=1}^{2} \frac{\omega_j}{k_j} [1 - \frac{\beta^2}{6} h^2(x)k_j^2]a_j(x)[1 - e^{-n\Lambda_j}]e^{i(k_j x - w_j t)} + c.c. \qquad (1.29)$$

With U in hand, the mass-transport velocity in the boundary layer is easily computed. First note that if U is averaged over a fundamental period

$t_1 = 2\pi/\omega_1$, we obtain zero. This is the usual state of affairs for small-amplitude waves, in which there are no first-order currents. For the drift velocity one must look to the next order, as Stokes (1847) already appreciated. At the second order the average over a period t_1 gives the lowest-order approximation of the mass-transport velocity u_m, and is not usually zero. In fact, Longuet-Higgins has determined its value generally, under assumptions weaker than those in force here. In our variables, this has the form

$$u_m(x,n,t) = 4 \int_t^{t+t_1} [\frac{\partial u_b}{\partial x}(x,t) \int_0^\tau U(x,n,s)ds]d\tau$$

$$+ 3\int_0^n \int_t^{t+t_1} \{[\frac{\partial U}{\partial x}(x,n',\tau) - \frac{\partial u_b}{\partial x}(x,t)] \int_0^\tau \frac{\partial U}{\partial n}(x,n',s)ds\}d\tau dn',$$

(1.30)

(see Longuet-Higgins, 1953, formula (169)). Using the previously obtained expressions for U and u_b, the mass-transport velocity, u_m, may be evaluated explicitly as

$$u_m(x,n) = \sum_{j=1}^2 \frac{\omega_j}{k_j} |a_j(x)|^2[1 - \frac{\beta^2}{6} h^2(x)k_j^2]^2 \tilde{D}_j(\delta),$$

(1.31)

where

$$\tilde{D}_j(\delta) = 5 - 8e^{-\mu_j}\cos(\mu_j) + 3e^{-2\mu_j},$$

with

(1.32)

$$\mu_j = n(\frac{\beta\omega_j}{2})^{1/2} = \frac{1}{\delta} z(\frac{\beta\omega_j}{2\delta})^{1/2}.$$

Note that U_m is independent of the fast time variable t. In obtaining (1.31) we have ignored the small quantities $h'(x)$ and $a_j'(x)$ in favour of the order-one terms retained above (and, as before, the terms multiplied by β^2 arising from dispersion are inconsistently kept). This would be mechanical if we had maintained the distinction between x and the 'slow' spatial variable X introduced in Section 1.1.

The goal in view now is to determine how the depth changes with time. In the present, very simplified model the temporal evolution of the depth will be intimately connected with the mass-transport velocity, u_m. As determined above, u_m depends directly on the vertical coordinate n in the bottom fluid layer, and on the horizontal coordinate x and time through a_1, a_2, and h. Moreover, a_1 and a_2 depend on time only through variation of h over time. Introduce another, temporal variable T which is such that significant changes in the bed profile take place over time scales T of order one. From the observational data we know that the time variable t representative of a wave period is essentially infinitesimal with regard to T. The depth h is now explicitly allowed to depend on time, but only through the very slow variable T, which is viewed as independent of t. Because $h = h(x,T)$, the amplitudes a_1 and a_2 also depend on T, as does the mass-transport velocity u_m. Of course ζ and u have a dependence on T induced by that of a_1, a_2, and h.

Consider a short time interval, say in which t varies by order one, $[T,T+\Delta T]$, and a small spatial interval $[x,x+\Delta x]$. Let $\rho(n)$ denote the assumed form of sediment density in the lower fluid. (For simplicity, ρ is taken to be identically zero outside the lower layer.) A typical distribution, having some theoretical justification (cf. Raudkivi, 1976), is the exponential form, $\rho(n) = \rho_0 e^{-\gamma n}$. A cruder assumption is to simply take the density to be constant in the lower layer, say set at some appropriate average value, $\rho(n) \equiv \bar{\rho}$. In any case, define

$$M(x,T) = \int_0^\infty \rho(n)u_m(x,n,T)dn. \tag{1.33}$$

Because of the assumption that the suspended sediment follows the fluid motion, the quantity

$$\int_T^{T+\Delta T}[M(x,\tau) - M(x+\Delta x,\tau)]d\tau \tag{1.34}$$

is proportional to the total mass accumulation in the region $I = \{x': x < x' < x + \Delta x\}$ during the interval $[T,T+\Delta T]$. Since the concentration of suspended sediment is time and space invariant, this mass accumulation must be reflected in the differential amount of sediment on the bed, namely

$$\rho_0 \int_x^{x+\Delta x}[h(x',T+\Delta T) - h(x',T)]dx', \tag{1.35}$$

where ρ_0 is the average density of the deposited sediments (the fact that the layer of sediment near the surface of the bed is a saturated porous medium is ignored here). Let K be the constant of proportionality to which allusion was made above (1.34). Then our mass balance has the form

$$K \int_T^{T+\Delta T} [M(x,\tau) - M(x+\Delta x, \tau)] d\tau \tag{1.36}$$

$$= \rho_0 \int_x^{x+\Delta x} [h(x',T+\Delta T) - h(x', T)] dx'.$$

Upon dividing by $\Delta x \, \Delta T$ and taking the limit formally as Δx and ΔT approach zero, the instantaneous version of the mass balance appears, namely

$$\frac{\partial h}{\partial T} = \frac{K}{\rho_0} \frac{\partial M}{\partial x} = K \int_0^\infty \frac{\rho(n)}{\rho_0} \frac{\partial u_m}{\partial x} (x,n,T) dn. \tag{1.37}$$

Again for the sake of simplicity we shall henceforth take $\rho(n) \equiv \bar{\rho}$, though such an assumption is not needed to proceed. Absorbing the dimensionless quantity $\bar{\rho}/\rho_0$ into K, and setting

$$U_m(x,T) = \frac{1}{\delta} \int_0^\delta u_m(x,n,T) dn, \tag{1.38}$$

(2.37) becomes

$$\frac{\partial h}{\partial T} = \tilde{K} \frac{\partial U_m}{\partial x} \qquad \text{where} \quad \tilde{K} = \delta K. \tag{1.39}$$

The variable U_m is readily evaluated in terms of a_1 and a_2 to be

$$U_m(x,T) = \sum_{j=1}^{2} \frac{\omega_j}{k_j} |a_j(x,T)|^2 [1 - \frac{\beta^2}{6} h^2(x,T) k_j^2]^2 D_j, \tag{1.40}$$

where

$$D_j = 5(1 - \frac{1}{2v_j}) - 3 \frac{e^{-2v_j}}{2v_j} + 4 \frac{e^{-v_j}}{v_j} (\cos v_j - \sin v_j)$$

and

$$v_j = (\frac{\beta \omega_j}{2})^{1/2}, \quad \text{for } j = 1,2. \tag{1.41}$$

The reader will note that after $\tilde{K}$ replaces K in (1.39), the boundary layer depth no longer appears explicitly in our formulas. In effect, the timescale T can be rescaled to eliminate δ from the model equations, a potentiality that has been exercised throughout the remainder of this paper.

The simple law (1.39) combined with the Lau-Barcilon equations (1.21) and supplemented by the original bottom configuration,

$$h(x,0) = H(x), \qquad (1.42)$$

for $0 \leqslant x \leqslant M$, serves to determine the bed profile for $T > 0$.

In summary, the model suggested here has the general form

$$\frac{\partial a_1}{\partial x} = A_1(a_1, a_2, h),$$

$$\frac{\partial a_2}{\partial x} = A_2(a_1, a_2, h), \qquad (1.43)$$

$$\frac{\partial h}{\partial T} = H(a_1, a_2, h),$$

where the functions A_1 and A_2 are defined in formulae (1.18), (1.21), and (1.22), and H is defined in (1.39), (1.40), and (1.41). The system (1.43) is supplemented by the auxilliary conditions

$$a_1(0,T) = a_1^{\,0} ,$$

$$a_2(0,T) = a_2^{\,0} , \qquad (1.44)$$

for $T \geqslant 0$, as in (1.19), corresponding to a fixed, incoming, deep-water wavetrain, and

$$h(x,0) = H(x), \qquad (1.45)$$

as in (1.42), corresponding to a given initial bed profile. Whilst several somewhat drastic simplifications have been made in the construction of this model, we shall see that the model still retains what appears to be certain of the fundamental elements of wave-bottom interaction. Moreover, the model seems to have predictive power, even in its present, preliminary stage of development.

2. Equilibrium Configurations

Since one of our principle aims is to explain in some measure the formation of stable bed structures in near-shore zones, it is natural to consider what kind of time-invariant solutions are possessed by the model developed in the last Section. Before embarking on a tudy of equilibrium solutions, it is useful to understand certain general aspects of the ecuations more fully.

As in the derivation, attention is given first to the situation wherein the bed is fixed. In this case we are concerned only with the Lau-Barcilon equations (1.21). As noticed by Armstrong et al. (1962), this system of equations admits a conservation law, namely

$$E = \frac{|a_1(x)|^2}{\alpha Q_1} + \frac{|a_2(x)|^2}{\alpha Q_2} .$$
(2.1)

That is, for all $x > 0$, $E(x) = E(0)$, and so E is determined by the initial data (1.19) for the incoming wavefield.

Consider first that the bed is flat and horizontal, so that $h(x) \equiv 1$ and $f(x) \equiv 0$. The system (1.21) simplifies to

$$a_1'(x) + i\alpha Q_1 e^{i\Delta k x} a_1^*(x) a_2(x) = 0,$$

(2.2)

$$a_2'(x) + ia Q_2 e^{-i\Delta k x} a_1^2(x) = 0,$$

for $0 < x < M$, with

$$a_1(0) = a_1^0, \qquad a_2(0) = a_2^0.$$

In this form the equations admit an exact solution, given already by Armstrong et al. (1962, see also Mei and Ünlüata, 1972). The salient results are merely reported here. Temporarily define new variables as follows:

$$E_0 = \frac{|a_1^0|^2}{\alpha Q_1} + \frac{|a_2^0|^2}{\alpha Q_2} ,$$
(2.3)

$$w(x) = \frac{|a_1(x)|}{(\alpha E_0 Q_1)^{1/2}} \quad , \qquad v(x) = \frac{|a_2(x)|}{(\alpha E_0 Q_2)^{1/2}} \quad , \tag{2.4}$$

and

$$\tilde{x} = \alpha Q_1 (\alpha Q_2 E_0)^{1/2} x. \tag{2.5}$$

View v and w as functions of $\tilde{x}$ rather than x. Then the conservation of energy expressed in (2.1) becomes

$$w^2(x) + v^2(x) = 1. \tag{2.6}$$

The solution of (2.2), expressed in the variables given in (2.3)-(2.5), is

$$v^2(\tilde{x}) = v_a^2 + (v_b^2 - v_a^2)\operatorname{sn}^2[(v_c^2 - v_a^2)^{1/2}(\tilde{x} - \tilde{x}_a); \gamma], \tag{2.7}$$

$$w^2(x) = 1 - v^2(x),$$

where sn denotes the Jacobian elliptic function, $v_a^2 < v_b^2 < v_c^2$ are the three real roots of the cubic equation

$$0 = P(v^2) = (1-v^2)v^2 - [\Gamma + \tfrac{1}{2}\Delta Q(v^2 - v^2(0))],$$

with

$$\Delta Q = \frac{\Delta k}{\alpha Q_1 (\alpha Q_2 E_0)^{1/2}} \quad ,$$

$$\Gamma = w^2(0)v(0) \cos(\theta(0)),$$

and the modulus γ of sn has the value

$$\gamma = \frac{v_b^2 - v_a^2}{v_c^2 - v_a^2} \quad {}^{1/2}.$$

The parameters governing this solution, Γ, ΔQ, $v(0)$, $w(0)$, $\theta(0)$, and $\tilde{x}_a$ are inter-related in easily determined ways, so that only three are free, and these comprise a constrained set within themselves. The exact forms of these constraints is not important here, but what is interesting is that the first and second harmonics both vary periodically in space with period

$$\tilde{L}_\lambda = \frac{2K}{(v_c^2 - v_a^2)^{1/2}}$$

in the $\tilde{x}$ variable, where K is the complete elliptic integral

$$K = K(\gamma) = \int_0^{\pi/2} (1 - \gamma^2 \sin^2(y))^{-1/2} dy \ .$$

In the original variables, the period of a complete cycle of exchange of energy between the two harmonics is

$$L_\lambda = \frac{2K}{(v_c^2 - v_a^2)^{1/2} \alpha Q_1 (\alpha Q_2 E_0)^{1/2}} \ .$$

This distance will be called the _repetition length_ of the system (the same parameter is named variously by other authors). This sort of periodic variation of the first two harmonics certainly accords qualitatively with what is observed in the laboratory (see again Figure 3(a)). An incoming wave train alluded to earlier that is especially interesting for our purposes is when $a_1^0 = 1/2$, $a_2^0 = 0$. The above formulae then simplify considerably:

$$\Gamma = v_a = x_a = 0,$$

$$v_b^2 = [1 + (\tfrac{\Delta Q}{4})^2]^{1/2} - \tfrac{\Delta Q}{4} , \qquad v_c = \frac{1}{v_b} ,$$

$$\gamma = \frac{v_b}{v_c} = v_b^2 , \qquad \tilde{L}_\lambda = 2v_b K(v_b^2).$$

In the form just given, the system's predictions have been set against a considerable range of laboratory data (see Mei and Ünlüata, 1972). The results of this comparison show the repetition length, L_λ, to be rather well predicted, especially for larger values of ΔQ. since the repetition length seems to be a major factor as regards the spacing of bar-like structures in the bed, the reliability of the model in this aspect is welcome.

If the bed is spatially variable, an analytical solution is not available. However, for a given bottom configuration $h(x)$ it is an easy task to accurately approximate numerically the solution a_1 and a_2 of (1.21). The system of equations is not stiff, and a straightforward application of the standard, fourth-order correct, Runge-Kutta discretization procedure gives very satisfactory results (cf. Isaacson and Keller, 1966). It is found that while the harmonic amplitudes $A_j(x) = |a_j(x)|$, $j = 1,2$, are not typically periodic over a gently-varying bed, they neverthless present a pattern of rhythmical exchange of energy. Examples of this phenomenon are pictured in Figure 6, where a fixed, incident

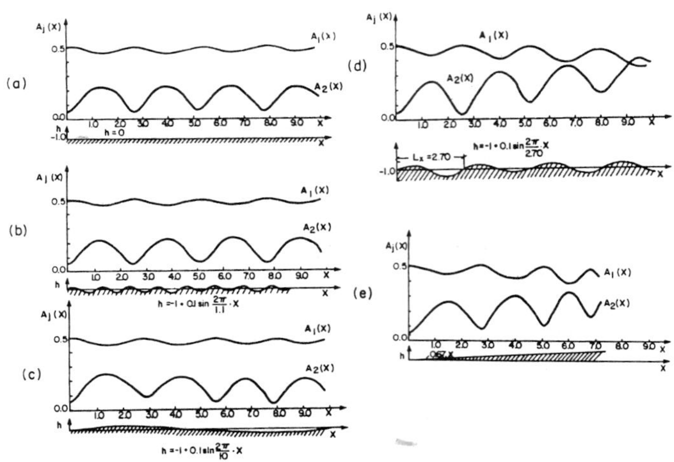

Figure 6. Rhythmical exchange of energy between the harmonic amplitudes $A_j(x) = |a_j(x)|$, $j = 1,2$, of an incident, sinusoidal wavetrain propagating over gently varying bed topographies: (a) a flat bed; (b) a wavy bed with corrugation-length scale shorter than the repetition length, L_λ; (c) a wavy bed with corrugation-length scale longer than the repetition length, L_λ; (d) wavy bed "tuned" to the repetition lenght, L_λ; (e) development of the wavetrain over a slightly sloped bed.

sinusoidal wavetrain is shown impinging on several, different slowly-varying bed topographies. Figure 6a shows the waves' response to a flat bed, and incidentally a nice display of the repetition that occurs over a flat bed. In 6b and 6c, the surface wave meets a wavy bed, but in both cases the length scale of the bed is quite different from the flat-bed repetition length. These bottom profiles draw about the same response as a flat bed. It is otherwise in 6d where the wavy bed is 'tuned' in length to the flat-bed repetition length, resulting in a substantial increase in the mean energy of the second harmonic. In 6e, an interesting example is presented where the wave encounters a gentle slope. Because the basic pattern of variation of A_1 and A_2 is similar over a wide range of topographies, it is tempting to use the term <u>repetition length</u> to mean the distance between successive minima of A_2, even in situations where this quantity varies with distance toward the shoreline. This terminology has proved to be of practical utility, and consequently it will be employed henceforth.

Turning now to equilibrium solutions, these are obtained from the full model by setting $\partial h/\partial T \equiv 0$ in (1.43), and solving the resulting system of equations. Since $\partial h/\partial T \equiv 0$, h is a function of x alone, and hence so are a_1 and a_2. Moreover, from (1.39), $\partial U_m/\partial x \equiv 0$, whence there is a constant Λ such that

$$U_m(x,T) \equiv \Lambda. \tag{2.8}$$

The constraint (2.8), when coupled with the Lau-Barcilon system (1.21), comprises the equations for a steady state, namely,

$$a_1'(x) + i\,\epsilon f_1 f(x) a_-(x) + i\alpha Q_1 e^{i\Delta kx} a_1^*(x) a_2(x) = 0,$$

$$a_2'(x) + i\,\epsilon f_2 f(x) a_2(x) + i\alpha Q_2 e^{-i\Delta kx} a_1^2(x) = 0, \tag{2.9}$$

with

$$\sum_{j=1}^{2} \frac{\omega_j}{k_j} |a_j(x)|^2 [1 - \frac{\beta^2}{6} h^2(x) k_j^2]^2 D_j = \Lambda.$$

It is convenient in our analysis to write a_1 and a_2 in the form

$$a_j(x) = A_j(x) e^{i\phi_j(x)}, \tag{2.10}$$

for $j = 1,2$, where $A_j = |a_j|$ as before. Also let

$$\theta(x) = 2\phi_1(x) - \phi_2(x) - x\,\Delta k \ . \tag{2.11}$$

The existence of the conservation law (2.1) implies that the four real differential equations represented in complex form in (1.39) may be integrated once, as Armstrong $\underline{et.al.}$ (1962) have noticed. The resulting system is

$$\frac{dA_1}{dx} = - \alpha Q_1 A_1 A_2 \sin(\theta),$$

$$\frac{dA_2}{dx} = \alpha Q_2 A_1^{\ 2} \sin(\theta), \tag{2.12}$$

$$\frac{d\theta}{dx} = - \Delta k - \epsilon f(x)(2f_1 - f_2) + \alpha(Q_2 \frac{A_1^{\ 2}}{A_2} - 2Q_1 A_2)\cos(\theta).$$

The equations (2.12) are just the Lau-Barcilon equations in another guise. For steady configurations, they are still constrained by the last equation in (2.9),

$$\sum_{j=1}^{2} \frac{\omega_j}{k_j} A_j^{\ 2}[1 - \frac{\beta^2}{6} h^2(x)k_j^{\ 2}]^2 D_j = \Lambda, \tag{2.13}$$

and the conservation law

$$\sum_{j=1}^{2} \frac{A_j^{\ 2}}{Q_j} = \alpha E, \tag{2.14}$$

where E is a constant. Because of (2.13) and (2.14), A_1, A_2, θ, and h are not independent. For example, if A_1 is known, then A_2 is determined from (2.14), h is then obtained from (2.13), and θ is found using the last equation in (2.12). Thus in principle the system describing steady states may be further reduced to two equations in two unknowns, say A_2 and θ. All this is predicated upon knowledge of the various parameters in the problem, α, β, ω_i, k_i, Q_i, and f_i, for $i = 1,2$. These in turn are all determined once α and β are specified. (In the variables defined in (1.2), the fundamental wavelength is 1. Consequently $k_1 = 2\pi$ and ω_1, $\omega_2 = 2\omega_1$, and k_2 are determined by the dispersion relation, (1.11).) The amplitude and wavelength parameters, α and β respectively, can take any values consistent with the order of magnitude assumptions posited earlier. The values of E and Λ depend upon how a_1, a_2, and h have been normalized. If h_0 is taken to be the value of the initial depth at $x = 0$, then

$$h(0) = 1. \tag{2.15}$$

Moreover, because u and ζ are order-one quantities when referred to the variables as they appear here, it is natural to pose

$$A_1^{\,2}(0) + A_2^{\,2}(0) = \frac{1}{4}. \tag{2.16}$$

In addition the phase is normalized by the stipulation that

$$\theta(0) = 0. \tag{2.17}$$

The simplest steady-state solutions are those where A_1, A_2, θ, and h are all x-independent, say equal to $A_1^{\,0}$, $A_2^{\,0}$, θ_0, and H_0, respectively. Such a solution necessarily satisfies

$$\theta_0 = 0, \quad H_0 = 1, \quad f(x) \equiv 0,$$
$$\Delta k - 2\alpha Q_1 A_2^{\,0} + \alpha Q_2 \frac{(A_1^{\,0})^2}{A_2^{\,0}} = 0, \tag{2.18}$$

and

$$(A_1^{\,0})^2 + A_2^{\,0})^2 = \frac{1}{4}.$$

Combining the last two equations in (2.18) leads to a quadratic equation for $A_2^{\,0}$, say, from which one deduces immediately that

$$A_2^{\,0} = \frac{-\Delta k + [(\Delta k)^2 + \alpha^2 Q_2(2Q_1 + Q_2)]^{1/2}}{2\alpha(2Q_1 + Q_2)},$$

$$A_1^{\,0} = [\tfrac{1}{4} - (A_2^{\,0})^2]^{1/2},$$

$$\theta_0 = 0, \quad H_0 = 1.$$

The equilibrium configurations written in (2.19) are not especially interesting. In particular, the notion of a repetition length plainly has no validity in their regard. However, these solutions may be perturbed to produce time-independent solutions with more structure.

To construct more general equilibrim configurations, it is helpful to reduce the system (2.12) along the lines already indicated. The choice of A_2 and θ as the primary dependent variables is convenient for the purposes at hand. Because of (2.14),

$$A_1^{\,2}(x) = Q_1[\alpha E - \frac{A_2^{\,2}(x)}{Q_2}], \tag{2.20}$$

and so A_1 is determined as the positive square root of the right-hand side of (2.20). For brevity, write

$$A_1(x) = F(A_2(x)), \tag{2.21}$$

where

$$F(y) = \{Q_1[\alpha E - \frac{y^2}{Q_2}]\}^{1/2}.$$

Using this relation in the last two equations in (2.12), there appear

$$\frac{dA_2}{dx} = \alpha Q_2 F(A_2(x))^2 \sin\theta(x),$$

$$\frac{d\theta}{dx} = -\Delta k + \varepsilon f(x)(f_2 - 2f_1) - 2\alpha Q_1 A_2(x)\cos\theta(x) \tag{2.22}$$

$$+ \alpha[Q_2 \frac{F^2(A_2(x))}{A_2^{\,2}(x)}]\cos\theta(x).$$

To complete the reduction to the two unknowns A_2 and θ, the depth variation $\varepsilon f(x) = h(x) - 1$ appearing in the second equation must be related to A_2 and θ. This is accomplished using (2.13) in conjunction with (2.21). Briefly, write

$$0 = h^4(x) \sum_{j=1}^{2} \frac{\beta^4}{36} \omega_j k_j^{\,3} A_j^{\,2}(x)D_j - h^2(x) \sum_{j=1}^{2} \frac{\beta^2}{3} \omega_j k_j A_j^{\,2}(x)D_j$$

$$+ \sum_{j=1}^{2} \frac{\omega_j}{k_j} A_j^{\,2}(x)D_j - \Lambda \tag{2.23}$$

$$= Ah^4(x) - Bh^2(x) + C,$$

163

say. Whenever A_1 appears, use (2.21) to express it in terms of A_2. The resulting quadratic equation may be solved thusly;

$$h^2(x) = \frac{B - [B^2 - 4AE]^{1/2}}{2A} = D,$$

say, so expressing h^2 as a function of A_2. The negative sign is taken in the solution of (2.23) because h is to be order one. Taking the positive square root, ϵf is obtained as a function of A_2,

$$\epsilon f(x) = \left[\frac{B - [B^2 - 4AE]^{1/2}}{2A} - 1 \right]^{1/2} \tag{2.24}$$

$$= G(A_2(x)).$$

Combining (2.24) with (2.22) gives a system for A_2 and θ, written symbolically as,

$$\frac{dA_2}{dx} = K(A_2, \theta),$$

$$\tag{2.25}$$

$$\frac{d\theta}{dx} = L(A_2, \theta),$$

where

$$K(y,z) = \alpha Q_2 F(y)^2 \sin(z)$$

and

$$L(y,z) = -\Delta k + (f_2 - 2f_1)G(y) - 2\alpha Q_1 y \cos(z) + \alpha [Q_2 \frac{F(y)^2}{y}] \cos(z).$$

The plane autonomous system (2.25) is investigated in a neighborhood of the constant solutions A_1^0, A_2^0, $\theta_0 = 0$, $H_0 = 1$ given in (2.18). The point (A_2^0, θ_0) comprises a critical point of (2.25), and therefore the behavior of the system close by this solution may be inferred from the properties of the linearized system

$$\frac{d(\delta A_2)}{dx} = \nabla K(A_2^0, \theta_0) \cdot (\delta A_2(x), \delta \theta(x)),$$

$$\tag{2.26}$$

$$\frac{d(\delta\theta)}{dx} = \nabla L(A_2^0, \theta_0) \cdot (\delta A_2(x), \delta\theta(x)),$$

near the origin $(0,0)$. Here $\nabla K = (\partial K/\partial A_2, \partial K/\partial \theta)$ and similarly for ∇L. After a little manipulation, (2.26) is put into the simple form

$$\frac{d^2(\delta A_2)}{dx^2} = - \alpha Q_2 W_1 (A_1^{\,0})^2 \delta A_2(x), \qquad (2.27)$$

$$\frac{d\theta}{dx} = - W_1 \delta A_2(x),$$

where

$$W_0 = \frac{3A_1^{\,0} \{\frac{\omega_2}{k_2}(1 - \frac{\beta^2}{6} k_2^2)^2 D_2 - \frac{\omega_1}{k_1}(1 - \frac{\beta^2}{6} k_1^2)^2 \frac{Q_1 D_1}{Q_2}\}}{\beta^2 \{\omega_2 k_2 (1 - \frac{\beta^2}{6} k_2^2) D_1 (A_2^{\,0})^2 + \omega_1 k_1 (1 - \frac{\beta^2}{6} k_1^2) D_2 (A_1^{\,0})^2\}}$$

and

$$W_1 = W_0(2f_1 - f_2) + 4\alpha Q_1 + \alpha Q_2 \left(\frac{A_1^{\,0}}{A_2^{\,0}}\right)^2.$$

As $W_1 > 0$ in the range of parameter values that interest us, the solutions of (2.27) are periodic of period $2\pi/(A_1^{\,0}\sqrt{\alpha Q_2 W_1})$. Thus infinitesimal perturbations of the constant states show periodicity, a kind of infinitesimal version of the repetition length discussed earlier, only now the bottom also has a periodic structure.

In any case, the foregoing calculations imply that $(0,0)$ is a center for the linearized system (2.26), and the general theory of plane autonomous systems (cf. Coddington and Levinson, 1955, chapter 15) assures that $(A_2^{\,0}, \theta_0)$ is itself a center, or else a spiral point, of the full nonlinear system (2.25).

The possibility of $(A_2^{\,0}, \theta_0)$ being a spiral point is discarded using the following argument. Supposing the contrary, let $(A_2(x), \theta(x))$ be an orbit of (2.25) that spirals into the critical point $(A_2^{\,0}, \theta_0) = (A_2^{\,0}, 0)$, say as $x \to +\infty$. Then by a translation of the independent variable, the normalization (2.17) that $\theta(0) = 0$ may be enforced. Let $\alpha > 0$ be the first zero of $\theta(x)$ for $x > 0$. Such a point a exists since $(A_2(x), \theta(x))$ spirals around $(A_2^{\,0}, 0)$. Define new functions $(r(x), \phi(x))$ on $[-a,a]$ as follows:

$$r(x) = \begin{cases} A_2(x), & 0 \leqslant x \leqslant a, \\ A_2(-x), & -a \leqslant x \leqslant 0, \end{cases} \qquad \phi(x) = \begin{cases} \theta(x), & 0 \leqslant x \leqslant A, \\ -\theta(-x), & -a \leqslant x \leqslant 0. \end{cases} \qquad (2.28)$$

A close examination convinces one that $(r(x), \phi(x))$ is a solution of (2.25) on
the entire interval of its definition, and that $(r(a), \phi(a)) = (r(-a), \phi(-a))$.
Extending the functions (r, ϕ) to all x by asking that they be periodic of
period $2a$, we obtain a closed orbit, that is, a periodic solution of (2.25). It
follows that $(r(x), \phi(x)) = (A_2(x), \theta(x))$ for all x, since if two orbits intersect they are identical, by uniqueness of solutions to the initial-value problem.
This contradicts the assumption that (A_2, θ) spirals in to $(A_2^0, 0)$ and proves that
(A_2^0, θ_0) is a center.

Figure 7 shows the dependence of the infinitesimal repetition length, the
periods of the solutions of the linearized system (2.26), as a function of α and
β. Note that L_λ decreases with an increase in either α or β, but that L_λ
depends more strongly on β than on α. This squares with the trends observed in
the laboratory for the repetition length over a horizontal bed (see Figure 8).

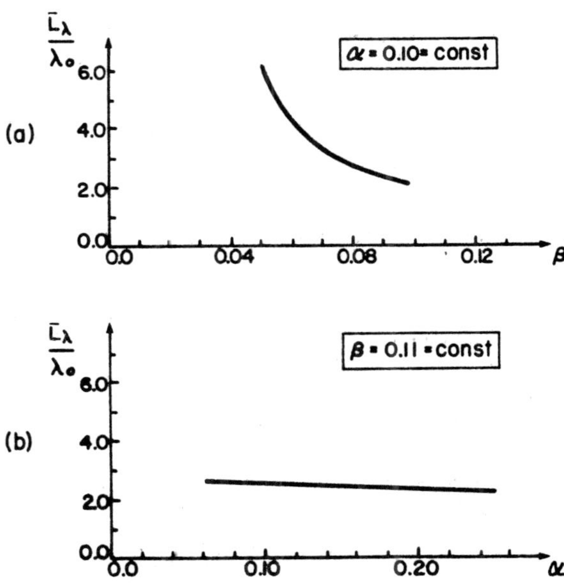

Figure 7. Dependence of the infinitesimal repetition length, L_λ, the periods of
the solution of the system (2.24), as a function of the wave parameters
α and β.

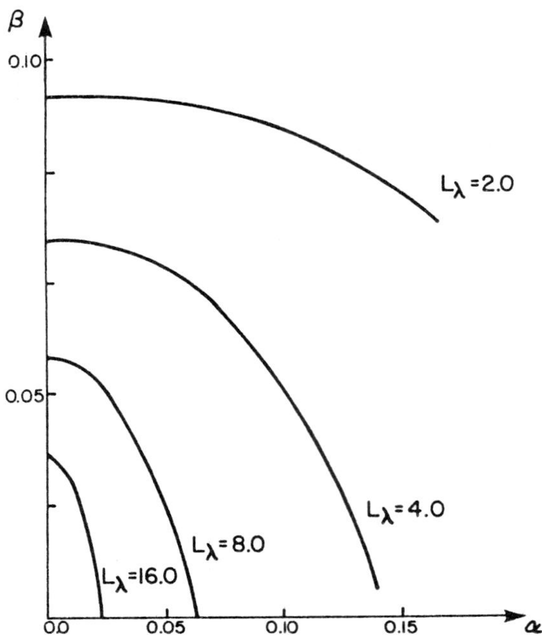

Figure 8.   Wavetank data for the repetition length $\overline{L}_\lambda/\lambda_0$ as a function of the wave parameters: (a) β with α = const, (b) α with β = const; (after Boczar-Karakiewicz and Bona, 1981).

The presence of periodic solutions to the linearized system is taken as an encouraging outcome, suggesting that it would be worthwhile to study the full, time-dependent version of the model. In this respect, the steady configurations can play another useful role, serving as 'exact' solutions on which to test the accuracy of a time-dependent numerical scheme. These developments are reported in the next Section.

3. Time-Dependent Configurations

The qualitative match of theory and experiment obtained in Section 2 regarding the steady-state solutions of the model suggested that the investigation of time-dependent solutions was worthwhile. In this section, a series of numerical integrations of the complete, time-dependent set of equations (1.43) will be described. In the present paper, only two types of initial bottom profiles are examined, namely, a ramp configuration (used in the wavetank experiments of Boczar-Karakiewicz et al. 1981, see Figure 1) and a flat bottom. In later stages of this investigation, more complex initial bed profiles will be introduced. Of especial interest are gently sloping beds which are typical of many different field situations.

A standard, forth-order correct, Runge-Kutta algorithm was used to integrate the first two equations in (1.43). A predictor-corrector scheme with a small amount of artificial damping was used to advance the bottom to the next time level. More precisely, the strong evidence of two, very disparate time scales, as described in Section 1, indicated that it would be propitious to decouple the temporal and spatial integrations. At a fixed time T, the wave harmonics the bed configuration $h(x,T)$ to be fixed. The updated bedform $h(x,T+\Delta T)$ is then determined from the third equation in (1.43). The process is then iterated. It gets its start from the given initial bed profile, of course.

The precise technique utilized is now described. The region of space and time of interest is the rectangle comprised of those points (x,T) for which $0 \leqslant x \leqslant M$ and $0 \leqslant T \leqslant T_0$. Consider the grid $(i\,\Delta x, j\,\Delta T)$ in (x,T) space where i and j are integers with $0 \leqslant i \leqslant N$, $0 \leqslant j \leqslant J$, $M = N\Delta x$, and $T_0 = J\Delta T$. For $x = i\,\Delta x$ and $T = j\,\Delta T$, let h^{ij} denote the approximate value of $h(x,T)$, and let f^{ij} denote the approximate value of $f(x,T)$, where f is the function defined in (1.8) that corresponds to h. Let a_1^{ij} and a_2^{ij} denote the approximate values of a_1 and a_2, respectively, at the same spatial and temporal point. Suppose at time T that h^{ij} is known for each value of i between 0 and N. Then values of a_1^{ij} and a_2^{ij}, for $0 \leqslant i \leqslant N$, are computed from equation (1.21) using the standard, forth-order correct, Runge-Kutta scheme (cf. Isaacson & Keller, 1966, Ch. 8). The boundary conditions explained in (1.19) are imposed in the form

$$a_1^{0j} = \bar{a}_1^{0} \quad \text{and} \quad a_2^{0j} = a_2^{0} \, ,$$

and used to initiate the numerical integration of (1.21). Once a_1^{ij}, a_2^{ij}, and h^{ij}, $0 < i < N$, are in hand, an approximation ϕ^{ij} to the function U_m at the points $x = i\Delta x$, $0 < i < N$, at time T is immediately available by application of formulas (1.40) and (1.41). A fourth-order approximation y^{ij} to $\partial U_m/\partial x$ at the points $x = i\Delta x$, $T = j\Delta T$, $0 < i < N$, is then obtained using standard, centered difference formulas, with appropriate modifications near the left- and right-hand endpoints of the spatial domain. An updated approximation $h^{i,j+1}$, $0 < i < N$, of the bottom profile is then obtained using a split-step technique consisting of the fourth-order correct Moulton's method (see, again, Isaacson & Keller, 1966) to predict a first version $\tilde{h}^{i,j+1}$ of the bottom followed by an averaging step

$$h^{i,j+1} = [\tilde{h}^{i-1,j+1} + \tilde{h}^{i,j+1} + \tilde{h}^{i+1,j+1}]/3.$$

The initial bottom profile, $h(x,0)$, is used to start Moulton predictor step for the case $j = 0$.

The spatial averaging that is featured in the method outlined above is used to add stability to the computational algorithm. It amounts to integrating the nonlinear parabolic equation

$$\frac{\partial h}{\partial T} = \frac{\gamma \Delta x}{3} \frac{\partial^2 h}{\partial x^2} + \frac{\partial \phi}{\partial x} \qquad \text{where} \quad \gamma = \lim_{\substack{\Delta x \to 0 \\ \Delta T \to 0}} \frac{\Delta x}{\Delta T} > 0. \qquad (3.1)$$

rather than the conservation law

$$\frac{\partial h}{\partial T} = \frac{\partial \phi}{\partial x} . \qquad (3.2)$$

The damping term deserves some comment. As far as the present article is concerned, this term is strictly artificial, and has been introduced to stabilize the temporal integration. However, we will argue in subsequent studies that a term something like this should appear when more detailed modelling of the sedimentation process is considered. As long as the solution of the differential equation (3.2) is smooth, then the error introduced by the damping term becomes negligible in the limit as the discretization parameters Δx and ΔT tend to zero appropriately.

The reliability of the integrations was checked in some detail. The temporal integration gave more trouble and the solutions did vary slightly (a maximum of about 10%) as ΔT was decreased, but their global behavior did not change significantly. As regards the spatial integration, a convergence study was performed with decreasing values of Δx . For such a study the exact, constant solutions of the steady-state equations (2.19) were used as test cases. For these solutions, the values of ζ_1 and ζ_2 must vary by quantities that are of order one so that A_1 and A_2 remain constant. Figure 9 shows a log-log plot of the absolute value of the maximum error in A_1 and A_2 over the spatial interval (0,10). The ordinate gives the percentage error and the abscissa shows the number of intervals used in the spatial integration. Notice that at a point lying between $\Delta x = 1/16$ and $\Delta x = 1/32$ numerical roundoff appears to prohibit any further increase in the accuracy of the integration. The slope of the dotted line shows how accuracy should behave in the fourth order scheme. This behavior is observed in the range where $\Delta x > 1/16$.

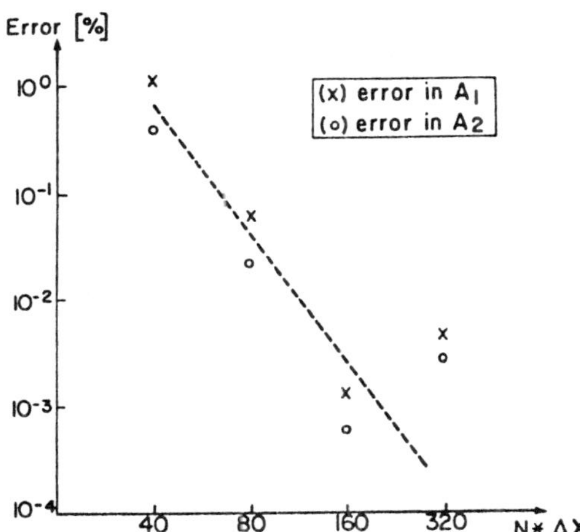

Figure 9. A convergence study of the numerical scheme for integrating the model equations (2.19). The predicted convergence rate for the integration algorithm used should be $O(\Delta x^4)$.

Figure 10 depicts several stages of the evolution of an initial ramp bottom profile as well as the associated wave field's harmonic components. The incident wave field is characterized by $\alpha = 0.1$ and $\beta = 0.08$, values that correspond closely to values of these parameters that obtain in actual flume experiments. The depth changes relatively rapidly in the earlier stages of the integration, but the entire system settles down to an equilibrium configuration in due course (shown at $T = 640\Delta T$). Notice in Figure 10 that the left-hand side of the (x,ζ) domain, where the waves enter the system, appears to reach its final configuration first. This corresponds to what is observed in wavetank experiments (see again Figure 1 and the article of Boczar-Karakiewicz <u>et al.</u> 1981).

A sample of the long term equilibrium solutions of our model system is presented in Figure 11. In these examples, the bottom configuration was completely flat at $T = 0$, and incident wave field was given four different pairs of parameter values ($\alpha = 0.05$ and 0.15, $\beta = 0.07$ and 0.09). Notice that the equilibrium repetition lengths appearing in Figure 11 vary with α and β in the same way that they do in the graph in Figure 7. The repetition length decreases with increasing α and β, and depends much more strongly upon β than α.

The equilibrium states of Figures 10 and 11 are all qualitatively similar. In more extensive studies we have found that a comparatively wide range of initial bottom configurations evolves into equilibrium bottom configurations that are periodic in x with a well defined repetition length that characterizes both the exchange of energy between the first and second harmonic components and the bottom variation. Again, this is the same state of affairs that obtains in wavetank experiments.

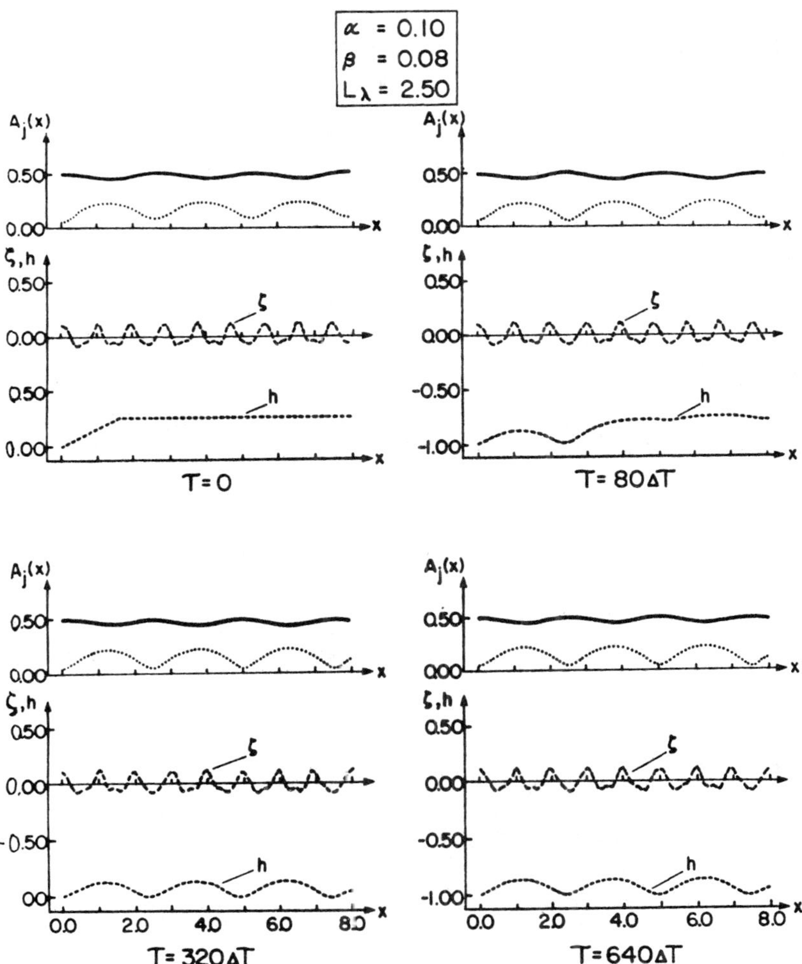

Figure 10. Final state solution (at $T = 640\Delta T$) of the time dependent model equations starting at $T = 0$ from an initial ramp bottom profile. The horizontal coordinate gives distances in units of the wavelength, the incident wave parameters are $\alpha = 0.10$ and $\beta = 0.08$; - connotes the value of $A_1(x)$ and --- connotes the value of $A_2(x)$.

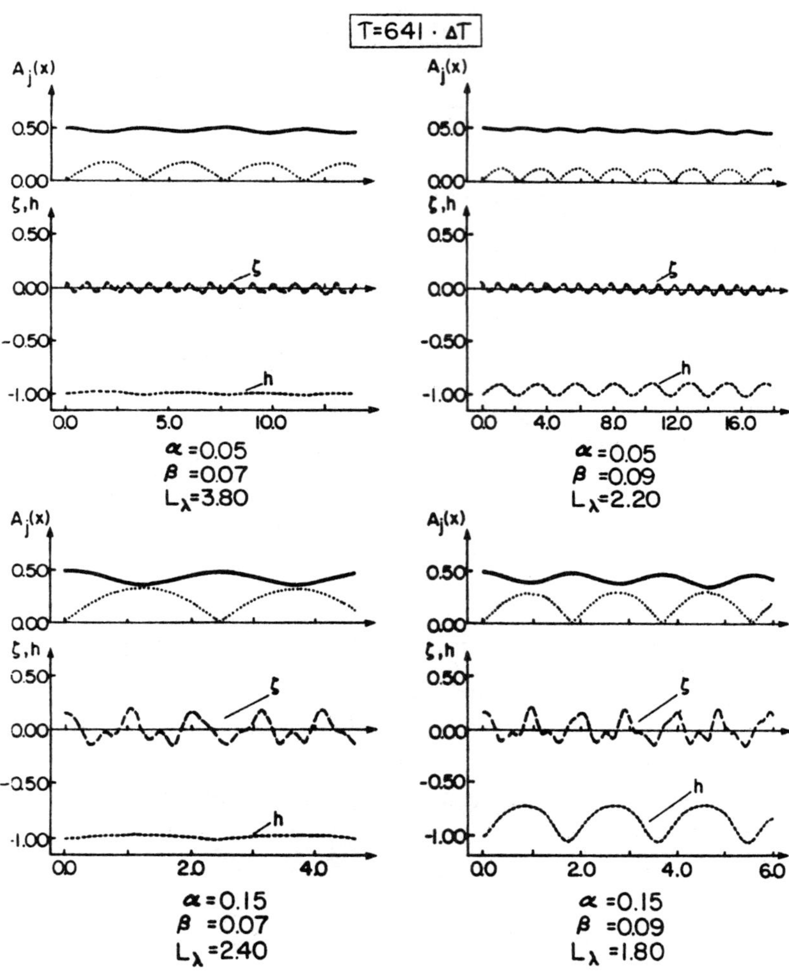

Figure 11. Final state solution (at $T = 641 \Delta T$) of the time dependent model equations starting at $T = 0$ from an initial flat bottom profile. The horizontal coordinate gives distances in units of the wavelength, the incident wave parameters are $\alpha = 0.05$ and 0.015; $\beta = 0.07$ and 0.09; L_λ denotes the corresponding repetition length; — connotes the values of $A_1(x)$ and --- connotes the values of $A_2(x)$.

4. Conclusions

It is a well known fact that the interaction of shallow water waves and bottom sediment may lead to the formation of equilibrium bar and trough configurations. Here we have used a simple, two-dimensional model to relate the change over time of a bottom profile to the spatial development of the wavefield's first and second harmonics. The resulting set of equations has been investigated both for equilibrium solutions and for time-dependent solutions. It has been found that time-dependent solutions may evolve toward an attractor which is an equilibrium solution.

Although data from wavetank experiments does not yet permit realistic quantitative comparison with our theory, the length scales and qualitative features predicted by our model agree tolerably well with laboratory findings. Preliminary experience with the model as regards field situations also holds up promise for its utility (cf. Boczar-Karakiewicz and Bona 1981, Boczar-Karakiewicz _et al_. 1984). Further investigation of the model's predictive power in both laboratoy and especially field situations seems warranted.

It is worth emphasis that the model developed here has been kept as simple as possible, and that there are many ways in which it might be improved. First, reflection is important on many real beaches, and this should certainly be taken into account in modelling the wave field. Secondly, dissipation of the wave field is important in some regimes to which the model might apply, and this too should be incorporated into the model. At a more sophisticated level, there is no reason why Boussinesq or Korteweg-de Vries type equations could not be used in modelling the wavefield, rather than passing to the two-harmonic approximation. One could even imagine retaining the full Euler equations if need be. More complex models for the boundary layer and the associated mass transport could be incorporated (cf. Bagnold 1963, Drew 1979, or Bowen 1980). And, one need not fix the sediment distribution in the fluid, but rather determine this locally according to the wave field. Finally, it would be worthwhile extending the model in the other spatial dimension, so allowing for some gradual variation along a coastline.

We expect to report on various of these theoretical and practical issues at a later stage.

References

Abbott, M.B. 1960. Boundary layer effects in estuaries. J. Mar. Res., 18: 83-100.

Amick, C.J. 1984. Regularity and uniqueness of solutions of the Boussinesq systems of equations. J. of Diff. Equ., 54: 231-247.

Armstrong, J.A., N. Bloombergen, J. Ducuing and P.S. Pershan. 1962. Interaction between light waves in a nonlinear dielectric. Phys. Rev., 127: 168-177.

Bagnold, R.A. 1963. Mechanics of marine sedimentation. In: The Sea, Hill, M.N. (ed.) Wiley-Interscience, New York, 3: 507-528.

Benjamin, T.B., B. Boczar-Karakiewicz and W.G. Pritchard. (To appear). Formation of sand bars by waves. Part 1: Flow over fixed beds.

Benjamin, T.B., J.L. Bona and J.J. Mahony. 1972. Model equations for long waves in nonlinear dispersive systems. Phil. Trans. Roy. Soc. Lond., Series A, 272: 47-78.

Boczar-Karakiewicz, B., B. Paplinska and J. Winiecki. 1981. Formation of sand bars by surface waves in shallow water. Laboratory experiments. Rozprawy Hydrotechniczne, 43: 111-125.

Boczar-Karakiewicz, B. and J.L. Bona. 1981. Uber die riffbildung an sandigen küsten durch progressive wellen. Mitteilungen des Leichtweiß-Instituts für Wasserbau der Technischen Universität Braunschweig, 70: 380-420.

Boczar-Karakiewicz, B., G. Drapeau and B.F. Long. 1984. Modélisation des barres sableuses littorales de la partie nord des Iles-de-la-Madeleine. Sciences et Techniques de l'Eau, 17: 35-39.

Bona, J.L., W.G. Pritchard and L.R. Scott. 1981. An evaluation of a model equation for water waves. Phil. Trans. Roy. Soc. Lond., Series A, 302: 457-510.

Bona, J.L. and V.A. Dougalis. 1980. An initial- and boundary-value problem for a model equation for propagation of long waves. J. Math. Anal. Appl., 75: 501-522.

Bona, J.L. and R. Smith. 1976. A model for the two-way propagation of water waves in a channel. Math. Proc. Camb. Phil. Soc., 79: 167-182.

Boussinesq, J. 1871. Théorie de l'intumescence liquide appelée onde solitaire ou de translation se propageant dans un canal rectangulaire. Comptes Rendus, 72: 755-759.

Bowen, A.J. 1980. Simple models of nearshore sedimentation; beach profiles and longshore bars. In: The Coastline of Canada, McCann, S.B. (ed.), Geological Survey of Canada, paper 80-10: 1-11.

Bowen, A.J. and D.L. Inman. 1971. Edge waves and crescentic bars. J. Geophys. Res., 76: 8662-8671.

Carter, T.G., P.L.F. Liu and C.C. Mei. 1973. Mass transport by waves and offshore sand bedforms. J. Waterways, Harbours, Coastal Eng., 99: 165-184.

Coddington, E.A. and N. Levinson. 1955. Theory of ordinary differential equations. McGraw-Hill, New York.

Drew, D.A. 1979. Dynamic model for channel bed erosion. J. Hydraulics Division A.S.C.E., 105: 721-735.

Druet, Cz, S.R. Massel, and R. Zeidler. 1972. The structure of wind waves in the southern coastal zone of the Baltic Sea. Rozprawy Hydrotechniczne, 30: 312-318 (in Polish).

Elgar, S. and R.T. Guza. 1985. Shoaling gravity waves: comparison between field observations, linear theory and a nonlinear model. J. Fluid Mech., 158: 47-70.

Greenwood, B. and R.G.D. Davidson-Arnott. 1979. Sedimentation and equilibrium in wave-formed bars: A review and case study. Can. J. Earth Sci., 16: 312-332.

Hammack, J.L. 1973. A note on tsunamis: their generation and propagation in an ocean of uniform depth. J. Fluid Mech., 60: 769-799.

Hammack, J.L. and H. Segur. 1974. The Korteweg-de Vries equation and water waves. Part 2. Comparison with experiments. J. Fluid Mech., 65; 289-314.

Isaacson, E. and H.B. Keller. 1966. Analysis of numerical methods. John Wiley, New York.

Johns, B. 1970. On the mass transport induced by oscillatory flow in a turbulent boundary layer. J. Fluid Mech., 43: 177-185.

Johns, B. 1975. The form of the velocity profile in a turbulent shear wave boundary layer. J.Geophys. Res., 80: 5109-5112.

Johns, B. 1977. Residual flow and boundary shear stress in the turbulent bottom layer beneath waves. J. Phys. Oceanogr. 7: 733-738.

Johnson, R.S. 1973. On the development of a solitary wave moving over an uneven bottom. Proc. Camb. Phil. Soc., 73: 183-203.

Kakutani, T. 1971. Effect of an uneven bottom on gravity waves. J. Phys. Soc. Japan, 30: 272-276.

Kortweg, D.J. and G. de Vries. 1895. On the change of form of long waves advancing in a rectangular channel, and on a new type of long stationary waves. Phil. Mag., 5: 422-443.

Lau, J. and A. Barcilon. 1972. Harmonic generation of shallow water waves over topography. J. Phys. Oceanogr., 2: 405-410.

Lau, J. and B. Travis. 1973. Slowly varying Stokes waves and submarine longshore bars. J. Geophys. Res., 78: 4489-4497.

Longuet-Higgens, M.S. 1953. Mass transport in water waves. Phil. Trans. Roy. Soc. Lond., Series A, 245: 535-581.

Longuet-Higgens, M.S. 1981. Oscillating flow over steep sand ripples. J. Fluid Mech., 107: 1-35.

Madsen, O.S. 1971. On the generation of long waves. J. Geophys. Res., 76: 8672-8683.

Madsen, O.S. and C.C. Mei. 1969. The transformation of a solitary wave over an uneven bottom. J. Fluid Mech., 39: 781-791.

McClain, C.R., N.E. Huang and L.J. Pietrafesa, 1977. Application of a radiation type boundary condition to the wave, porous bed problem. J. Phys. Oceanogr., 7: 823-835.

Mei, C.C. and B.L. Mehauté. 1966. Note on the equations of long water waves over an uneven bottom. J. Geophys. Res., 71: 393-400.

Mei, C.C. and U. Ünlüata. 1972. Harmonic generation of shallow waves. In: Waves on beaches and resulting sediment transport, Meyer, R.E., (ed.), Academic Press, New York: 181-202.

Peregrine, D.H. 1972. Equations for water waves and the approximation behind them. In: Waves on beaches and resulting sediment transport, Meyer, R.E. (ed.), Academic Press, New York: 95-121.

Peregrine, D.H. 1967. Long waves on a beach. J. Fluid Mech., 27: 815-825.

Raudkivi, A.J. 1976. Loose boundary hydraulics. 2nd edition, Pergamon Press, Oxford.

Russel, R.C.H. and J.D.C. Osorio. 1958. An experimental investigation of drift profiles in a closed channel. Proc. 6th Conf. Coast. Eng., London: 171-193.

Schonbek, M.E. 1981. Existence of solutions for the Boussinesq system of equations, J. Diff. Equa., 52: 325-352.

Sleath, J.F.A. Measurements of bed load in oscillatory flow. J. Waterway Port Coast. Ocean Div., 4: 291-322.

Stokes, G.G. 1847. On the theory of oscillatory waves. Trans. Camb. Phil. Soc. Collected papers, 1: 197-229.

Svendsen, I.A. and J. Buhr-Hanson. 1978. On the deformation of periodic long waves over a gently sloping bottom. J. Fluid Mech., 87: 433-448.

Thornton, E.B., J.J. Galvin, F.L. Bub and D.P. Richardson. 1976. Kinematics of breaking waves. Proc. 15th Conf. Coast. Eng., Honolulu: 460-476.

Thornton, E.B. and G. Schaeffer. 1978. Probability density functions of breaking waves. Proc. 16th Conf. Coast. Eng., Hamburg: 507-519.

Van de Graff, I. and W.M.K. Tilmans. 1980. Sand transport by waves. Proc. 17th Coast. Eng. Conf., Sydney: 1140-1157.

Whitham, G.B. 1974. Linear and nonlinear waves. John Wiley & Sons, New York.

Winther, R. 1982. Finite element method for a version of the Boussineq equation. SIAM J. Num. Anal., 19: 561-570.

Zabusky, N.J. and C.J. Galvin. 1971. Shallow-water waves, the Korteweg-de Vries equation and solitons. J. Fluid Mech., 47: 811-824.

Zenkovitch, V.P. 1967. Processes of coastal development. Interscience, New York.

CLASSICAL PIEZOELECTRICITY - IS THE THEORY COMPLETE?

Peter J. Chen

Sandia National Laboratories
Albuquerque, New Mexico 87185

Abstract

In this paper, we cite experimental evidences which show that the classical theory of piezoelectricity is not complete in describing the responses of ferroelectric ceramics. There are also some preliminary evidences that the theory as applied to piezoelectric single crystals needs more thorough investigations.

1. Introduction

The classical linear constitutive relations describing the responses of piezoelectric materials are of the form

$$
\begin{aligned}
T_{ij} &= C_{ijkm} S_{km} - e_{kij} E_k, \\
D_i &= e_{ikm} S_{km} + \varepsilon_{ik} E_k
\end{aligned}
\tag{1.1}
$$

where C_{ijkm}, e_{ijk} and ε_{ij} are the elastic, piezoelectric and dielectric constants. T_{ij} is the stress, D_i is the electric displacement, E_i is the electric field given by

$$
E_i = -\phi_{,i}
\tag{1.2}
$$

where ϕ is the electric potential, and S_{ij} is the strain, defined by

$$
S_{ij} = \frac{1}{2} (u_{i,j} + u_{j,i})
\tag{1.3}
$$

with u_i being the mechanical displacement. To the preceding equations we must adjoin the field equations, viz. balance of linear momentum

$$
T_{ij,j} = \rho_0 \ddot{u}_i
\tag{1.4}
$$

where ρ_0 is the reference mass density; and Gauss' law for a charge free body

178

$$D_{i,i} = 0. \qquad\qquad (1.5)$$

Thus, there are 22 equations for the determination of 22 unknowns.

The system of equations is said to apply to single crystals as well as polycrystalline materials such as poled ferroelectric ceramics. With the x_3-axis being the poling direction, ferroelectric ceramics effectively have the symmetry of a hexagonal crystal in class C_{6v}, Tiersten [1], p. 54. The non-zero material constants are

$$c_{1111} = c_{2222}, \; c_{3333},$$
$$c_{1122}, \; c_{1133} = c_{2233}, \; c_{2323}$$

with

$$c_{1212} = \frac{1}{2}(c_{1111} - c_{1122});$$

$$e_{113} = e_{223}, \; e_{311} = e_{322}, \; e_{333};$$

and

$$\varepsilon_{11} = \varepsilon_{22}, \; \varepsilon_{33}.$$

A virgin or thermally depoled ferroelectric ceramic is generally regarded as isotropic in the macroscopic context <u>exhibiting no piezoelectric coupling</u> with

$$c_{1111} = c_{3333}, \; c_{1122} = c_{1133}, \; c_{1212} = c_{2323}, \; \varepsilon_{11} = \varepsilon_{33},$$

i.e. a virgin or thermally depoled ceramic is isotropic and presumably electro-mechanically <u>inactive</u>.

2. Some Background Information

It is common knowledge that when a piezoelectric specimen is driven by a small amplitude cyclic electric field "resonances" at particular frequencies can occur depending on the geometry of the specimen and its material properties. Because of the existence of resonances, resonators have been employed extensively in numerous applications as well as in the determination of material properties including, of course, those of poled ferroelectric materials; see, for example, the text Piezoelectric Ceramics [2], and Holland and EerNisse [3]. The conventional method of discerning the onset of "resonances" is to monitor the admittance (or impedance) which becomes large (or small) at such instances. The divergence of the admittance is equivalent to the divergence of the time rate of the change of the electric displacement, so that these resonances may be viewed as <u>electrical</u> resonances. It is generally believed that <u>mechanical</u> resonances of the specimen also occur simultaneously. This belief has been the basis of much theoretical and experimental work for many years. Perhaps, the best single source of the current practices is the IEEE Standard on Piezoelectricity [4].

Recent experimental observations have shown that the conventional belief and practice need not be completely valid. This is because mechanical resonances of quite large amplitudes can exist in piezoelectric and ferroelectric materials independent of any noticeable electrical disturbance. These resonances occur at frequencies which are much lower than those of the lower detectable electrical resonances and at intermediate frequencies between electrical resonances. Mechanical resonances, but <u>no</u> electrical resonances, are also detected in virgin and thermally depoled ferroelectric ceramic specimens, and these resonances also occur when the specimens are subsequently poled, Chen [5] and [6]. Examples of some of the specimens which have been examined are X-cut quartz, Z-cut $LiNbO_3$, ADP, slim loop ferroelectrics, PZT65/35, PZT7/65/35, $BaTiO_3$ ceramic, Clevite PZT8 and Channel 5500 ceramic.

3. Experimental Results

During the early stages of our work, the measurements were carried out using a single-beam displacement laser interferometer system described in the paper by Allensworth [7]. Briefly, we measure the axial mechanical displacements of selected surface points of cylindrical disc specimens, and the phase angles between the mechanical displacements and driving ac voltages. We also measure the charge across an integrating capacitor of large capacitance connected in series with specimens, and the phase angles between the electric displacements and the applied voltages. Time rate of change of the charge gives the current in the circuit. Recently, we have developed a dual-beam laser interferometer system with a single light source. This system permits us to measure the mechanical displacements of opposite points on the surfaces of the specimens which are held in nearly traction-free conditions. Some of the details of this system are given in the paper by Chen [8].

Initial announcements of the existence of electrically excitable mechanical resonances in ferroelectric ceramics are given by Chen [5] and [6]. Results concerning virgin, thermally depoled, and poled specimens are reported in these papers. It is shown that for virgin and thermally depoled specimens only mechanical resonances at particular frequencies are detected, and these resonances are not accompanied by any detectable electrical disturbance. Electrical resonances are, of course, detected for poled specimens and these resonances are always accompanied by mechanical resonances. However, mechanical resonances which are not accompanied by detectable electrical disturbances continue to be present. Certain characteristics of two of these resonances are reported in [9]. In a subsequent paper [10], an extensive experimental study has been carried out to show that the electrically excitable mechanical resonances are influenced by the domain structure of the specimens even though the specimens do not exhibit any remanent polarization.

Perhaps, the most interesting observation thus far is that virgin and thermally depoled ferroelectric materials such as PLZT7/65/35 and PZT65/35 exhibit a new electric to mechanical coupling phenomenon. The results concerning PLZT7/65/35 are reported in [8]. Along any diameter of a disc the mechanical displacements of both surfaces are parabolic and are in the positive direction of a small magnitude applied static electric field with no measureable axial strain. Similar results are valid when the direction of the applied static field is switched. These results cannot be explained using the theory of §1.

Because of this new electric to mechanical coupling phenomenon, the specimen vibrates flexurally when driven by cyclic electric fields, and flexural resonances occur at particular frequencies. The flexural resonant frequencies agree quite well with those predicted within the framework of the classical bending theory of thin plates, see e.g. [12], viz.

$$D \nabla^4 u_3 = -\rho_0 h \ddot{u}_3 \qquad (3.1)$$

where h is the plate thickness, and

$$D = \frac{E h^3}{12(1 - \nu^2)} \qquad (3.2)$$

with E being Young's modulus and ν being Poisson's ratio. The nature of the resonant mode shapes are, however, quite interesting. Those with <u>only</u> nodal rings are as predicted by the classical theory. On the other hand, those with nodal diameters (nodal rings may or may not be present) are not. The classical resonant mode shapes are exhibited by the quadrature components of the measured mechanical displacements when nodal diameters are present. The details of these results are given in the paper by Chen [11]. In Fig. 1, we exhibit examples of two such resonances of a virgin PLZT7/65/35 specimen of diameter 2.342×10^{-2} m and thickness 2.616×10^{-4} m. The contours of the amplitudes of the displacements are on the left, and the contours of the quadrature components are on the right. The units of the contours are 10^{-9} m. Clearly, the contours of quadrature components exhibit mode shapes with 2 nodal diameters and 1 nodal ring (10.405 kHz), and 3 nodal diameters (3.598 kHz) as predicted by the classical theory.

4. Closure

It is not difficult to conclude from the experimental results cited in §3 that the classical equations of piezoelectricity do not completely describe the responses of ferroelectric ceramics. There is evidence to support the existence of a new electric to mechanical coupling phenomenon which is not accounted for in the equations given in §1. This is particularly true for virgin or thermally depoled ferroelectric ceramics for which one can excite mechanical resonances electrically, and yet classical piezoelectricity states that these ceramics are isotropic exhibiting no electromechanical coupling. More detailed experiments

182

concerning piezoelectric single crystals will be conducted in the near future.

Acknowledgment

The continuing assistance of Dwight L. Allensworth is gratefully acknowledged. This work performed at Sandia National Laboratories was supported by the U.S. Department of Energy under contract #DE-AC04-DP00789.

References

1. H.F. Tiersten, Linear Peizoelectric Plate Vibrations. Plenum Press, New York, 1969.

2. Piezoelectric Ceramics, ed. by J. vanm Randeraat. Applications Book, published by Electronic Components and Materials Division, N.V. Philips, Eindhoven, 1968.

3. R. Holland and E.P. EerNisse, Accurate measurements of coefficients in a ferroelectric ceramic. IEEE Trans. Sonics Ultrason. SU-$\underset{\sim}{16}$, 173-181 (1969).

4. IEEE Standard on Piezoelectricity. IEEE Std 178-1978, The Institute of Electrical and Electronics Engineers, Inc. (1978).

5. P.J. Chen, Observation of the existence of electrically excited purely mechanical resonances in piezoelectric and ferroelectric materials. Il Nuovo Cimento $\underset{\sim}{2}$D, 1145-1155 (1983).

6. P.J. Chen, Electrically excitable purely mechanical resonances in piezoelectric and ferroelectric materials - geometrical considerations. Wave Motion $\underset{\sim}{5}$, 177-183 (1983).

7. D.L. Allensworth, Interferometer for the determination of strains due to domain switching in ferroelectrics. Rev. Sci. Instrum. 51, 1330-1334 (1980).

8. P.J. Chen, A new electromechanical coupling phenomenon in the electro-optic ceramic PLZT7/65/35. Il Nuovo Cimento $\underset{\sim}{4}$D, 280-292 (1984).

9. P.J. Chen, Certain characteristics of electrically excitable mechanical resonances in PZT65/35. Acta Mechanica $\underset{\sim}{51}$, 217-226 (1984).

10. P.J. Chen, Effects of domain structure of electrically excitable mechanical resonances in ferroelectric ceramics. Int. J. Solids Struct. $\underset{\sim}{20}$, 121-128 (1984).

11. P.J. Chen, Electrically excitable mechanical resonant mode shapes in the electro-optic ceramic PLZT7/65/35, Int. J. Solids Struct., forthcoming.

12. R.C. Colwell and H.C. Hardy, The frequencies and nodal systems of circular plates. Phil. Mag. S7, $\underset{\sim}{24}$, 1041-1055 (1937).

183

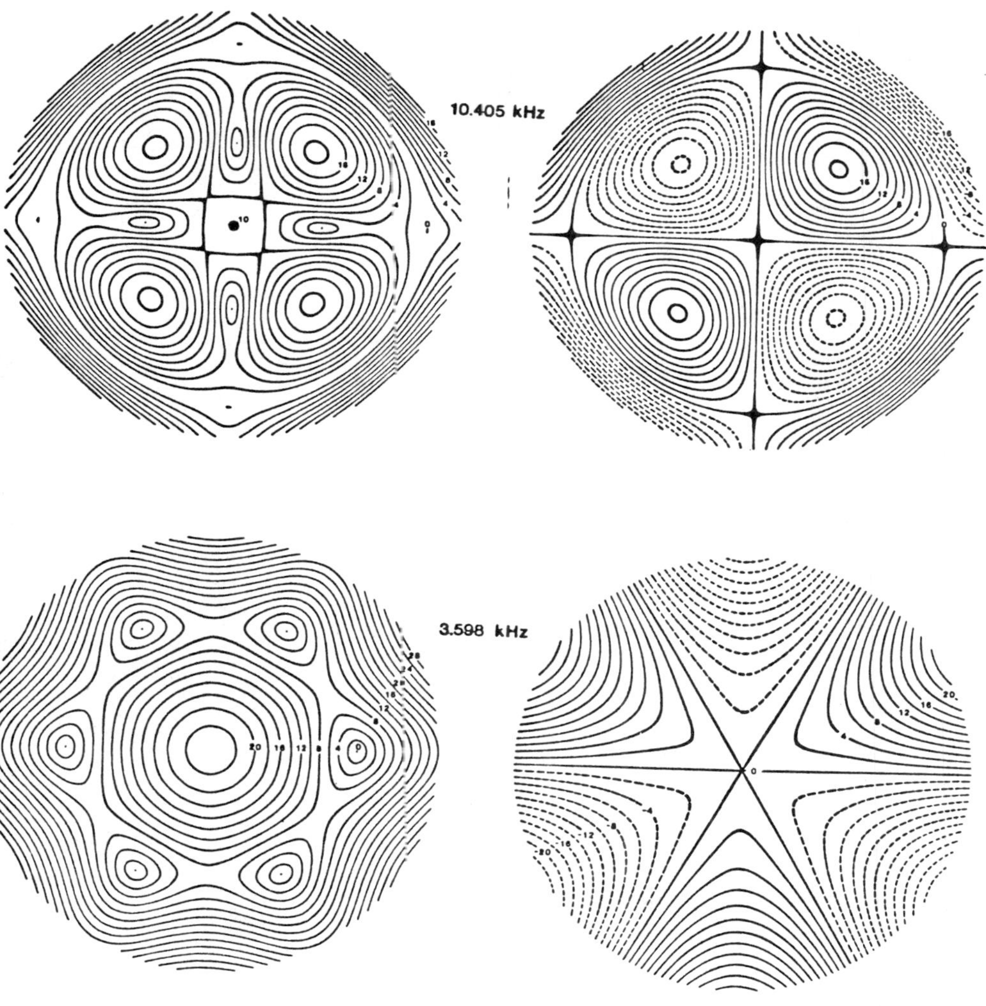

Figure Caption

Figure 1   Contours of amplitudes of mechanical displacements (figures on the
left) and contours of quadrature components (figures on the right).
The units of the contours are 10^{-9}m.

ON THE DYNAMICS OF A COLLISIONLESS PLASMA

Robert T. Glassey

Walter A. Strauss

Department of Mathematics
Indiana University
Bloomington, Indiana   47405

Department of Mathematics
Brown University
Providence, Rhode Island   02912

A plasma is, roughly speaking, a bunch of many fast-moving charged particles. Physicists are fond of saying that 99% of all matter in the universe is in the plasma state. The interiors of stars, the ionosphere, the solar wind, and fusion reactors such as tokamaks provide examples of plasmas. Under certain circumstances the kinetic effects of the particles dominate and collisions can be neglected. This is the case for the solar wind, which is very dilute, and for a fusion reactor after it is powered up. Then the only forces are the attraction and repulsion of charges.$^{(*)}$

Imagine several types of particles, indexed by $\alpha = 1,\ldots,N$; for example, electrons and ions. A single particle of species α, with mass m_α and charge e_α, would experience a force

$$m_\alpha \ddot{x} = e_\alpha (E + \frac{v}{c} \times B)$$

due to the electromagnetic field E, B. This equation is sufficient for most engineering applications but, since some of the particles are expected to travel at relativistically fast speeds, we need to replace it by its relativistic version

$$\dot{x} = \hat{v}_\alpha \equiv v(m_\alpha^2 + |v|^2/c^2)^{-1/2}$$

(1)

$$\dot{v} = e_\alpha (E + c^{-1}\hat{v}_\alpha \times B).$$

Here v is the momentum and $\hat{v}_\alpha$ the velocity of the particle. Note that $|\hat{v}_\alpha| < c$ as it should be.

---

*   Lecture given by the second author at the IMA Workshop on Dynamical Problems in Continuum Physics, June 3, 1985.

*   Research supported in part by NSF MCS83-19944 and NSF DMS84-20957.

Now let $f_\alpha(t,x,v)$ be the density of the particles of species α in phase space $(x,v) \in R^3 \times R^3$. Thus its integral over a box $\Omega_x \times \Omega_v$ in six-dimensional space yields the number of particles in Ω_x with momentum in Ω_v at time t. Conservation of mass immediately yields a partial differential equation for f_α,

$$
(2) \qquad \frac{\partial f_\alpha}{\partial t} + \hat{v}_\alpha \cdot \nabla_x f_\alpha + e_\alpha (E + \frac{1}{c} \hat{v}_\alpha \times B) \cdot \nabla_v f_\alpha = 0,
$$

known as the Vlasov equation. Note that (1) are the characteristic equations for (2). There is a Vlasov equation for each species α. The charge and current densities due to all the particles are

$$
(3) \qquad
\begin{aligned}
\rho(t,x) &= 4\pi \sum_\alpha \int e_\alpha f_\alpha \, dv, \\[2mm]
j(t,x) &= 4\pi \sum_\alpha \int \hat{v}_\alpha e_\alpha f_\alpha \, dv.
\end{aligned}
$$

The electromagnetic field $E(t,x)$, $B(t,x)$ satisfies the standard Maxwell equations

$$
(4) \qquad
\begin{aligned}
\frac{\partial E}{\partial t} &= c \, \nabla \times B - j & \nabla \cdot E &= \rho \\[2mm]
\frac{\partial B}{\partial t} &= -c \, \nabla \times E & \nabla \cdot B &= 0.
\end{aligned}
$$

(We have put the 4π factor in a nonstandard place for notational convenience.) The equations (2), (3), (4) comprise the relativistic Vlasov-Maxwell system (RVM). It is this system we wish to study, specifically the Cauchy problem with initial conditions

$$
f_\alpha(0,x,v) = f_{\alpha 0}(x,v), \quad E(0,x) = E_0(x), \quad B(0,x) = B_0(x)
$$

which satisfy the necessary constraints

$$
\nabla \cdot E_0(x) = \rho_0(x) = \rho(0,x), \quad \nabla \cdot B_0(x) = 0,
$$

and the natural assumption $f_{\alpha 0}(x,v) \geq 0$.

The most basic facts about RVM are the following. Each $f_\alpha(t,x,v)$ is constant on the characteristic curves defined by (1). Hence it is non-negative and bounded by an upper bound on its initial values, in any region which is reached by the characteristics. Secondly, the x and v derivatives in (2) can be moved past their coefficients (because $\hat{v}_\alpha$ has zero curl) and so the number of particles of each species

$$N_\alpha = \iint f_\alpha(t,x,v)dvdx$$

is conserved. Thirdly, if equation (2) is multiplied by $c^2(m_\alpha^2 + v^2/c^2)^{1/2}$, equations (4) are multiplied by E and B, and the results are added together, we obtain the conservation of energy

$$E = \int (\frac{1}{2} |E|^2 + \frac{1}{2} |B|^2)dx + \sum_c \iint c^2(m_\alpha^2 + v^2/c^2)^{1/2} f_\alpha \, dvdx.$$

It can be shown [2] that the <u>local</u> energy, the integral over a bounded set in x-space, tends to zero as $t \to \infty$.

A large part of the physics literature, and almost all the mathematical, assume zero magnetic field. In that case, $\nabla_x \cdot E = 0$ from the Maxwell equations, so that we can write $E = \nabla_x u$, u being the electric potential and $\Delta_x u = \rho$. Thus we simply have Poisson's equation coupled with Vlasov's which takes the form

$$\frac{\partial f_\alpha}{\partial t} + \hat{v}_\alpha \cdot \nabla_x f_\alpha + e_\alpha \nabla_x u \cdot \nabla_v f_\alpha = 0.$$

This is the relativistic Vlasov-Poisson system (RVP). The only vestige of relativity is the single coefficient $\hat{v}_\alpha$. With $\hat{v}_\alpha$ replaced by v we have the standard Vlasov-Poisson system (VP). The main known facts about the Cauchy problem for VP with nice initial data (smooth and small at infinity) are the existence of a weak solution [Arsenev], the existence of a smooth solution assuming radial or cylindrical symmetry [Batt, Wollman, Horst], and the existence of a smooth solution assuming small initial data [Bardos and Degond]. In general, a smooth solution is not known to exist for all time, and uniqueness is not known for the weak solutions. For these references, and others, see [3].

An interesting variant is obtained when the interior of a star is modeled by a single species of particles which interact only through their gravitational attraction. If u denotes the gravitational potential, then we have the system VP_-, the same as VP except that the sign of the nonlinear term is reversed. For VP_- the gravitational and kinetic energies have opposite signs. Assuming spherical symmetry, smooth solutions exist for all time [Batt]. However, if we take RVP_-, the system with $\hat{v}_\alpha$ and a reversed sign, then spherically symmetric solutions exist with nice initial data which blow up in a finite time [Glassey and Schaeffer]! This is a striking illustration of the subtlety of our problem.

Returning to the full Maxwell case, the mathematical literature is very sparse. Wollman [5] and Degond [1] have recently proved the existence of smooth solution of VM for a short time. Whether such a solution can be extended globally is an open question. (In the nonrelativistic version (VM), $\hat{v}_\alpha$ is simply replaced by v in equations (1), (2) and (3).) Since VM and RVM look like classical nonlinear conservation laws, one could reasonably expect the formation of shock waves. We have recently discovered some necessary conditions for the formation of shocks or any other singularity for RVM. The simplest such condition is that the $f_\alpha(t,x,v)$ could not have compact support in the v-variable. Recalling that v denotes the momentum, our result means that if no particle ever travels faster than 99% of the speed of light, then no singularity could ever be formed and a smooth solution would exist for all time. This is proved in [3]. Here is a sharper condition (see [4]).

<u>Theorem 1</u>. Let $f_{\alpha 0}(x,v)$ be C^1 functions with compact support and $E_0(x)$, $B_0(x)$ be C^2 functions satisfying the constraints mentioned above. Assume the <u>a priori</u> bound

$$(5) \qquad\qquad \int |v| f_\alpha(t,x,v)\,dv \leq C_T$$

for $v \in R^3$, $x \in R^3$, $t \in [0,T]$, valid for any α and any T and for any reasonable approximation of f_α. Then the Cauchy problem for RVM has a unique solution for all time.

Since f_α is automatically bounded, (5) is just a condition on its behavior for large v: there should not be too many very fast-moving particles. An equivalent way to state (5) is that the kinetic energy density (the inner integral in the second term in **E**) be bounded. A more "honest" theorem without any a priori assumption can be stated for a "cutoff" version of the system. Let $\zeta(|v|)$ be a smooth function of compact support. The ζVM system is equations (3) and (4) together with equation (2) modified by placing $\zeta(|v|)$ as a coefficient of the nonlinear term.

Theorem 2. Same as Theorem 1 for the system ζVM without any <u>a priori</u> assumption.

The idea of Theorem 2 is that the characteristics satisfy $\dot{v} = 0$ for large v. Since the $f_{\alpha 0}(x,v)$ vanish for large v, so do $f_\alpha(t,x,v)$ for all t. So we automatically have the situation of Theorem 1 except for a lot of error terms coming from the cutoff function.

We shall sketch the proof of Theorem 1 in case the $f_\alpha(t,x,v)$ vanish for large v. Note that the Vlasov characteristics (1) satisfy $|\dot{x}| = |\hat{v}_\alpha| < c$, as they ought in a relativistic theory. The assumption means that $|\dot{x}|$ is bounded away from c. That is, the Vlasov and the Maxwell characteristics are separated and the system is strictly hyperbolic except for the multiplicity inherent in Maxwell's equations.

Our proof makes no use of the conserved quantities N_α and E and is therefore valid even for equations with reversed signs like RVP_-. We shall estimate f_α, E and B in the C^1 norm, just the maxima of the absolute values of them and their first derivatives. Thus the result is in some sense very crude. We shall denote the C^0 norm (just the max) by $\| \ \|$. For the sake of exposition we shall subject the reader to several abuses of notation. We'll drop the subscripts α, imagining only one species. We'll let $m_\alpha = e_\alpha = c = 1$. We'll use D to denote any first derivative, $\|E\|$ to denote $\|E\| + \|B\|$, and $\|E\| + \|DE\|$ to denote the sum of the C^1 norms of E and B. Thus, in this crazy notation, we need to estimate $\|f\|$, $\|Df\|$, $\|E\|$, and $\|DE\|$. If this can be accomplished, the existence theorem is proved by the procedure, standard in PDE's, of appropriately approximating the equations, deriving the same estimates for the approximations, and passing to the limit.

First $\|f\|$ is already known to be bounded as was mentioned earlier. Suppose $D = \partial/\partial x_j$ and apply it to the Vlasov equation (2) to get

$$\{\partial_t + \hat{v} \cdot \nabla_x + (E + \hat{v} \times B) \cdot \nabla_v\}Df = -(DE + \hat{v} \times DB) \cdot \nabla_v f.$$

Therefore $Df(t,x,v)$ is expressible as an integral of the right-hand side over the characteristic curve which passes through the point (t,x,v) back to time zero. So we have an estimate

$$(6) \qquad \|Df(t)\| \leq c + c \int_0^t \|DE(\tau)\| \; \|Df(\tau)\| d\tau.$$

A similar estimate holds if $D = \partial/\partial v_j$.

Now we estimate the field E, B which satisfies the second-order equation

$$(\partial_t^2 - \Delta)E = -\nabla_x \rho - \partial_t j$$

and similarly for B. Recall that ρ and j are integrals of f over v-space. A crude estimate of this inhomogeneous wave equation in 3 dimensions only gives $\|E\|$ bounded in terms of $\|f\| + \|Df\|$, which is hopeless. Our final estimate is not _that_ crude. In fact, let's write $E(t,x)$ explicitly in terms of the initial data and an integral of $(-\nabla_x \rho - \partial_t j)$ over the backwards characteristic cone K with vertex at (t,x). Thus

$$(7) \qquad E(t,x) = E^{(0)}(t,x) + \int_K \int_{R^3} (\quad)Df \; dv \; dK,$$

where the parentheses indicate a kernel depending on space variables and on v, and D is a linear combination of time and space derivatives. Now there are 4 characteristic operators, one

$$S = \frac{\partial}{\partial t} + \sum_{k=1}^{3} \hat{v}_k \frac{\partial}{\partial x_k}$$

for Vlasov, and three

$$T_j = -\omega_j \frac{\partial}{\partial t} + \frac{\partial}{\partial x_j}$$

for Maxwell where ω_j denotes an angular variable. Since they are linearly independent, we can express Df as a linear combination of Sf, $T_1 f$, $T_2 f$ and $T_3 f$.

In equation (7) the T_j operators are tangential along the cone K, so an integration by parts along K makes them hit only the kernel. As for the term Sf, we use Vlasov's equation (2) to write it as $-\nabla_v \cdot [(E + \hat{v} \times B)f]$. Here an integration by parts in the v variable will make the ∇_v hit only the kernel. We end up with the representation

$$E = E^{(0)} + \int\int(\quad)f \, dv \, dK + \int\int(\quad)(E + \hat{v} \cdot B)f \, dv \, dK$$

These new kernels are harmless because the K-integration runs over a bounded set (the cone) while the v-integration runs over the support of f which is assumed to be bounded. Furthermore the factors $\ne$ in the integrals are bounded. Therefore

$$\|E(t)\| \leqslant c + c \int_0^t \|E(\tau)\| \, d\tau$$

so that $\|E\|$ is bounded by Gronwall.

Next we must estimate DE and DB. Think of D as $\partial/\partial x_j$. Differentiating (7), we have

$$DE = DE^{(0)} + \int\int(\quad)D^2f \, dv \, dK.$$

Now we have two derivatives on f, both of which we split into Sf and $T_j f$, and then integrate by parts either in v or in K. This process leads to an equation of the type

$$(8) \qquad DE = DE^{(0)} + \int\int\{(\quad)f + (\quad)Ef + (\quad)E^2f + (\quad)(DE)f\}dv \, dK$$

in our cryptic notation. But we already know that f and E are bounded. Therefore

$$(9) \qquad \|DE(t)\| \leqslant c + c\int_0^t \|DE(\tau)\| \, d\tau,$$

so that $\|DE\|$ is bounded. By (6), $\|Df\|$ is also bounded and the proof is complete.

There were several cheats in this sketch of a proof, but one was downright fraud. The kernels are not in fact bounded. Denoting the distance of a point of the cone K to the vertex (t,x) by r, the original kernel in (7) is of course $O(r^{-1})$. It is not bounded but is locally integrable on K. The difficulty occurs upon integrating by parts. Each time an operator T_j hits the kernel, it reduces the power of r by one. When this is done twice we get $O(r^{-3})$ which is no longer integrable on K. This situation occurs in only one term, the first integral term in (8), which has the form

$$\iint \frac{a}{r^3} f \, dv \, dK.$$

Luckily the average of a over the angular variables vanishes, so that we have a classical singular integral. When this difficulty is taken into account, we end up with

$$\|DE(t)\| \leq c + c \int_0^t \{\|DE(\tau)\| + \log^+ \|Df(\tau)\|\} d\tau$$

instead of (9). Using Gronwall and putting the result into (6), we get

$$\|Df(t)\| \leq c + c \int (\log^+ \|Df(\tau)\|) \|Df(\tau)\| d\tau.$$

Hence $\|Df\|$, and then $\|DE\|$, are bounded once again.

References

1. Degond, P., "Local existence of solutions of the Vlasov-Maxwell equations and convergence to the Vlasov-Poisson equations for infinite light velocity", to appear.

2. Glassey, R. and Strauss, W., "Remarks on collisionless plasmas", Contemp. Math. 28, Fluids and Plasmas (1984), 269-279.

3. ibid., "Singularity formation in a collisionless plasma could occur only at high velocities", Arch. Rat. Mech. Anal., 92 (1986), 59-90.

4. ibid., "High velocity particles in a collisionless plasma", to appear in Math. Meth. in Appl. Sci.

5. Wollman, S., "An existence and uniqueness theorem for the Vlasov-Maxwell system", Comm. Pure Appl. Math. 37 (1984), 457-462.

ACOUSTOELASTICITY

Joseph B. Keller

Departments of Mathematics and Mechanical Engineering
Stanford University
Stanford, California   94305

1. Introduction

Acoustoelasticity is the study of acoustic waves in stressed elastic
materials. Because the equations of elasticity are nonlinear, the propagation of
acoustic waves is affected by a preexisting stress. In particular, their speed of
propagation depends upon the stress, and this dependence is called the
acoustoelastic effect. One practical use of it is to determine the stress in a
body. In this application the speed of an acoustic wave is measured along a path
in the body. Then the stress is calculated from the relation between speed and
stress, which must be known. The resulting stress is an average value along the
path. One virtue of this method is that it permits the determination of stress
inside a body, where strain gauges cannot be used. Also, it can be used on opaque
bodies, such as metals, in contrast to photoelasticity, which is limited to
transparent bodies.

The theory of acoustoelasticity is complicated by the fact that metals and
many other solids are heterogeneous on a small scale. They are composed of
microcrystalline grains which are randomly oriented, and each grain is slightly
anisotropic. As a consequence, the solid is a heterogeneous medium which is
weakly anisotropic. Therefore acoustic waves in the solid are scattered by the
inhomogeneities.

If the granular structure is ignored, the theory of acoustoelasticity is much
simpler. Therefore we shall describe this homogeneous case first. In particular
we shall show how to calculate the accustoelastic effect and the acoustoelastic
coefficient tensor which occur in the expression for it.

Then we shall take account of graininess by employing the smoothing method.
In this method the medium is viewed as random with a known probability distribution. Then an equation for the mean wave is deduced and solved. From it the

speed and attention of the mean wave are calculated in terms of the statistical properties of the medium. Finally we shall indicate how these statistical properties can be evaluated in terms of the one and two point probability distributions of grain orientation.

The detailed analysis upon which the present account is based is given in a paper by L.L. Bonilla and J.B. Keller [1]. That paper also describes previous work on this subject and gives references to it. Therefore we shall present only the ideas and general form of the theory.

Before presenting the theory, we shall mention the relation of the present work to some previous work. First we note that the theory of acoustic waves in a pre-stressed solid is well known. It provides the basis for acoustoelasticity in the absence of graininess. The resulting waves are nondispersive and unattenuated in the usual theory, based upon nonlinear elasticity. Secondly we recall that the theory of elastic waves in heterogeneous grainy or random media has also been studied for many years, although not as extensively as the corresponding theory of electromagnetic waves. The simplest analysis, employing the single scattering or Born approximation, can be used to deduce the attenuation of waves. In a modified form it could also be used to determine the change in propagation speed due to inhomogeneities. Both of these results and more are accomplished by the smoothing method, which we employ.

The smoothing method is a procedure for determining the mean or coherent wave in a random medium, as well as higher moments and correlation functions of the wave field. It was applied to waves in random isotropic elastic media by Karal and Keller [2] and to waves in random anisotropic elastic media by Stanke and Kino [3]. The method is based upon the smallness of the random inhomogeneity. Therefore it automatically achieves closed sets of equations by restricting itself to effects of a given order of smallness. Other methods, which deal with inhomogeneities that are not small, generally employ some ad hoc closure hypothesis to get closed sets of equations.

In taking account of inhomogeneities in elastostatics, two common averaging procedures have been used traditionally. One, due to Voigt, is to determine the effective stiffnesses by averaging the nonuniform or random stiffness. (The stiffnesses are the coefficients which occur when stress is expressed in terms of strain. The compliances occur in the expression of strain in terms of stress.) The other method, due to Reuss, is to find the effective compliances by averaging the random compliances. Johnson [4] has applied both of them to a metal composed of randomly oriented anisotropic grains with cubic symmetry. Then he used the results to compute the acoustoelastic effect. The two results differ from one another, although for some metals they are not very different. The experimental results appear to lie between them, as might be expected.

The present analysis employing the smoothing method, avoids the necessity of arbitrarily choosing an averaging method. Instead once the goal of calculating the coherent wave is formulated, the appropriate averaged quantities are determined by the analysis.

2. Formulation and Solution for Homogeneous Media in an Elastic Body

Let $x(X,t)$ denote the position at time t of the material point X. We write the equations governing x is the form

$$F(x,\sigma,\alpha) = 0. \tag{2.1}$$

Here the parameter σ is a measure of the applied load and the parameter α is a measure of the imposed acoustic forcing. Thus we shall write $x(X,t,\sigma,\alpha)$ to indicate the dependence of the motion on σ and α.

When $\sigma = \alpha = 0$ we assume that the body is undeformed with $x(X,t,0,0) = X$, so (2.1) yields

$$F(X,0,0) = 0. \tag{2.2}$$

Now when $\alpha = 0$, (2.1) becomes

$$F(x,\sigma,0) = 0. \tag{2.3}$$

Differentiating (2.3) with respect to σ at $\sigma = 0$ yields

$$F_x(X,0,0)x_\sigma = -F_\sigma(X,0,0). \tag{2.4}$$

The solution of (2.4) is

$$X_\sigma(X,t,0,0) = -F_X^{-1}(X,0,0)F_\sigma(X,0,0). \tag{2.5}$$

The preceding results determine two terms in the Taylor expansion of X:

$$x(X,t,\sigma,0) = x(X,t,0,0) + \sigma x_\sigma(X,t,0,0) + O(\sigma^2)$$
$$= X - \sigma F_X^{-1}(X,0,0)F_\sigma(X,0,0) + O(\sigma^2). \tag{2.6}$$

This result gives the deformed configuration of the body to first order in the load parameter σ.

Returning to our derivation, we differentiate (2.1) with respect to α at $\alpha = 0$ to get

$$F_X[x(X,t,\sigma,0),\sigma,0]x_\alpha(X,t,\sigma,0) = -F_\alpha[x(X,t,\sigma,0),\sigma,0]. \tag{2.7}$$

This is the equation for the acoustic motion $x_\alpha(X,t,\sigma,0)$ around the stressed or deformed state. Thus this is the basic equation of acoustoelasticity. We shall examine its homogeneous form, obtained by setting $F_\alpha = 0$, which describes free acoustic waves. For simplicity we shall assume that the deformed configuration (2.6) is independent of time, that it is a homogeneous deformation, and that the body is unbounded and homogeneous.

Under these conditions the operator F_X in (2.7) is translationally invariant in t and X, so it has periodic plane wave solutions. Therefore we seek x_α in the form

$$x_\alpha(X,t,\sigma,0) = Be^{i(k \cdot X-\omega t)} \tag{2.8}$$

Then we can write the homogeneous form of (2.7) as

$$\left[e^{-(ik \cdot X-\omega t)}F_X[x(X,\sigma,0)]e^{i(k \cdot X-\omega t)}\right]B = 0. \tag{2.9}$$

The coefficient of the vector B in (2.9) is a matrix which is independent of X and t because of the assumed homogeneity of the medium and of the deformation. We shall call it the acoustic matrix and write it as

$$A(k,\omega,\sigma) = e^{-i(k \cdot X-\omega t)}F_X[x(X,\sigma,0),\sigma,0]e^{i(k \cdot X-\omega t)} \tag{2.10}$$

In order that (2.9) have a solution $B \neq 0$, the determination of A must vanish:

$$\det A(k,\omega,\sigma) = 0 \tag{2.11}$$

This is the dispersion equation relating k and ω. It is convenient to set $K = |k|$ and $\hat{k} = k/K$. From the form of F_x it follows that each entry of the matrix A is real and homogeneous of degree one in K^2 and ω^2. Therefore (2.11) is a cubic equation in the ratio $(K/\omega)^2$. When K/ω is real it is the phase velocity c, so (2.11) is a cubic in c^2. Let us denote its roots $\pm c_i(\sigma)$, $i = 1,2,3$. The dependence of c upon σ is the acoustoelastic effect. For σ small we can write

$$c_i(\sigma) = c_i(0) + \sigma c_{i\sigma}(0) + O(\sigma^2). \tag{2.12}$$

The derivative $c_{i\sigma}(0)$ can be found from (2.11).

Since the term $\sigma c_{i\sigma}(0)$ is linear in σ, it is also linear in the deformation gradient of the deformed configuration at $\sigma = 0$. Thus we can write

$$\frac{c_i(\sigma) - c_i(0)}{c_i(0)} = \sigma \frac{c_{i\sigma}(0)}{c_i(0)} + O(\sigma^2) \equiv \sigma P^i \frac{\partial(x - X)}{\partial X} + O(\sigma^2). \tag{2.13}$$

The coefficient P^i in (2.13) is a second rank tensor which we call the acoustoelastic tensor for waves of type i. It can be found from (2.11). Once $c_i(\sigma)$ is determined, the vector F is given as a null vector of the matrix A, as (2.9) shows.

These calculations complete the analysis of free acoustic waves in a homogeneous medium which undergoes homogeneous deformation. They are not presented in [1], which deals only with grainy or heterogeneous media. However the results for this case can be obtained from those of [1] by setting $\varepsilon = 0$, since ε is a measure of the random inhomogeneity.

It is to be noted that the phase speeds c_i are independent of ω, so the waves are nondispersive. Furthermore K^2 is real, so the waves are not attenuated. If the undeformed medium is isotropic, the speed $c_i(0)$ in (2.12) is independent of the propagation direction $\hat{k}$. Then one of the speeds is the compressional wave speed and for it $B = \hat{k}$, while the other two both equal the shear wave speed and for them $B \cdot \hat{k} = 0$. However the correction term $\sigma c_{i\sigma}(0)$ in (2.12) does depend upon $\hat{k}$ in general because the deformation makes the medium anisotropic.

To calculate $c_{i\sigma}(0)$ it suffices to expand F_x in (2.10) to first order in σ. Thus the solution (2.6) for x is adequate for this purpose. In fact it is not necessary to evaluate the determinant in (2.11). Instead it is possible to solve (2.9), which is just $A(k,\omega,\sigma)B = 0$, for both B and K/ω to first order in σ by the first order perturbation method.

Finally we note that (2.13) can also be written in terms of the stress instead of the deformation gradient. Then a different acoustoelastic tensor $P' = PS$ occurs, where S is the compliance tensor, i.e. the inverse of the stiffness or elasticity tensor.

3. Heterogeneous Media

When graininess is taken into account, the medium is no longer homogeneous. Instead its properties vary with X on the small scale of the grains. The reason is that each grain is weakly anisotropic, and different grains are oriented in different directions. Consequently the medium is weakly anisotropic, and the anisotropy varies rapidly with X.

To represent this behavior of the medium, we shall write its elasticity tensor C in the form

$$C(X,\varepsilon) = C^I + \varepsilon\, c(X,\mu) \qquad (3.1)$$

Here C^I is the large constant isotropic part of C, and $\varepsilon c(X,\mu)$ is the small inhomogeneous anisotropic part. The small parameter ε is a measure of this inhomogeneity.

We consider the rapidly varying function $c(X,\mu)$ to be a random function. By this we mean that it depends upon a random variable μ which has a probability density $dP(\mu)$. In terms of this probability we define the average of any function $f(\mu)$ by

$$<f(\mu)> = \int f(\mu)dP(\mu). \qquad (3.2)$$

Now we can write the requirement that $c(X,\mu)$ be statistically stationary, to

199

second order, in the form

$$<c(X,\mu)> = <c> = \text{constant}, \tag{3.3}$$

$$<c(X,\mu)c(Y,\mu)> = R(X - Y). \tag{3.4}$$

The function R in (3.4) is the two-point correlation tensor of c. These conditions (3.3) and (3.4) express the fact that the statistical properties of c
are independent of position in the medium.

When C has the form (3.1), the equation (2.1) governing x has the more
complicated form

$$F(x,\sigma,\alpha,\varepsilon,X,\mu) = 0. \tag{3.5}$$

The symbols in (3.5) have the following significance: σ is a measure of the
applied stress, α is a measure of the acoustic wave amplitude and ε is a
measure of the random inhomogeneity, which depends upon position X and upon the
random variable μ. As a consequence the solution $x(X,t,\sigma,\alpha,\varepsilon,\mu)$ is also a random function.

Differentiation of (3.5) with respect to α at $\alpha = 0$ yields the linear
equation

$$F_X(x,\sigma,0,\varepsilon,X,\mu)x_\alpha(X,t,\sigma,0,\varepsilon,\mu) = -F_\alpha(x,\sigma,0,\varepsilon,X,\mu) \tag{3.6}$$

This equation governs the acoustic motion around the deformed state. In order to
use it we must first calculate this state, just as we did in the previous case.

We assume that when $\sigma = 0$ and $\alpha = 0$, the medium is undeformed so x = X.
Then differentiation of (3.5) with respect to σ at $\sigma = 0$ and $\alpha = 0$ yields

$$F_X(X,0,0,\varepsilon,X,\mu)x_\sigma(X,0,0,\varepsilon,\mu) = -F_\sigma(X,0,0,\varepsilon,X,\mu). \tag{3.7}$$

We have omitted the argument t since x_σ is independent of t by assumption.
From (3.7) we see that x_σ is random. In terms of it we have

$$x(X,\sigma,0,\varepsilon,\mu) = X + \sigma x_\sigma(X,0,0,\varepsilon,\mu) + O(\sigma^2) \tag{3.8}$$

Now we can use (3.8) for x in (3.6) to obtain the equation for x_α.
We see that both (3.6) and (3.7) are linear equations involving random coef-

ficients. Such equations are called stochastic equations, and we must now consider how to treat them. Before doing so, however, we observe that the right sides F_α and F_σ are both zero in the present case because the stress and the acoustic disturbance are both applied at the boundary of the medium, not in the interior. Therefore α and σ do not occur in the equations (2.1) and (3.5). Furthermore we see that the operator in (3.7) is obtained from that in (3.6) by setting $\sigma = 0$ and setting the time derivatives equal to zero. Thus results for (3.7) can be obtained from those from (3.6) by these specializations.

In view of these considerations, we must examine the homogeneous form of the linear stochastic equation (3.6). In the next section we shall describe a method for doing this. However since the coefficients in (3.6) do not involve t, it is convenient to write x_α in the form

$$x_\alpha(X,t) = U(X)e^{-i\omega t}. \tag{3.9}$$

Then (3.6) becomes an equation for $U(X)$ with coefficients involving ω.

4. The Smoothing Method

Let us consider a linear stochastic equation of the form

$$(M + \varepsilon V)U = g. \tag{4.1}$$

Here M is an invertible non-random linear operator, V is a random linear operator, g is a given non-random function and U is the unknown solution. Since V is random, U is random also. We seek an equation for the mean value $\langle U \rangle$ of the solution. To find it we shall take advantage of the small parameter ε multiplying V.

We can write the soluton of (4.1) formally as

$$U = (M + \varepsilon V)^{-1}g = (1 + \varepsilon M^{-1}V)^{-1}M^{-1}g. \tag{4.2}$$

Averaging (4.2) yields

$$\langle U \rangle = \langle (1 + \varepsilon M^{-1}V)^{-1} \rangle M^{-1}g. \tag{4.3}$$

From (4.3) we can write

$$M<(1 + \epsilon M^{-1}V)^{-1}>^{-1}<U> = g. \tag{4.4}$$

This is an equation for $<U>$, but it is not too useful as it stands.

To make it more useful, we use the binomial theorem to expand $(1 + \epsilon M^{-1}V)^{-1}$ in powers of ϵ. Then we average each term and expand the inverse of the result in powers of ϵ again. In this way we get

$$[M + \epsilon<V> - \epsilon^2<V'M^{-1}V'> + O(\epsilon^3)]<U> = g. \tag{4.5}$$

Here $V' = V - <V>$ denotes the fluctuating part of V.

The result (4.5) yields the first order smoothing approximation to the equation for $<U>$ when we omit $O(\epsilon^3)$. This equation contains only smooth or average quantities, and they involve only the first and second moments of V. To make it more explicit, we shall express M^{-1} as an integral operator with kernel $G(X,Y)$. Thus G is the Green's function which satisfies the equation

$$MG(X,Y) = I\delta(X - Y). \tag{4.6}$$

When M is translationally invariant then $G = G(X - Y)$ is also translationally invariant.

By using G in (4.5) and omitting $O(\epsilon^3)$, we obtain the equaton

$$(M + \epsilon<V>)<U(X)> - \epsilon^2 \int <V'(X)G(X - Y)V'(Y)><U(Y)>dY = g(X). \tag{4.7}$$

When $M(X,\nabla)$ and $V(X,\nabla)$ are differential operators, (4.7) is an integrodifferential equation for $<U>$.

We now assume that $M(\nabla)$ is translationally invariant and that V is statistically stationary to second order. Then $<V(X,\nabla)> = <V(\nabla)>$ is independent of X and $<V'(X,\nabla_X)V'(Y,\nabla_Y)> = R(X - Y,\nabla_X,\nabla_Y)$. We next set $g(X) = 0$ in (4.7) and seek a periodic plane wave solution of the resulting homogeneous equation. Thus we write

$$<U(X)> = Be^{ik \cdot X} \tag{4.8}$$

and substitute (4.8) into (4.7) with $g = 0$. Then we multiply the result by $e^{-ik \cdot X}$ to get

$$[M(ik) + \varepsilon<V(ik)> - \varepsilon^2 \int e^{-ik\cdot(X-Y)} <V'(X,\nabla_X)G(X - Y)V'(Y,ik)>dY]B = 0. \quad (4.9)$$

The mean value in the integrand in (4.9) can be written in the following more convenient form if G and $V'(Y,ik)$ commute:

$$<V'(X,\nabla_X)G(X - Y)V'(Y,ik)> = <V'(X,\nabla_X)V'(Y,ik)>G(X - Y)$$
$$= R(X - Y,\nabla_X,ik)G(X - Y). \quad (4.10)$$

Then the integral in (4.9) is just the Fourier transform of the right side of (4.10) with respect to $X - Y$, evaluated at the transform variable k. We shall write it as $\widehat{RG}(k)$, and then (4.9) becomes

$$[M(ik) + \varepsilon<V(ik)> - \varepsilon^2\widehat{RG}(k)]B = 0. \quad (4.11)$$

When G and V' are matrices or tensors which do not commute, we can still interchange their order provided that we introduce components with subscripts, and use them in writing (4.10) and (4.11).

The result (4.11) is of the form $A(k)B = 0$ where the acoustic matrix A is given by the bracketed expression in (4.11). Thus A must be singular, and therefore k must satisfy the dispersion equation $\det A(k) = 0$. Then B must be a null vector of $A(k)$. In general k will be complex, so the plane waves will be attenuated.

When we apply the result (4.11) to (3.6) with $F_\alpha = 0$, the operator M and V involve the frequency ω and the small parameter σ. Therefore $A = A(k,\omega,\sigma)$ as in the previous case. We can again write $K = |k|$ and solve the dispersion equation for K in terms of ω, $\hat{k}$ and σ by expanding to first order in σ. Alternatively we can solve (4.11) for K and B to first order in σ by the perturbation method. The results for $\sigma = 0$ describe wave propagation in the unstrained medium, and exhibit dispersion and decay due to scattering by the inhomogeneities.

5. Grain Statistics

Suppose that the medium consists of anisotropic grains, all of which have the same second and third order elasticity tensors relative to fixed axes in each grain. Let Γ denote the Euler angles between the axes of a grain and the laboratory axes. We introduce the rotation matrix $\sigma(\Gamma)$ which transforms the grain axes to the laboratory axes. Then by using $\sigma(\Gamma)$, we can express the elasticity tensors of each grain in terms of the laboratory axes.

Now we introduce the probability $p(\Gamma,X)$ that a grain at X has Euler angles Γ, and the probability $p(\Gamma_1,\Gamma_2,X_1,X_2)$ that grains at X_1 and X_2 have Euler angles Γ_1 and Γ_2. For a statistically homogeneous medium $p(\Gamma,X)$ is independent of X and $p(\Gamma_1,\Gamma_2,X_1,X_2) = p(\Gamma_1,\Gamma_2,X_1 - X_2)$. In terms of these two probabilities we can calculate the mean and the second moment of the elasticity tensors of the medium because they are expressible in terms of $\sigma(\Gamma)$.

To facilitate this calculation we let $W(X_1,X_2) = W(|X_1 - X_2|)$ be the probability that X_1 and X_2 are points in the same grain. We assume that $\Gamma_1 = \Gamma_2$ when X_1 and X_2 are in the same grain, and that Γ_1 and Γ_2 are independent when X_1 and X_2 are in different grains. Then $p(\Gamma_1,\Gamma_2,X_1,X_2)$ can be expressed in terms of $p(\Gamma)$ and W in the form

$$p(\Gamma_1,\Gamma_2,X_1,X_2) = \delta(\Gamma_1-\Gamma_2)p(\Gamma_1)W(|X_1-X_2|)+p(\Gamma_1)p(\Gamma_2)[1-W(|X_1-X_2|)].$$

These considerations were introduced in [1] to calculate means and second moments when the individual grains have cubic symmetry. It was also assumed that the medium is transversely isotropic about a fixed axis so that $p(\Gamma)$ is symmetric about that axis. Most of the calculations were based on the work of L.L. Bonilla [5], who utilized invariance arguments to obtain simple explicit results.

References

1. Luis L. Bonilla and Joseph B. Keller, Acoustoelastic effect and wave propagation in heterogeneous weakly anisotropic materials, J. Mech. Phys. Solids, 33, 241-261, 1985.

2. F.C. Karal and Joseph B. Keller, J. Math. Phys. 5, 537, 1964.

3. F.E. Stanke and G.S. Kino, J. Acoust. Soc. Am., to appear.

4. G.C. Johnson, J. Nondestructive Eval. 3, 1, 1982.

5. Luis L. Bonilla, J. Mech. Phys. Solids, 33, 227, 1985.

ONE DIMENSIONAL FINITE AMPLITUDE PULSE PROPAGATION
IN ELECTROELASTIC SEMICONDUCTORS

Matthew F. McCarthy

National University of Ireland
University College
Galway, Ireland

Abstract

The propagation of high frequency finite amplitude pulses in electroelastic
semiconductors is examined, using the techniques of modulated simple wave theory.
The differential equations which govern the propagation of high frequency pulses
are derived and their consequences are examined. Small amplitude high frequency
pulses are examined in detail and the influence of the biasing electric field on
the evolutionary behaviour of such waves is discussed in depth. The behaviour of
pulses of finite amplitude which propagate into a region which is initially in a
homogeneous steady state is also examinec and the behaviour of such pulses is
studied in a number of particular situations. In particular, the conditions under
which a finite amplitude pulse may become saturated are determined.

I. Introduction

The propagation of acoustic waves in piezoelectric semiconductors has long
attracted attention. The primary reason for this is the well known fact that the
application of a d.c. voltage above a certain threshold value across a
piezoelectric semiconductor results in an oscillatory response [1-3]. As the
voltage is increased considerably beyond the threshold value, a growing wave front
has been found to form, the sharpness of which increases with increasing voltage
[2,3], ultimately resulting in the formation of a surface of discontinuity and a
consequent spiked response. The region of large deformation and electric field
behind the wave is called an acoustoelectric domain. Previous analyses of
acoustic wave propagation have for the most part been based entirely on linear
theory [4,5,6], or else they have been based on perturbation analyses of materials

in which nonlinearities are solely of electronic origin and whose electroelastic response is linear [7-12]. These analyses hold for weakly nonlinear situations and are certainly not valid for highly nonlinear cases in which both the deformation and electric field are large.

Recently, a general nonlinear rotationally invariant theory of electroelastic semiconductors was presented by deLorenzi and Tiersten [13]. Subsequently McCarthy and Tiersten [14,15] studied the propagation and growth of acceleration waves in such media. Their analysis indicated that for electric fields above a threshold value the amplitude of the acceleration wave would increase without bound and become a shock. A detailed analysis of the behaviour of these shocks was subsequently presented [16].

In this paper it is shown how our earlier results on one dimensional acceleration waves [14] may be generalized by applying the concepts of modulated simple wave theory [17-21] to the study of finite amplitude one dimensional pulses in nonlinear electroelastic semiconductors of the type described in [13]. In particular, we seek to determine the manner in which the propagation of such pulses is influenced by (i) the electroelastic and semiconduction properties of the material, (ii) the amplitude of the pulse and (iii) the conditions prevailing ahead of the pulse.

In Section 2 we briefly review the equations which govern the one dimensional motion of electroelastic semiconductors. The properties of acceleration waves propagating in piezoelectric semiconductors which are initially in a uniform state are reviewed in Section 3, and the influence of the applied electric field on the evolutionary behavior of the amplitudes of such waves is discussed in detail. In particular, it is noted that, when the attenuation length of the material is much greater than the distance traversed by an acceleration wave, semiconduction effects may be neglected and, as far as the propagation of acceleration waves is concerned, the electroelastic semiconductor behaves like an elastic dielectric. This suggests that, when the wavelengths associated with any dynamical disturbances or the distances traversed by such disturbances are small compared with the magnitude of the attenuation length, conduction and semiconduction effects may be neglected and the disturbances then propagate as simple waves. The properties of these waves are examined and are compared with those of acceleration waves. Naturally, we can only expect to be able to model the propagation of a pulse in an electroelastic semiconductor by a simple wave in exceptional circumstances, and even then our solution can only be expected to be valid over distances which are exceptionally small when compared with the natural attenuation length for acceleration waves. This observation prompts us to attempt to model the propagation of a pulse in terms of modulated simple wave theory in Section 4. An iterative method is developed, based on the assumption that the pulse length is short when compared with the attenuation length of the material. At the first iteration it is shown that, in a pulse of finite amplitude, the velocity of the lattice continuum, the strain of the lattice continuum and the electric field are related in precisely the same manner as they would be in a nonconducting electroelastic dielectric. The gradient of the free electronic charge density is also determined as a function of strain at each wavelet of the pulse and it is shown that, at the first iteration, the electric displacement, the electric potential and the free electronic charge density are unaltered by the passage of the pulse. The variation of the strain (and hence the velocity, electric field, etc.) on each wavelet is determined by a first order ordinary differential equation, as are the effects of pulse distortion.

The results derived in Section 4 are applied in Section 5 to the study of high frequency small amplitude pulses [19,21,22,23]. Here we derive an asymptotic expansion valid as $\omega \to \infty$, where ω is the ratio of the attenuation length of the medium to the length of the pulse. The first order terms in the expansion are studied in Section 5.1. The manner in which the strain, and hence the velocity of points of the lattice continuum, the electric field, the gradient of the free electronic charge density, the stress and the conduction current vary on each wavelet of the pulse as it traverses the material is determined and the equation which describes the effects of amplitude dispersion, distortion and shock formation is derived and its implications are examined. The second order theory is treated in Section 5.2 and, in particular, expressions for the changes in the electric displacement, free electronic charge density and electric potential, all of which vanish in the first order solutions, are derived. It is shown that the value of these quantities on any wavelet when it has reached a particular point are influenced by the value of the strain on each of the wavelets which have already passed through this particular point. The second order term in the strain as well as other quantities such as lattice point velocity, electric field, etc. is derived and, on any particular wavelet, it is fuond that these quantities are influenced by the precursor wavelets which have already passed through the point at which the wavelet under study is currently located. However, it is noted that, even at second order, the criterion for shock formation on any wavelet is not influenced by precursor wavelets.

An effort is made in Section 6 to generalize some of the results of Section 5 to the case of finite amplitude pulses. The nonlinear differential equation which governs the variation of the strain on any particular wavelet of a finite ampli- tude pulse is presented, and many of the expressions derived in Section 5 are generalized to the case of finite amplitude pulses. However, a detailed discussion of the implications of the results of Section 6 is deferred until Section 7 where pulse propagation in electroelastic semiconductors which are ini- tially homogeneously strained and subjected to a uniform electric field is exa- mined. Section 7 divides in three subsections in which (i) the consequences of Section 5 are examined in detail, (ii) the propagation of "weak" shocks is studied and (iii) the implications of our results for pulses of finite amplitude which propagate into a uniform region are presented. First of all, the implications of our results for high frequency small amplitude waves are examined when the electroelastic response of the material is linear, and the consequences of our results are examined in detail when the conduction current is given by the usual "nonlinear" constitutive equation of semiconductor physics [24]. For this par- ticular type of material we show that the applied electric field exerts a critical influence on the manner in which the strain varies on each wavelet. In par- ticular, we show that the strain will increase exponentially whenever the applied electric field exceeds a certain critical value. We observe however that, when this begins to happen, the small amplitude theory ceases to be valid and we must then treat the pulse as one of finite amplitude. The special forms which the results of Section 5 assume are also examined in detail in Section 7 and many results which generalize our earlier work [14] on acceleration waves are noted. The possibility of shock formation is treated and the propagation of shock waves which are weak in the sense of Whitham [25] is discussed. Once more the critical role played by the applied electric field in determining the behaviour of the amplitude of such waves is noted. Finally, the behaviour of finite amplitude pulses which propagate into a material which is initially at rest in a uniform state is examined. It is shown that the variation of the amplitude of the strain on each wavelet of the pulse is determined by a Bernoulli equation. The effect of amplitude attenuation is noted and it is shown that if the applied electric field exceeds a certain critical value then the strain on each wavelet of the pulse tends to a fixed limit as the pulse traverses the material. Thus, the material becomes saturated; a phenomenon that has been noted in a number of earlier works [24]. Finally, it is shown that, contrary to the impression given in some of the literature, see e.g., [8], saturation may occur in media whose electroelastic response is nonlinear even though the constitutive equation which governs the semiconduction response of the material is linear.

2. Equations for Electroelastic Semiconductor

In the general nonlinear rotationally invariant theory of electroelastic semiconductors developed by deLorenzi and Tiersten [13], the one dimensional, i.e. purely longitudinal, motion of a body may be described by the mapping

$$y = y(X,t) \tag{2.1}$$

which gives the position of y at time t of the point whose reference coordinate is X. Any sufficiently smooth motion of such a material will be governed by the balance laws [14]

$$\partial_X \Sigma = \rho_0 \dot{v}, \quad \partial_X p^e = \mu^e(\mathcal{E} + \mathcal{E}^e) \tag{2.2}$$

$$\partial_x D = \bar{\mu}, \quad \underline{\mathcal{E}} = -\partial_x \phi, \quad \partial_x \mathcal{J} + \dot{\bar{\mu}} = 0 \tag{2.3}$$

where

$$\Sigma = \tau + T - p^e, \quad T = \varepsilon_0 E^2/2,$$

$$\mathcal{E} = FE, \quad \mathcal{E}^e = FE^e, \quad \bar{\mu} = \mu F = \mu^e F + \bar{\mu}^r(X), \tag{2.4}$$

$$F = \partial_x y, \quad \mathcal{J} = \mu^e(v^e - v).$$

In these equations, τ and T represent the usual mechanical and free space Maxwell stresses, respectively; v^e, v, E, E^e, D and $\mathcal{J}$ represent the velocity of the free electronic fluid, velocity of the solid lattice, electric field, local electric field experienced by the free electronic fluid, electric displacement and electric current, respectively; and μ^e, p^e, ρ_0, ϕ, ϕ^e, μ and μ^r represent the free electronic charge density, the free electronic pressure, mass density of the lattice in the reference configuration, electrical potential, free electronic chemical potential, net charge density and residual charge density, respectively, and ε_0 is the permittivity of free space. The motion of any nonmaterial surface of discontinuity with respect to the reference coordinates is given by

$$X = \tilde{X}(t) \tag{2.5}$$

and at this surface the integral forms of the balance laws (2.2), (2.3) yield the relations [14]

$$[\Sigma] = -\rho_0 c[v], \quad [p^e] = 0, \quad [D] = 0, \tag{2.6}$$

$$[\phi] = 0, \quad c[\bar{\mu}] = [\mathcal{J}] \tag{2.7}$$

where

$$c = c(t) = \frac{d\tilde{X}}{dt}(t) > 0 \tag{2.8}$$

is the intrinsic velocity of the surface of discontinuity which is a measure of the speed of propagation of the discontinuity surface with respect to the reference coordinates of material points and

$$[\phi] = [\phi](t) = \phi^-(t) - \phi^+(t) \tag{2.9}$$

where ϕ^- and ϕ^+ are the limiting values of $\phi(x,t)$ immediately ahead of and just behind the wavefront.

The constitutive equations associated with the balance laws (2.2), (2.3) take the forms [13]

$$\tau = \hat{\tau}(F,\mathcal{E}), T = \varepsilon_0 F^{-2} \varepsilon^2/2, \ p^e = (\mu^e)^2 \partial_\mu e(\varepsilon^e),$$

$$\phi^e = \hat{\varepsilon}_\mu e(\mu^e \varepsilon^e), \quad \varepsilon^e = \hat{\varepsilon}^e(\mu^e), \tag{2.10}$$

$$D = \varepsilon_0 F^{-1} \mathcal{E} + \hat{P}(F,\mathcal{E}) = \hat{D}(F,\mathcal{E}), \tag{2.11}$$

$$\mathcal{J} = \hat{\mathcal{J}}(F,\mathcal{E},\mu^e,\mathcal{E}^e). \tag{2.12}$$

where ε^e is the internal energy per unit charge of the free electronic fluid. It is clear from $(2.4)_1$ and (2.10) that we have

$$\Sigma = \hat{\Sigma}(F,\mathcal{E},\mu^e). \tag{2.13}$$

Since $(2.2)_2$ may be recast in the form

$$\mathcal{E} + \mathcal{E}^e = \frac{1}{\mu^e}(\partial_{\varepsilon^e} p^e)g, \ g = \partial_x \mu^e, \tag{2.14}$$

we may eliminate $\mathcal{E}^e$ from (2.12) and rewrite this relation in the form

$$= \hat{\mathcal{J}}(F, , \mu^e, g). \tag{2.15}$$

The ensuing analysis may be made more transparent by introducing a new notation through the definitions

$$\lambda = F - 1, \ u = \mu^e, \ g = \partial_x u. \tag{2.16}$$

The motion of the body is now governed by the equations

$$\alpha_1 \partial_x \lambda + \alpha_2 \partial_x \mathcal{E} + \alpha_3 g = \dot{v}, \tag{2.17}$$

$$\partial_x v = \dot{\lambda}, \tag{2.18}$$

$$\beta_1 \partial_x \lambda + \beta_2 \partial_x \mathcal{E} = uF + \mu^r(X), \tag{2.19}$$

212

$$\gamma_1 \partial_x \lambda + \gamma_2 \partial_x \mathcal{E} + \gamma_3 g + \gamma_4 \partial_x g = \dot{u}\lambda + \dot{u}F, \qquad (2.20)$$

where

$$\alpha_1 = \rho_0^{-1} \partial_F \hat{\Sigma}, \quad \alpha_2 = \rho_0^{-1} \partial_{\mathcal{E}} \hat{\Sigma}, \quad \alpha_3 = \rho_0^{-1} \partial_{\mu^e} \hat{\Sigma},$$

$$\beta_1 = \partial_F \hat{D}, \quad \beta_2 = \partial_{\mathcal{E}} \hat{D} \qquad (2.21)$$

$$\gamma_1 = D_F \hat{g}, \quad \gamma_2 = D \, \hat{g}, \quad \gamma_3 = \partial_{\mu^e} \hat{g}, \quad \gamma_4 = \partial_g \hat{g}$$

and we assume that

$$\alpha_1 > 0, \quad \beta_2 > 0, \quad \alpha_2 \neq 0, \quad \beta_1 \neq 0,$$

$$\alpha_3 \neq 0, \quad \gamma_4 \neq 0. \qquad (2.22)$$

Finally, we assume that all the coefficients are $C^{(1)}$ functions of their arguments.

3. Acceleration Waves and Simple Waves

In this section we review briefly the properties of acceleration waves which propagate in piezoelectric semiconductors and we show that, in certain limiting situations, the behaviour of such acceleration waves is similar to that of simple waves in nonconducting elastic dielectrics. Detailed analyses of the properties of acceleration waves in electroelastic semiconductors have been presented elsewhere [14,15], and only those results which have particular relevance to our present study will be quoted here.

Suppose that the one dimensional motion under study contains an acceleration wave which is located at the material point $X = \tilde{X}(t)$ at time t and has intrinsic velocity c. The discontinuities in $\dot{v}, \dot{F}$ and $\partial_x F$ across this wave are related as follows [14]:

$$a = [\dot{v}] = -c[\dot{F}] = c^2[\partial_x F]. \qquad (3.1)$$

We have shown in [14] that

$$c^2 = \zeta_1(F, \mathcal{E}) = \alpha_1 - \alpha_2 \beta_1 / \beta_2 \tag{3.2}$$

and, when the constitutive equations (2.11) and (2.13) are inverted and rewritten as

$$\mathcal{E} = \tilde{\mathcal{E}}(F, D), \Sigma = \tilde{\Sigma}(F, D, \mu^e), \tag{3.3}$$

it follows that we have

$$c^2 = \zeta_1(F, D) = \rho_0^{-1} \partial_F \tilde{\Sigma}. \tag{3.4}$$

Further, when a standard analysis, the details of which will be found in [14], is carried out it is found that if the medium ahead of the wave is in a steady, i.e. time independent, uniform state the amplitude a of the wave satisfies the Bernoulli equation

$$\frac{da}{dt} = -\xi_0 a + \beta_0 a^2, \tag{3.5}$$

where

$$\xi_0 = \frac{1}{2c_0} \left(\frac{\alpha_2}{\beta_2} + \frac{\alpha_3}{\gamma_4} c_0 \right) \mu^e + \frac{1}{2c_0} \frac{\alpha_3}{\gamma_4} \left(\frac{\gamma_2 \beta_1}{\beta_2} - \gamma_1 \right), \tag{3.6}$$

$$\beta_0 = -\frac{1}{2c_0^3} \left\{ \alpha_{11} - 2 \frac{\alpha_{12} \beta_1}{\beta_2} + \frac{\alpha_{22} \beta_1^2}{\beta_2^2} - \alpha_2 \frac{\beta_{11}}{\beta_2} + 2 \frac{\alpha_2 \beta_{12} \beta_1}{\beta_2^2} - \alpha_2 \frac{\beta_{22} \beta_1^2}{\beta_2^3} \right\}$$

$$= -\frac{1}{2c_0^3} \zeta_{11} = -\frac{1}{2c_0^3 \rho_0} \partial_F^2 \tilde{\Sigma}, \tag{3.7}$$

with

$$\alpha_{11} = \partial_F^2 \hat{\Sigma} / \rho_0, \quad \alpha_{12} = \partial_F \partial \hat{\Sigma} / \rho_0, \quad \alpha_{22} = \partial^2 \hat{\Sigma} / \rho_0,$$

$$\beta_{11} = \partial_F^2 \hat{D}, \quad \beta_{12} = \partial_F \partial \hat{D} \quad \beta_{22} = \partial^2 \hat{D}. \tag{3.8}$$

It is to be noted that all coefficients occuring in (3.6) and (3.7) are evaluated by setting F, $\mathcal{E}$ and μ^e equal to the constant values F_0, $\mathcal{E}_0$ and μ_0^e which these quantities have immediately ahead of the acceleration wave.

The Bernoulli equation (3.5) admits the solution

$$\frac{a(t)}{a_c} = \frac{\nu \exp[-\xi_0(t-t_1)]}{1-\nu\{1-\exp[-\xi_0(t-t_1)]\}} = \frac{\nu \exp[-\overline{\xi}_0(X-X_1)]}{1-\nu\{1-\exp[-\overline{\xi}_0(X-X_1)]\}} \tag{3.9}$$

where $\nu = a(0)/a_c$, $a_c = \xi_0/\beta_0$ is the _critical jump in acceleration_, $X = c_0 t$, $\overline{\xi}_0 = \xi_0/c_0$ and $a(0)$ is the value at $a(t)$ when $t = t_1$. The quantities $\tau_r = \xi_0^{-1}$ and $L_a = c_0\xi_0^{-1}$ are the "_relaxation time_" and "_attenuation length_" for one dimensional acceleration waves which propagate into the uniform state described above. Clearly, the values of a_c, τ_r and L_a depend on the nature of the electroelastic semiconductor under study, as well as the electromechanical conditions which describe the steady uniform conditions prevailing ahead of the wave [14]. While the expressions (3.9) have the same form as those which determine the behaviour of one dimensional acceleration waves which propagate into homogeneously deformed materials with memory [26], the similarity is superficial since the sign of ξ_0 may be positive or negative depending on the electrical conditions which prevail ahead of the wave. In particular, it is to be noted that the amplitudes of some waves which decay to zero when $\xi_0 > 0$ may become unbounded if $\xi_0 < 0$. In the limit when $|X - X_1| \ll |L_a|$, so that the magnitude of the "attenuation length" is very much greater than the distance traversed by the wave, (3.9) reduces to

$$a(X) = \frac{a(0)}{1 + \dfrac{a(0)\,\zeta_{11}}{2c_0^4}\,(X - X_1)} \tag{3.10}$$

which has the same form as the expression given earlier for the amplitude of a longitudinal acceleration wave in an elastic material by Green [27], and is precisely the same as that derived by McCarthy [28] in his study of acceleration waves in elastic dielectrics.

When the wavelengths associated with any dynamical disturbances are small compared with the magnitude of the attenuation length, (3.10) suggests that conduction and semiconduction effects may be neglected and the motion of the material will then be governed by the equations

$$\alpha_1 \partial_x \lambda + \alpha_2 \partial_x = \dot{v}$$
$$\partial_x v = \dot{\lambda} \qquad (3.11)$$
$$\beta_1 \partial_x \lambda + \beta_2 \partial_x \mathcal{E} = 0.$$

Equations (3.11) admit solutions $v(\alpha)$, $\lambda(\alpha)$, $\mathcal{E}(\alpha)$ which describe simple waves [29] in which the wavelets $\alpha(X,t) = $ const. propagate at speed

$$c = c(\alpha) = -\dot{\alpha}/\partial_x \alpha. \qquad (3.12)$$

The functions $v(\alpha)$, $\lambda(\alpha)$, (α) satisfy the system of ordinary differential equations

$$\begin{bmatrix} c & \alpha_1 & \alpha_2 \\ 1 & c & 0 \\ 0 & \beta_1 & \beta_2 \end{bmatrix} \begin{bmatrix} \partial_\alpha v \\ \partial_\alpha \lambda \\ \partial_\alpha \varepsilon \end{bmatrix} = 0 \qquad (3.13)$$

and, since the determinant of the coefficients of equations (3.13) must vanish, c is given once more by (3.2).

Suppose that λ is prescribed at some fixed point $X = X_1$ so that $\lambda(X_1,t) = k(t/\tau_p)$, where k will be taken to represent a pulse if $k(y) = 0$ whenever $y \notin [0,1]$. The label $\alpha(X,t)$ of each wavelet may be chosen so that $\alpha(X_1,t) = t/\tau_p$ and then we have $\lambda(X,t) = k(\alpha)$ where

$$X - X_1 = c(\alpha)(t - \tau_p \alpha). \qquad (3.14)$$

The functions $v(\alpha)$ and $\mathcal{E}(\alpha)$ are related to $\lambda(\alpha)$ by the equations

$$\frac{dv}{c} = -d\lambda = \frac{\beta_2}{\beta_1} d\mathcal{E} \qquad (3.15)$$

and are found by solving these equations and choosing the constants of integration in an appropriate fashion.

When equation (3.14) is differentiated with respect to t, we find that, on any wavelet $\alpha = $ constant, the acceleration is given by

$$a(X,t) = v'(\alpha)\dot{\alpha} = v'(\alpha)/\{\tau_p - (X - X_1)c'(\alpha)/c^2(\alpha)\}$$
$$= v'(\alpha)/\{\tau_p + (X - X_1)\zeta_{11}(\alpha)v'(\alpha)/2c(\alpha)^4\}. \tag{3.16}$$

The similarity between formulae (3.10) and (3.16) should be clear. Of course (3.16) holds throughout the wave and not just at the wavefront. If we set $\alpha = 0$ and replace $v'(0)$ by $\tau_p a(0)$ in (3.16) we find that the acceleration on the wavelet $\alpha = 0$ is given by (3.10). Furthermore, the differential equation (3.5) may be recast in the form

$$\frac{d}{dt}\, v'(0) = -\omega^{-1}v'(0) + \frac{\beta_0}{\tau_p}\, v'(0)^2 \tag{3.17}$$

with $\omega = \tau_r/\tau_p = L_a/L_p$, $\bar{t} = t/\tau_p$, where $L_p = c\tau_p$ is the reference length of the pulse under study. In the limit when $|\omega| \gg 1$ the first term on the right hand side of (3.17) is negligible and may be omitted. The assumption that $|\omega| \gg 1$, so that the first term on the right hand side of (3.17) may be dropped, is clearly equivalent to assuming that the conductivity and semiconduction properties of the material have a negligible effect on the propagation of the pulse.

4. Modulated Simple Waves

We seek to describe the mode of propagation of a disturbance propagating into the region $X > 0$, which is supposed to be in a time independent steady state, in terms of the variation in λ as the pulse passes the point $X = 0$. Thus, we seek to solve equations (2.17)-(2.20) subject to the initial conditions $v = 0$, $\lambda = \lambda_0(X)$, $\mathcal{E} = \mathcal{E}_0(X)$, $u = u_0(X) = \mu_0^e(X)$ for $t = 0$ and $X > 0$ and the boundary condition[*]

$$\lambda(0,t) = k(t/\tau_p), \quad 0 < t < \tau_p. \tag{4.1}$$

In the limit when $|\omega| = |\tau_r/\tau_p| \gg 1$, (4.1) describes a short duration or high frequency pulse.

We have seen in the previous section that the propagation of the pulse may be

---

[*] The method which is described here may also be applied to the study of the propagation of high frequency time periodic disturbances in which case the function $g(\cdot)$ will be a periodic function of period τ_p.

described in terms of simple waves in the limiting situation which arises when the distance travelled by the pulse is significantly less than the attenuation length of the medium. This prompts us to describe the pulse in terms of modulated simple waves [17-19,21] when the effects of conduction and semiconduction may not be ignored. In such a description, the wave is modulated by the dissipative mechanisms present in the material and solutions of this type may be conveniently described in terms of the characteristic variable $\alpha = \alpha(X,t)$ and X. If $t = T(X,\alpha)$ denotes the arrival time of the wavelet α at X and if we write $h(X,t) = \bar{h}(X,\alpha)$ then

$$\dot{h} = \bar{\ell}^{-1} \partial_\alpha \bar{h}, \quad \partial_X h = \partial_X \bar{h} - s\bar{\ell}^{-1} \partial_\alpha \bar{h}, \tag{4.2}$$

where

$$\ell = \partial_\alpha T, \quad s = \partial_X T. \tag{4.3}$$

$\ell = \bar{\ell}(X,\alpha)$ is the <u>incremental arrival time</u> at position X and $s = c^{-1} = \bar{s}(x,\alpha))$ is the <u>slowness</u> of the wavelet x. It follows from (4.3) that we have the compatibility condition

$$\partial_X \bar{\ell} = \partial_\alpha \bar{s} \tag{4.4}$$

and if α is chosen so that $\alpha = t/\tau_p$ at $X = 0$ equations (4.1)-(4.3) together imply that

$$\bar{\lambda}(0,\alpha) = k(\alpha), \quad \bar{\ell}(0,\alpha) = \tau_p; \quad 0 < \alpha < 1. \tag{4.5}$$

The motion of the electroelastic semiconductor is determined by equations (2.2), (2.3), or equivalently, by equations (2.17)-(2.20) and, when formulae (2.21) and (4.2) are employed, these equations may be rewritten as

$$\partial_\alpha \bar{v} + s(\alpha_1 \partial_\alpha \bar{\lambda} + \alpha_2 \partial_\alpha \bar{\mathcal{E}} + \alpha_3 \partial_\alpha \bar{u}) = \rho_0^{-1} \ell \partial_X \bar{\Sigma},$$

$$s \partial_\alpha \bar{v} + \partial_\alpha \bar{\lambda} = \ell \partial_X \bar{v},$$

$$s\beta_1 \partial_\alpha \bar{\lambda} + s\beta_2 \partial_\alpha \bar{\mathcal{E}} = \ell(\partial_X \bar{D} - \bar{u} F - \mu^r), \tag{4.6}$$

$$(s\gamma_1 - u)\partial_\alpha \bar{\lambda} + s\gamma_2 \partial_\alpha \bar{\mathcal{E}} + (s\gamma_3 - F)\partial_\alpha \bar{u} + s\gamma_4 \partial_\alpha \bar{g} = \ell \partial_X \bar{q},$$

while from the definitions $v = y(X,t)$ and $q = \partial_X u(X,t)$ and formulae (4.2) we have the relations

$$\partial_\alpha \bar{y} = \ell \bar{v}, \qquad \partial_\alpha \bar{u} = s^{-1} \ell (\partial_X \bar{u} - \bar{g}). \tag{4.7}$$

Equations (4.4), (4.6) and (4.7) describe the motion of the electroelastic semiconductor for any choice of $s = s(X,\alpha)$. When the body is in a steady state, the right hand sides of (4.6) and (4.7) vanish identically. Thus, since we assume that the region ahead of the pulse is in a steady state, equations (4.6) furnish us with a set of homogeneous linear equations which relate $\partial_\alpha \bar{v}$, $\partial_\alpha \bar{\lambda}$, $\partial_\alpha \bar{\mathcal{E}}$ and $\partial_\alpha \bar{g}$ at the wavefront $\alpha = 0$ where, in view of $(4.7)_2$, $\partial_\alpha \bar{y}$, $\partial_\alpha \bar{u} = 0$. Of course, since conditions are steady ahead of the pulse, the first three of equations (4.6) reduce at the wavefront $\alpha = 0$ to equations (3.13) which apply throughout the simple wave which may propagate in the elastic dielectric to which our material reduces when dissipative effects are ignored and conditions are uniform ahead of the pulse. When conditions are steady but not uniform ahead of the pulse, it will have the same structure as that of a simple wave, i.e., at the head of the pulse, , λ and u will still be related by equations (3.13). Clearly, the right hand sides of (4.6) and (4.7) will only begin to change at any point X when the leading edge $\alpha = 0$ of the pulse has reached the point i.e. all changes are induced by the pulse. The representation of the pulse by a simple wave at point "very close" to the front should thus yield a reasonably accurate description of what is actually happening at these points, as has been noted in a number of other situations [19,21]. The accuracy of this representation improves as τ_p, or equivalently, the pulse length decreases.

Equations (4.6), (4.7) are valid for any choice of s. Let us choose $s = c^{-1}$ so that c is determined by formula (3.2). Next, equations $(4.6)_{2,3}$ and $(4.7)_2$ may be rewritten in the forms

$$\partial_\alpha \bar{v} = s^{-1} \{ \ell \partial_X \bar{v} - \partial_\alpha \bar{\lambda} \},$$

$$\partial_\alpha \bar{\mathcal{E}} = (s \beta_2)^{-1} \{ \ell (\partial_X \bar{D} - \bar{u}F - \mu^r) - s \beta_1 \partial_\alpha \bar{\lambda} \}, \tag{4.8}$$

$$\partial_\alpha \bar{u} = s^{-1} \ell (\partial_X \bar{u} - g),$$

and when these expressions are substituted into equation (4.6) we arrive at the equation

$$\rho_0^{-1} \partial_x \overline{\Sigma} - s^{-1} \partial_x \overline{v} - \alpha_2 \beta_2^{-1} (\partial_x \overline{D} - \overline{u}\overline{F} - \mu^r) - \alpha_3 (\partial_x \overline{u} - g) = 0. \tag{4.9}$$

Equation (4.9) is the _transport equation_ and it should be noted that it contains no derivatives with respect to α.

The form of the transport equation (4.9) may seem rather surprising at first sight, since it does not appear to contain any reference to the current which we expect to play an important role in determining the evolutionary behaviour of the pulse. However, it follows from equations (2.3)$_1$ and (2.3)$_3$ that

$$\mathcal{J} + \dot{D} = \mathcal{J}_0 \tag{4.10}$$

where

$$\mathcal{J}_0 = \hat{\mathcal{J}}(F_0, \mathcal{E}_0, \mu_0^e, \partial_x \mu_0^e) \tag{4.11}$$

is the steady current which flows in the region ahead of the pulse. When formula (4.2)$_1$ is used, equation (4.10) yields the relation

$$\partial_\alpha \overline{D} = \beta_1 \partial_\alpha \overline{\lambda} + \beta_2 \partial_\alpha \overline{\mathcal{E}} = \ell(\mathcal{J}_0 - \overline{\mathcal{J}}) \tag{4.12}$$

which is, of course, equivalent to (4.6)$_3$. If (4.12) is now used, instead of (4.8)$_2$, in order to eliminate $\partial_\alpha \overline{\mathcal{E}}$ from (4.6), we find that

$$\rho_0^{-1} \partial_x \overline{\Sigma} - s^{-1} \partial_x \overline{v} - s\alpha_2 \beta_2^{-1} (\mathcal{J}_0 - \overline{\mathcal{J}}) - \alpha_3 (\partial_x \overline{u} - g) = 0. \tag{4.13}$$

Of course, equation (4.13) is simply another form of the transport equation (4.9).

The motion of the body is now described by equations (4.4), (4.8) and the transport equation either in the form (4.9) or (4.13). Once these equations have been solved, $\overline{y}(X, \alpha)$ and $\mu^e = \overline{u}(X, \alpha)$ follow on integration of equations (4.7). The electrical potential $\phi(X, t) = \overline{\phi}(X, \alpha)$ may be obtained by integrating the equation

$$\partial_\alpha \overline{\phi} = s^{-1} \ell(\partial_x \overline{\phi} + \mathcal{E}) \tag{4.14}$$

which follows equation $(2.3)_2$ and $(4.2)_2$. The rate at which energy flows past any point X is given by the formula [30]

$$Q = \Sigma v - (\phi + \phi^e)\dot{g} - \phi \dot{D} \tag{4.15}$$

and this quantity may be calculated at any point once $\bar{y}(X,\alpha)$, $\bar{u}(X,\alpha)$ and $\bar{\phi}(X,\alpha)$ are known. The function $T(X,\alpha)$ may be obtained by integrating $(4.3)_2$ and thus all the functions needed to describe the motion of the electroelastic material may be expressed in terms of X and t.

The foregoing system of equations may be solved iteratively in the manner described by Parker [18] and others with $v, \lambda, ,u$ and g specified on the front $\alpha = 0$ and $\lambda(0,t) = k(t/\tau_p)$ prescribed at $X = 0$. To a first approximation, conditions in the pulse may be determined by neglecting the right hand sides of equations $(4.6)_4$ and (4.8) so that, at the first iteration, we have

$$\partial_\alpha \bar{v} = -s^{-1} \partial_\alpha \bar{\lambda}, \quad \partial_\alpha \bar{\mathcal{E}} = (-\beta_1/\beta_2) \partial_\alpha \bar{\lambda},$$

$$\partial_\alpha \bar{g} = -\gamma_4^{-1} \{\gamma_1 - \gamma_2 \beta_1/\beta_2 + \frac{\bar{u}}{s}\} \partial_\alpha \bar{\lambda}, \quad \partial_\alpha \bar{u} = 0. \tag{4.16}$$

It follows from (4.12) and (4.14) that at the first iteration

$$\partial_\alpha \bar{D} = 0, \quad \partial_\alpha \bar{\phi} = 0, \tag{4.17}$$

and we note that $(4.17)_1$ is equivalent to $(4.16)_2$. It is now clear from $(4.16)_4$ and (4.17) that, at the first iteration, we have

$$\bar{u} = u_0^e(X), \quad \bar{D} = D_0(X), \quad \bar{\phi} = \phi_0(X) \tag{4.18}$$

so that, at the first iteration, the density of free electronic charge, the electric displacement and the electric potential are not altered by the pulse. Next, an elementary calculation shows that

$$\delta_1 = \tilde{\delta}_1(F,D) = \partial_F \tilde{\mathcal{E}}(F,D) = -\beta_1/\beta_2 = \delta_1(\bar{\lambda},X), \tag{4.19}$$

while equation (3.4) may be used to express s as a function of $\bar{\lambda}$ and X:

$$s = \mathcal{S}(\bar{\lambda},D_0(X)) = \mathcal{S}(\bar{\lambda},X). \tag{4.20}$$

Thus, the first two of equations (4.16) serve as a pair of ordinary differential equations which relate $\partial_\alpha \bar{v}$ and $\partial_\alpha \bar{\mathcal{E}}$ to $\partial_\alpha \bar{\lambda}$ and $\bar{\lambda}$ at the point X:

$$\partial_\alpha \bar{v} = -\mathcal{J}(\bar{\lambda},X)^{-1}\partial_\alpha \bar{\lambda}, \quad \partial_\alpha \bar{\mathcal{E}} = \delta_1(\bar{\lambda},X)\partial_\alpha \bar{\lambda}. \tag{4.21}$$

Once $\bar{v}$ and $\bar{\mathcal{E}}$ have been determined by solving (4.21), subject to appropriate initial conditions, $(4.16)_3$ serves as a differential equation which relates $\partial_\alpha \bar{g}$ to $\bar{\lambda},\bar{g}$ and $\partial_\alpha \bar{\lambda}$. It is important to appreciate [17] that equations (4.21) are approximate and cannot, in general, be integrated to obtain formulae giving $\bar{v}$ and $\bar{\mathcal{E}}$ as functions of $\bar{\lambda}$ and X which are uniformly valid for all time. The terms which have been neglected in arriving at equations (4.21) and $(4.16)_3$ will, in general, ultimately produce first order contributions to $\bar{v}, \bar{\mathcal{E}}, \bar{g}$ as well as to $\bar{\phi}$ and $\bar{u}$. However, as has been noted in [17], these contributions are negligible whenever the wave is a "high frequency" periodic disturbance or pulse. Equations (4.21) may now be integrated at constant X, subject to the initial conditions which are specified on the wavelet $\alpha = 0$ and, when $(4.16)_3$ is used, this will result in relations of the form

$$\bar{v} = \hat{v}(\lambda,X), \quad \bar{\mathcal{E}} = \hat{\mathcal{E}}(\lambda,X), \quad \bar{g} = \hat{g}(\lambda,X) \tag{4.22}$$

and these formulae, when used in conjunction with equations (4.18), determine the detailed structure of the wave at each point X. Substitution from $(4.18)_1$ and (4.22) into (4.13) yields the first order ordinary differential equation

$$2c^2\partial_X \bar{\lambda} + \rho_0^{-1}\partial_X \hat{\Sigma} - c\partial_X \hat{v} - s\alpha_2\beta_2^{-1}(\hat{q}_0 - \hat{q}) - \alpha_3(\partial_X \mu_0^e - \hat{g}) = 0, \tag{4.23}$$

where we have used the notation $\bar{\Sigma} = \hat{\Sigma}(\bar{\lambda},X), \bar{v} = \hat{v}(\bar{\lambda},X), \bar{g} = \hat{g}(\bar{\lambda},X)$ and it is to be noted that all the coefficients in (4.23) are functions of $\bar{\lambda}$ and X only. The first order ordinary nonlinear differential equation (4.23) determines the variation of $\bar{\lambda}(x,\alpha)$ with X at each wavelet $\alpha = $ constant. It may be integrated subject to the boundary condition $\bar{\lambda}(0,\alpha) = k(\alpha)$ to give $\bar{\lambda}(X,\alpha)$ and hence $\bar{v}(X,\alpha), \bar{\mathcal{J}}(X,\alpha), \bar{g}(X,\alpha), \bar{s}(X,\alpha) = c^{-1}$. Equation (4.4) may then be integrated with respect to X to yield $\ell(X,\alpha)$ and hence $t = T(X,\alpha)$, the arrival time at X of the wavelet which left $X = 0$ at $t = \tau_p \alpha$. Finally, integration of (4.7), yields $y = y(X,\alpha)$.

Once $y(X,\alpha)$ and $T(X,\alpha)$ have been calculated, the first approximation of the simple wave description of the pulse has been determined. Higher order corrections to these results may be found in a straightforward manner [31]. At each iteration the structure of the wave is determined by equations (4.6), (4.7) and (4.14). Equation (4.23) serves as the transport equation for $\bar{\lambda}(X,\alpha)$ which determines the manner in which the amplitude of the pulse is attenuated, while equation (4.4) governs amplitude dispersion.

Before discussing the implications of the foregoing analysis in greater detail, we analyse the propagation of small amplitude rapid pulses.

5. High Frequency Small Amplitude Pulses

It is well known [19, 21-23] that the applicability of many of the results of the theory of acceleration waves may be extended to high frequency small amplitude pulses. Our primary objective in this section is to extend the results which we presented earlier for acceleration waves [14] to cover the propagation of small amplitude finite rate pulses in electroelastic semiconductors.

In order to clarify the limitations of our analysis, it is essential to make the terms "high frequency" and "small" precise. With this objective in mind, we introduce dimensionless variables X^*, y^*, t^*, Σ^*, D^*, $\vec{\mu}^*$, $\mathcal{E}^*$, ϕ^*, $\mathcal{J}^*$ through the definitions

$$X = L_a X^*, \quad y = L_a y^*, \quad t = \tau_r t^*, \quad \Sigma = \rho_0 \bar{c}_0^2 \Sigma^*,$$

$$D = -\bar{\mu}^e \bar{c}_0 \tau_r D^*, \quad u = -\bar{\mu}^e u^*, \quad \bar{\mu} = \bar{\mu}^e \vec{\mu}^*, \quad \mathcal{E} = \bar{E}_0 \mathcal{E}^*, \quad \phi = \bar{E}_0 L_a \phi^*, \qquad (5.1)$$

$$\mathcal{J} = -\bar{\mu}_0^e \bar{c}_\theta \mathcal{J}^*.$$

In the definitions (5.1) L_a, τ_r and $\bar{c}_0$ are respectively the attenuation length, relaxation time and speed of propagation of a one dimensional acceleration wave which is propagating into a region which is initially undeformed, is subject to a uniform electric field $\bar{E}_0$ and in which the mass density and charge density of the free electronic fluid have the constant values ρ_0 and $\overline{\mu}_0^e\,(<0)$, respectively. Inspection shows that, with the exception of (4.13), the governing equations are invariant under these transformations and, consequently, retain their present form when the asterisks are dropped from the dimensionless variables. The dimensionless form of (4.13) is obtained by setting $\rho_0 = 1$ in the original equation. The equations must now be solved subject to the initial conditions $\lambda(X,0) = \lambda_0(X)$, $\mathcal{E}(X,0) = \mathcal{E}_0(X)$, $\mu^e(X,0) = u_0(X)$ and the boundary condition

$$\lambda(0,t) = k(\omega t), \quad 0 < \omega t < 1, \tag{5.2}$$

where $\omega = L_a/c_0\tau_p$. Let us write $k(\xi) = \omega\epsilon f(\xi)$, where $\epsilon > 0$ and $f(\xi)$ is normalized so that $\max|f(\xi)| = 1$, $\xi\epsilon[0,1]$. We assume that $\omega \gg 1$, while $a = \omega^2\epsilon$ is finite so that we are dealing with <u>small amplitude high frequency finite acceleration pulses</u>, [19], for which

$$\lambda(0,t) = \frac{a}{\omega} f(\omega t), \quad 0 < \omega t < 1. \tag{5.3}$$

We now seek solutions which formally satisfy equations (4.4), $(4.7)_2$, (4.8), (4.12), (4.13) and (4.14) in the limit as $\omega \to \infty$, a remaining finite but not necessarily small. We content ourselves with solutions of these equations of the form

$$\overline{h} = \overline{h}(X,\alpha) = h_0(X) + \frac{1}{\omega} h^{(1)}(X,\alpha) + \frac{1}{\omega^2} h^{(2)}(X,\alpha), \tag{5.4}$$

where $\overline{h}$ represents one of the variables, $\overline{\lambda},\overline{v},\overline{\mathcal{E}},\overline{u},\overline{g},\overline{\ell},\overline{c},\overline{s},\overline{\Sigma},\overline{\mathcal{J}},\overline{D}$ or $\overline{\phi}$ and where the functions $\lambda_0(X)$, $u_0(X)$, $\mathcal{E}_0(X)$ and $g_0(X)$ are solutions of the equations

$$\partial_X\Sigma_0 = 0, \quad \partial_X D_0 = u_0(X) + \mu_r(X), \quad \partial_X\mathcal{J}_0 = 0, \tag{5.5}$$

and hence Σ_0, D_0 and $\mathcal{J}_0$ are defined by

$$\Sigma_0 = \hat{\Sigma}(F_0, \mathcal{E}_0, u_0), \quad D_0 = \hat{D}(F_0, \mathcal{E}_0), \quad \mathcal{J}_0 = \hat{\mathcal{J}}(F_0, \mathcal{E}_0, u_0, g_0). \tag{5.6}$$

We also note that when v, ℓ and s are represented by an expansion of type (5.4) then $v_0 = \ell_0 = 0$ while $s_0 = 1/c_0$.

The representations (5.4) may be substituted into the governing equations and, when coefficients of powers of ω^{-1} are equated, we obtain differential equations which govern the first and second order terms in the expansions (5.4). These equations are presented in the appendix and their consequences are examined here.

5.1 First Order Solution

We wish to solve equations (A.1)-(A.7) and (A.9) subject to the initial conditions $v^{(1)} = D^{(1)} = \mathcal{E}^{(1)} = u^{(1)} = g^{(1)} = \phi^{(1)} = \Sigma^{(1)} = \mathcal{J}^{(1)} = 0$ on $\alpha = 0$. If we choose α so that $\omega t = \alpha$ at $X = 0$, then the equations must be solved subject to the boundary condition

$$\lambda^{(1)}(0,\alpha) = af(\alpha). \tag{5.7}$$

It follows from equations (A.2), (A.4) and (A.6) that

$$D^{(1)}(X,\alpha) = u^{(1)}(X,\alpha) = \phi^{(1)}(X,\alpha) = 0, \tag{5.8}$$

so that, when terms of order $1/\omega^2$ are neglected, the electric displacement, the free electronic charge density and the electric potential are not altered by the passage of the pulse. Equations (5.8) are, of course, consistent with our earlier results as contained in equations (4.18). Next, when equations (A.1)-(A.3) and (A.9) are solved subject to the appropriate initial conditions we find that

$$\begin{aligned}
v^{(1)} &= -c_0\lambda^{(1)}(X,\alpha), & \mathcal{E}^{(1)} &= \delta_1\lambda^{(1)}(X,\alpha) \\
g^{(1)} &= \pi_1\lambda^{(1)}(X,\alpha), & \mathcal{J}^{(1)} &= c_0 u_0 \lambda^{(1)}(X,\alpha)
\end{aligned} \tag{5.9}$$

where

$$\delta_1 = \delta_1(X) = -\beta_1^{(0)}/\beta_2^{(0)}, \quad \pi_1 = \pi_1(X) = -\{\gamma_1^{(0)} - \gamma_2^{(0)}\beta_1^{(0)}/\beta_2^{(0)} - c_0 u_0\}\gamma_4^{(0)-1}.$$

$$\tag{5.10}$$

In (5.10), as well as in the remainder of this paper, whenever a function m is written as either $m(0)$ or m_0 it is to be assumed that the function is evaluated at the steady state values of its arguments, thus if

$m = \hat{m}(F, ,u,g) = \tilde{m}(F,D,u,g)$ then $m_0 = m^{(0)} = \hat{m}(F_0,\mathcal{E}_0,u_0,g_0) = \tilde{m}(F_0,D_0,u_0,g_0)$.

In order to be able to use the transport equation (A.7) as a differential equation for $\lambda^{(1)}(X,\alpha)$, we need to express $\Sigma^{(1)}$ as a function of $\lambda^{(1)}$. When $(3.3)_2$ is used, an elementary calculation shows that

$$\Sigma^{(1)} = \zeta_1^{(0)}\lambda^{(1)} = c_0^2\lambda^{(1)}. \tag{5.11}$$

Substitution from (5.8), (5.9) and (5.11) into equation (A.7) yields the linear transport equation for $\lambda^{(1)}(X,c)$ as

$$c_0^2\partial_X\lambda^{(1)} + \{ \tfrac{3}{4} \partial_X(c_0^2) + c_0\xi_0\}\lambda^{(1)} = 0, \tag{5.12}$$

where ξ_0 has been defined in equation (3.6). Equation (5.12) may be recast in the form

$$\partial_X\{c_0^{3/2}\lambda^{(1)} \exp[\int_0^X \xi_0(Y)/c_0(Y)]\} = 0, \tag{5.13}$$

and, when (5.7) is used, the solution of this equation is seen to be

$$\lambda^{(1)}(X,\alpha) = af(\alpha)M^{(1)}(X) \tag{5.14}$$

where $M^{(1)}(X)$ is the <u>amplitude modulation factor</u>

$$M^{(1)}(X) = \frac{c_0^{3/2}(0)}{c_0^{3/2}(X)N^{(1)}(X)} \tag{5.15}$$

with

$$N^{(1)}(X) = \exp\{\int_0^X [\xi_0(Y)/c_0(Y)]dY\}. \tag{5.16}$$

In order to complete our first order description of the propagating pulse, we need to determine the incremental arrival time $\ell^{(1)}(X,\alpha)$ by solving equation (A.5). Since $s = c^{-1}$, we have

$$s^{(1)} = -c^{(1)}/c_0^2 = K\lambda^{(1)}, \quad K = -\zeta_{11}^{(0)}/2c_0^3 \tag{5.17}$$

where ζ_{11} has been defined in (3.7). It now follows from (5.14), (5.17) and (A.5) that

$$\partial_X \ell^{(1)} \, K \partial_\alpha \lambda^{(1)} = aK(X)M^{(1)}(X)f'(\alpha). \tag{5.18}$$

Since $\ell^{(1)}(0,\alpha) = 1$, equation (5.18) may be integrated to yield the formula

$$\ell^{(1)}(X,\alpha) = 1 - a\Psi(X)f'(\alpha) \tag{5.19}$$

where

$$\Psi(X) = -\int_0^\alpha K(Y)M^{(1)}(Y)dY \tag{5.20}$$

is the <u>distortion factor</u> [21]. Since $\ell = \partial_\alpha T$ it follows, when terms of order $1/\omega^2$ are neglected, that the arrival time $T(X,\alpha)$ of the wavelet $\alpha = $ constant at the point X is given by the formula

$$\omega\left[T(X,\alpha) - \int_0^X \frac{ds}{c(s)}\right] = \alpha - a\Psi(X)f(\alpha), \tag{5.21}$$

and this formula determines the positions of the characteristics implicitly.

It is clear from (5.17) that the slowness of any particular wavelet is influenced by the amplitude of the signal carried by the wavelet. Thus, the pulse profile distorts as it propagates, since different portions of the pulse propagate at different speeds. This phenomenon is known as <u>amplitude dispersion</u> [19,31] and is a purely nonlinear phenomenon. It is also obvious from (5.14) and (5.21), that, in the present approximation, conditions at any point $X > 0$ at the passage of the wavelet $\overline{\alpha}$ are determined solely by conditions previously encountered by the wavelet $\overline{\alpha}$ and, in particular, are independent of the conditions prevailing at the wavelet $0 < \alpha < \overline{\alpha}$ which have already passed the point X.

When equations (4.3), (5.9)$_1$, (5.14) and (5.21) are employed, we find, when terms of order $1/\omega^2$ are neglected, that the acceleration is given by the forumla

$$a^{(1)}(X,\alpha) = \dot{v}^{(1)} = \frac{-ac_0(X)M^{(1)}(X)f'(\alpha)}{1 - a\Psi(X)f'(\alpha)}. \tag{5.22}$$

Direct calculation shows that the variation of $a^{(1)}(X,\alpha)$ along each wavelet is governed by the Bernoulli equation

$$\partial_X a^{(1)} = \overline{\beta}a^{(1)}(a^{(1)} - a_c), \tag{5.23}$$

where

$$a_c = a_c(X) = -\partial_X \{ \ell n[c_0(X)M^{(1)}(X)] \}/ \bar{\beta},$$
$$\bar{\beta} = \bar{\beta}(X) = K(X)/c_0(X). \tag{5.24}$$

Of course, (5.23) is just the equation which governs the propagation of an acceleration wave whose initial amplitude is $a(0,\alpha)$. While (5.23) reduces to
equation (3.5) when we set $\alpha = 0$ and assume that conditions ahead of the pulse
are initially uniform, it should be emphasized that (5.23) applies throughout the
pulse, i.e., it gives the variation with X of the acceleration on any wavelet in
the pulse. The implications of (5.23) for acceleration waves have been examined
in detail by Bailey and Chen [32] and their results generalize at once to apply to
the situation under study here. We later examine the implications of equation
(5.23) for pulses which propagate into electroelastic semiconductors which are in
a uniform state.

The acceleration $a^{(1)}(X,\alpha)$ becomes unbounded if the denominator of (5.21)
vanishes, i.e., if the equation

$$1 - a\Psi(X)f'(\alpha) = 0 \tag{5.25}$$

has solutions $X_s > 0$ and $\alpha_s \geqslant 0$. This usually heralds the formation of a shock
at the point X_s. The implications (5.25) and the propagation of weak shock waves
are examined in detail below when we study pulse propagation into regions which
are initially in a uniform state.

5.2 Second Order Solution

In order to gain a deeper insight into the manner in which pulses propagate,
we determine the final terms in the expansions of type (5.4). Thus, we seek to
solve equations (A.10)-(A.18) subject to the initial conditions that all coefficients of order $1/\omega^2$ in the expansions (5.4) vanish on the wavelet $\alpha = 0$ and
the boundary condition $\lambda^{(2)}(0,\alpha) = 0$.

When the first order solutions (5.9) and (5.14) are used, it follows from
equations (A.11), (A.14) and (A.16) that

$$D^{(2)} = -c_0 u_0 a M^{(1)} \mathcal{Y}, \quad u^{(2)} = -c_0 \pi_1 a M^{(1)} \quad , \quad \phi^{(2)} = c_0 \delta_1 a M^{(1)} \tag{5.26}$$

where

$$\mathcal{Y} = (X, \alpha) = \int_0^\alpha f(\xi)d\xi - \frac{1}{2} a \Psi(X) f^2(\alpha) \tag{5.27}$$

may be called the hysteresis function. Notice that when a given wavelet $\alpha = \alpha_1$, say, reaches a point X, the function $\mathcal{Y}$ is determined not only by the conditions previously encountered by the wavelet α_1, as manifested through the distortion factor $\Psi(X)$, but also by the values which the signal function $f(\cdot)$ had on all wavelets $0 < \alpha < \alpha_1$ which have already passed through the point X. Thus, it follows from (5.26) that the purely second order fields induced by the pulse at the wavelet α at the point X are determined not only by the conditions previously experienced by the wavelet α but by the conditions which prevailed at all other wavelets which have already passed the point X.

When equations (A.10), (A.12) and (A.13) are integrated subject to the appropriate initial conditions we find that

$$v^{(2)} = -c_0 \lambda^{(2)} + V_1 \mathcal{Y} + V_{11} f^2,$$
$$\mathcal{E}^{(2)} = \delta_1 \lambda^{(2)} + \mathcal{E}_1 \mathcal{Y} + {}_{11} f^2, \tag{5.28}$$
$$g^{(2)} = \pi_1 \lambda^{(2)} + G_1 \mathcal{Y} + G_{11} f^2,$$

where

$$V_1 = -a c_0 \partial_x (c_0 M^{(1)}), \quad \mathcal{E}_1 = -a c_0 u_0 M^{(2)} / \beta_2^{(0)},$$

$$G_1 = a\{c_0 \partial_x (c_0 u_0 M^{(1)}) + \mu_2^{(0)} u_0 M^{(1)} + (\mu_3^{(0)} - c_0 F_0)c_0 \pi_1 M^{(1)}\}/\mu_4^{(0)},$$

$$V_{11} = \frac{1}{2} a^2 c_0^2 K M^{(1)^2}, \quad \mathcal{E}_{11} = \frac{1}{2} a^2 \delta_{11}^{(0)} M^{(1)^2}, \tag{5.29}$$

$$G_{11} = -\frac{1}{2} a^2 \{(\mu_1^{(0)} + \pi_1 \mu_{14}^{(0)}) + (\mu_{14}^{(0)} + \pi_1 \mu_{44}^{(0)})\}M^{(1)^2}/\mu_4^{(0)},$$

and, in order to avoid even more lengthy algebraic expressions in (5.29), we have introduced the notations

229

$$\delta_{11} = \partial_F^2 \tilde{\mathcal{E}}, \quad \mu_1 = \partial_F \tilde{\mathcal{J}}, \quad \mu_2 = \partial_D \tilde{\mathcal{J}}, \quad \mu_3 = \partial_{\mu^e} \tilde{\mathcal{J}},$$
$$\mu_4 = \partial_g \tilde{\mathcal{J}}, \quad \mu_{14} = \partial_F \partial_g \tilde{\mathcal{J}}, \quad \mu_{44} = \partial_g^2 \tilde{\mathcal{J}} \tag{5.30}$$

where $\tilde{\mathcal{J}} = \tilde{\mathcal{J}}(F, \tilde{\mathcal{E}}(F,D), \mu^e, g) = \tilde{\mathcal{J}}(F,D,u,g)$. Equation (A.18) may now be integrated to yield the formula

$$\mathcal{J}^{(2)} = c_0 u_0 \lambda^{(2)} + J_1 \mathcal{G} + J_{11} f^2, \tag{5.31}$$

where

$$J_1 = a\{c_0 \partial_x(u_0 c_0 M^{(1)}) - c_0^2 \pi_1 F_0 M^{(1)}\}, \quad J_{11} = -\tfrac{1}{2} a^2 u_0 c_0^2 K M^{(1)2}, \tag{5.32}$$

and when (5.9), (5.14) and (5.26) are used in (A.22) we find that

$$\Sigma^{(2)} = c_0^2 \lambda^{(2)} + \Sigma_1 \mathcal{G} + \Sigma_{11} f^2 \tag{5.33}$$

where

$$\Sigma_1 = -a\{\delta_1^{(0)} \alpha_2^{(0)} u_0 + \alpha_3^{(0)} \pi_1 c_0\} M^{(1)}, \quad \Sigma_{11} = \tfrac{1}{2} a^2 \zeta_{11}^{(0)} M^{(1)2}. \tag{5.34}$$

Substitution from equations (5.9), (5.14), (5.26), (5.28), (5.31) and (A.17) into (A.19) now yields the second order linear transport equation

$$\partial_x\{c_0^{3/2} N(x) \lambda^{(2)}\} = P_1(X) \quad (X,\alpha) + P_2(X) f^2(\alpha) \tag{5.35}$$

where

$$2c_0^3 P_1 = -N\{c_0 \partial_x \Sigma_1 - c^2 \partial_x V_1 + \alpha_2^{(0)} J_1/\beta_2^{(0)} + c_0 a \alpha_3^{(0)} \partial_x(c_0 \pi_1 M^{(1)}) + c_0 \alpha_3^{(0)} G_1\}, \tag{5.36}$$

$$2c_0^3 P_2 = N\{-c_0 \partial_x \Sigma_{11} + c_0^2 \partial_x V_{11} - \alpha_2^{(0)} J_{11}/\beta_2^{(0)} - c_0 \alpha_3^{(0)} G_{11}$$

$$- \tfrac{1}{2} a[c_0 \Sigma_1 - c_0^2 V_1 + a c_0^2 \alpha_3^{(0)} \pi_1] K M^{(1)}$$

$$+ K c_0^2 a^2 [c_0 \partial_x(c_0 M^{(1)}) - \alpha_2^{(0)} u_0 M^{(1)}/\beta_2^{(0)}]$$

$$- c_0 u_0 a^2 M^{(1)2} [\beta_2^{(0)}(\alpha_{12}^{(0)} + \delta_1^{(0)} \alpha_{22}^{(0)}) + \alpha_2^{(0)}(\beta_{12}^{(0)} + \delta_1^{(0)} \beta_{22}^{(0)})]/\beta_2^{(0)2}\}.$$

It follows from equation (5.35) that

$$\lambda^{(2)}(X,\alpha) = Q_1(X) \int_0^\alpha f(\xi)d\xi + Q_2(X)f^2(\alpha) \tag{5.37}$$

where

$$c_0^{3/2}(X)N(X)Q_1(X) = \int_0^X N(Y)P_1(Y)dY,$$

$$c_0^{3/2}(X)N(X)Q_2(X) = \int_0^X N(Y)\{P_2(Y) - \tfrac{1}{2} a\Psi(Y)P_1(Y)\}dY. \tag{5.38}$$

In order to complete our second order description of the propagating pulse, we need to determine $\ell^{(2)}(X,\alpha)$ by solving equation (A.15). In view of equation (A.21), we need to evaluate $c^{(2)}$ in order to calculate $s^{(2)}$. We find that

$$c^{(2)} = C_1 \int_0^\alpha f(\xi)d\xi + C_{11}f^2(\alpha) \tag{5.39}$$

where

$$2c_0 C_1 = \zeta_{11}^{(0)}Q_1 - u_0 a\zeta_{12}^{(0)}M^{(1)}$$

$$2c_0 C_{11} = \zeta_{11}^{(0)}Q_2 + (\zeta_{111}^{(0)} - \zeta_{11}^{(0)}/2c_0^2)a^2 M^{(1)2}/2 \tag{5.40}$$

$$+ \tfrac{1}{2} u_0 a^2 \zeta_{12}^{(0)} \psi M^{(1)}$$

and here

$$\zeta_{111} = \partial_F^3 \widetilde{\Sigma}(F,D,u), \quad \zeta_{12} = \partial_D \partial_F \widetilde{\Sigma}(F,D,u). \tag{5.41}$$

When we substitute from (5.17) and (5.39) into (A.21) it follows that

$$s^{(2)} = S_1 \int_0^\alpha f(\xi)d\xi + S_{11}f^2(\alpha) \tag{5.42}$$

where

$$S_1 = -C_1/c_0^2, \quad S_{11} = \{\frac{1}{4c_0^5} \zeta_{11}^2 a^2 M^{(1)2} - \frac{1}{c_0^2} C_{11}\}. \tag{5.43}$$

Equations (5.42) and (A.15) now imply that

$$\partial_X \ell^{(2)} = S_1(X)f(\alpha) + 2S_{11}(X)ff' \tag{5.44}$$

and, since $\ell^{(2)}(0,\alpha) = 0$, equation (5.44) may be integrated to yield the formula

$$\ell^{(2)}(X,\alpha) = (L_1 + L_{11}f')f, \tag{5.45}$$

where

$$L_1 = \int_0^X S_1(Y)dY, \quad L_{11} = 2\int_0^X S_{11}(Y)dY. \tag{5.46}$$

Integration of the relation $\ell = \partial_\alpha T$ yields the formula

$$\omega \left\{ T(X,\alpha) - \int_0^X \frac{ds}{c(s)} \right\} = \alpha - a\,\psi f(\alpha) + \frac{1}{\omega}\left\{ L_1 \int_0^\alpha f(\zeta)d\zeta + \frac{1}{2} L_1 f^2(\alpha) \right\} \tag{5.47}$$

for the determination of the characteristics. It is to be noted that the characteristic curve of the wavelet α is influenced by the signal carried by the wavelets α_1 s.t. $0 \leqslant \alpha_1 \leqslant \alpha$.

In view of (4.8), the acceleration on the wavelet α when it reaches the point X is given by the formula

$$a(X,\alpha) = \frac{-c_0 \partial_\alpha \lambda^{(1)} - \omega^{-1}\{c_0 \partial_\alpha \lambda^{(2)} + c^{(1)}\partial_\alpha \lambda^{(1)} - c_0 \ell^{(1)}\partial_X v^{(1)}\}}{\ell^{(1)} + \ell^{(2)}/\omega} \tag{5.48}$$

and, of course, $a(X,\alpha)$ becomes unbounded if the denominator of (5.48) vanishes, i.e. if the equation

$$1 - a\,\psi(X)f'(\alpha) + \omega^{-1}\{L_1(X) + L_{11}(X)f'(\alpha)\}f(\alpha) = 0 \tag{5.49}$$

has solutions $X_s > 0$ and $\alpha_s \geqslant 0$. We note that the criterion for shock formation on the wavelet α is independent of the conditions prevailing on the precursor wavelets.

It is instructive to compare the first and second order solutions. First of all, to a first order in a high frequency pulse the strength and structure of the signal carried by the wavelet $\bar{\alpha}$ when it reaches the point X are determined by the conditions prevailing at the point $X = 0$ when the wavelet $\bar{\alpha}$ passed, as well as the steady state conditions encountered by the wavelet during its passage between the points $X = 0$ and X. In particular, conditions in the wavelet $\bar{\alpha}$ at the point X are not influenced by the precursor wavelets $0 < \alpha < \bar{\alpha}$ which have already passed through the point X. Furthermore, correct to first order, no changes are induced in the free charge density, electric potential or electric displacement at the point X. To second order, we find that when the wavelet $\bar{\alpha}$ reaches the point X, the structure of the signal is considerably changed and depends, inter alia, on the wavelets $0 < \alpha < \bar{\alpha}$ which have already passed through the point X. Second order changes are induced in the free charge density, the electric potential and electric field. It is interesting to note that while the second order changes as exhibited by equations (5.26), (5.28), (5.31), (5.33) and (5.37) depend on the function $\mathcal{Y}\,(X,\bar{\alpha})$ and on the wavelet $\bar{\alpha}$ at the point X and thus are influenced by the precursor wavelets $0 \leqslant a < \bar{\alpha}$, the incremental time of arrival $\ell(x,\alpha)$ even at the second order is not influenced by the wavelets which have already passed through the point X.

6. Pulses of Finite Amplitude

In the previous section, we obtained a number of concrete results by restricting our attention to small amplitude, finite rate high frequency pulses. By carrying out a perturbation analysis, we obtained first and second order solutions and, in particular, we demonstrated how second order effects modify the predictions of first order theory. Our objective here is somewhat different in that we seek to explore the predictions of Section 4 without placing any restrictions on the amplitude of the pulse which propagates. We restrict our analysis by examining the implications of the equations of Section 4 at the first iteration. Thus, our results may be regarded as generalizations of those of Section 5.1. The analysis presented here closely parallels that presented in somewhat different contexts by a number of authors [17,19,21,23].

The structure of the pulse is determined by equations (4.16) and (4.17) to be that of a simple wave and we have

$$v = \hat{v}(\overline{\lambda}, X), \quad \mathscr{E} = \hat{\mathscr{E}}(\overline{\lambda}, X), \quad g = \hat{g}(\overline{\lambda}, X),$$
$$\overline{\lambda} = \overline{\lambda}(X, \alpha), \quad \jmath = \mu_0^e(X), \quad D = D_0(X), \quad \phi = \phi_0(X), \tag{6.1}$$

where

$$\partial_\lambda \hat{v} = -c, \quad \partial_\lambda \hat{\mathscr{E}} = \delta_1, \quad \partial_\lambda \hat{g} = \pi_1 \tag{6.2}$$

and the functions $\hat{v}, \hat{\mathscr{E}}$ and $\hat{g}$ satisfy the initial conditions

$$\hat{v} = 0, \quad \hat{\mathscr{E}} = \mathscr{E}_0(X), \quad \hat{g} = g_0(X) \tag{6.3}$$

when $\alpha = 0$. The manner in which the stress and the conduction current are influenced by the pulse becomes more transparent by noting the formulae

$$\partial_\lambda \hat{\Sigma} = \zeta_1, \quad \partial_\lambda \hat{\jmath} = c\mu_0^e. \tag{6.4}$$

The variation of the value of $\overline{\lambda}$ associated with a given wavelet α of the pulse as the pulse traverses the material is determined by equation (4.23) which may be recast in the form

$$\partial_X \overline{\lambda} = \mathscr{F}(\overline{\lambda}, X), \tag{6.5}$$

where

$$2c^2 = -\rho_0^{-1}\partial_X \hat{\Sigma} + c\partial_X \hat{v} + s\alpha_2\beta_2^{-1}(\mathscr{J}_0 - \hat{\jmath}) + \alpha_3(\partial_X \overline{u}_3 - \hat{g}). \tag{6.6}$$

Once the functions introduced in (6.1) are known, the right hand side of (6.6) may be expressed in terms of $\overline{\lambda}$ and the known initial conditions ahead of the pulse so that $\mathscr{F} = \mathscr{F}(\overline{\lambda}, X)$ is then a known function of $\overline{\lambda}$ and X only. Obviously, the nonlinear transport equation (6.5) is much more complicated than the first order linear transport equation (5.12) of which it is, of course, a generalization. In general, $\mathscr{F}$ will not be a linear function of the pulse amplitude $\overline{\lambda}$ so that, in contrast to the result derived in Section 5.1, the attenuation of the signal associated with each wavelet α will depend on the amplitude of the signal. This phenomenon is usually called <u>amplitude attenuation</u> and has been discussed in a number of situations [17,19,31]. An example of the effect of amplitude attenuation in electroelastic semiconductors is discussed in the next section. The solution of equation (6.5) may be expressed in the form

$$\overline{\lambda}(X,\alpha) = \Lambda(X,k(\alpha)) \tag{6.7}$$

where Λ is the solution of the initial value problem

$$\frac{d\Lambda}{dX} = \mathcal{F}(\Lambda,X), \quad \Lambda(0,k) = k \tag{6.8}$$

and thus the amplitude on any wavelet $\overline{\alpha}$ which is instantaneously at the point X is determined only by the amplitude of the signal on the wavelet $\overline{\alpha}$ at the point $X = 0$ and the conditions encountered by the pulse during its transit from $X = 0$ to the point X. In particular, the precursor wavelets $0 < \alpha < \overline{\alpha}$ do not influence the amplitude of the signal on the wavelet $\overline{\alpha}$. It is clear from equation (6.6) that $\mathcal{F}$ has the same form throughout the pulse and hence that the nonlinear transport equation (6.8) has the same form at each wavelet of the pulse.

The expression (5.22) for the first order value of the acceleration in a small amplitude finite rate pulse may also be generalized to the case when the amplitude of the pulse if finite. The slowness s at any wavelet may be written

$$s = c^{-1} = s(\overline{\lambda},X) = \mathcal{S}(X,k) \tag{6.9}$$

and since $s = \partial_X T$ and $\alpha = t/\tau_p$ at $X = 0$ we have

$$T(X,\alpha) = \alpha\tau_p + \int_0^X \mathcal{S}(Y,k)dY. \tag{6.10}$$

It follows from (4.3) and (6.11) that

$$\ell(X,\alpha) = \tau_p - \varrho(X,k)k'(\alpha) \tag{6.11}$$

where the nonlinear distortion factor ϱ is given by

$$\varrho(X,k) = -\int_0^X \partial_k \mathcal{S}(Y,k)dY. \tag{6.12}$$

Equations $(4.8)_1$, (6.7), (6.8) and (6.11) together imply that

$$a(X,\alpha) = \frac{-c\,\partial\Lambda(X,k)/\partial k\,k'(\alpha)}{\tau_p - \varrho(X,k)k'(\alpha)} + c\{\partial_X\hat{v} - c\,\mathcal{F}\}. \tag{6.13}$$

Notice, in particular, that (6.13) shows that the arrival time of any wavelet α at X depends only on the signal associated with the wavelet α and is not influenced by the precursor wavelets. Likewise the shock formation criterion $\ell(X,\alpha) = 0$ depends only on the signal carried by the wavelet α at $X = 0$ and the point X.

Next, let $\alpha^* \in [0,1]$ be the least value of α for which $k'(\alpha)$ does not vanish. It follows from (6.13) that the acceleration a suffers a jump $[a]$ across the wavelet α^* and

$$[a] = \frac{-c \, \partial \Lambda(x,k)/\partial k \; k'(\alpha^*)}{\tau_p - \hat{X}(X,k)k'(\alpha^*)} \; . \tag{6.14}$$

Thus, the wavelet α^* is an acceleration wave. It is clear from (6.14) that $[a]$ may become unbounded, thus indicating the possibility of shock formation at the wavelet α^*.

Thus far we have confined our attention to the implications of our theory for finite amplitude pulses at the first iteration. We now generalize formulae (5.26) for pulses of finite amplitude. Firstly, we recall equations (4.12) and (4.14):

$$\partial_\alpha \overline{D}(X,\alpha) = \ell(X,\alpha)\{\mathcal{J}_0(X) - \overline{\mathcal{J}}(X,\alpha)\}$$

$$\partial_\alpha \overline{\phi}(X,\alpha) = c(X,\alpha)\ell(X,\alpha)\{\overline{\mathcal{E}}(X,\alpha) - \mathcal{E}_0\}. \tag{6.15}$$

Of course $\mathcal{E}(X,\alpha) = \hat{\mathcal{E}}(\lambda(X,\alpha),X)$ is known at the first iteration and $\mathcal{J}(X,\alpha) = \hat{\mathcal{J}}(\lambda(X,\alpha),X)$ may be calculated by substituting from (6.1) into the constitutive equation (2.15). When (6.11) is used, equations (6.15) yield the formulae

$$D = D_0(X) + \tau_p \int_0^\alpha \{\mathcal{J}_0 - \hat{\mathcal{J}}(\Lambda(X,\xi),X)\}d\xi$$

$$- \int_0^\alpha \{\mathcal{J}_0 - \hat{\mathcal{J}}(\Lambda(X,\xi),X)\}\hat{X}(X,k(\xi))k'(\xi)d\xi, \tag{6.16}$$

and

$$\phi = \phi_0(X) + \tau_p \int_0^\alpha \hat{c}(\Lambda(X,\xi),X)\{\hat{\mathcal{E}}(\Lambda(X,\xi),X) - \mathcal{E}_0(X)\}d\xi$$

$$-\int_0^\alpha \hat{c}(\Lambda(X,\xi),X)\{\hat{\mathcal{E}}(\Lambda(X,\xi),X) - \mathcal{E}_{\Lambda_0}(X)\}\hat{\chi}(X,k(\xi))k'(\xi)d\xi. \qquad (6.17)$$

These relations generalize formulae $(5.26)_{1,3}$ for pulses of finite amplitude. The change in the free electronic charge density induced by the finite amplitude pulse may be calculated by substituting from (6.16) into $(2.3)_1$ using formulae $(4.2)_3$.

7. Pulses Propagating into Uniform Regions

In our original study of acceleration waves, [4], we found that our results simplified considerably when the region ahead of the body ahead of the wave was in a homogeneous time independent state. Motivated by this observation, we now turn our attention to the study of pulses which propagate into an electroelastic semi-conductor which is homogeneously deformed, has a uniform charge density of free electronic fluid and is subjected to a uniform electric field at all points ahead of the pulse. If we choose the steady homogeneous configuration ahead of the pulse as our reference configuration, then the governing equations of Sections 4-6 must be solved subject to the initial conditions

$$\overline{\lambda}(X,0) = 0, \quad \overline{}(X,0) = \mathcal{E}_0, \quad \overline{u}(X,0) = u_0, \quad \overline{g}(X,0) = 0, \qquad (7.1)$$

where $\mathcal{E}_0$ and u_0 are constants.

Let us begin by considering the implications of the results of Section 5.1 for high frequency small amplitude pulses. It follows from equations (5.14)-(5.16) that when the wavelet α reaches the point X the associated pulse amplitude is given by

$$\lambda^{(1)}(X,\alpha) = af(\alpha)e^{-\overline{\xi}_0 X}, \quad \overline{\xi}_0 = \xi_0/c_0, \qquad (7.2)$$

and when this result is used in (5.15), (5.16) and (5.20) the expression for the distortion factor reduces to

$$\Psi(X) = -K_0(1 - e^{-\overline{\xi}_0 X})/\overline{\xi}_0. \qquad (7.3)$$

Thus, formulae (5.19) and (5.21) become

$$\ell^{(1)}(X,\alpha) = 1 + aK_0 f'(\alpha)(1 - e^{-\overline{\xi}_0 X})/\overline{\xi}_0, \qquad (7.4)$$

and

$$\omega(T - X/c_0) = \alpha + aK_0 f(\alpha)(1 - e^{-\overline{\xi}_0 X})/\overline{\xi}_0, \qquad (7.5)$$

respectively. Equations (A.5) and (7.4) now imply that, correct to order $1/\omega$, the slowness at the wavelet α is

$$s(X,\alpha) = 1/c_0 + \frac{a}{\omega} K_0 f(\alpha)e^{-\overline{\xi}_0 X} \qquad (7.6)$$

Expressions for $v^{(1)}$, $\mathcal{E}^{(1)}$, $g^{(1)}$ and $\mathcal{J}^{(1)}$ follow on substitution from (7.2) into the formulae (5.9).

In studying the behaviour of pulses in electroelastic semiconductors, which are initially in a homogeneous steady state, it is instructive to begin by following the approach adopted in much of the literature [4,24] and consider first the case wherein K_0 vanishes, which, of course, corresponds to a linear electroelastic response of the material. In such materials nonlinear effects of electronic origin which are associated with the semiconduction properties of the medium are the sole source of nonlinearity. The distortion factor ψ now vanishes, the incremental arrival time ℓ has the constant value and the characteristic curves associated with each wavelet of the pulse are parallel straight lines. In view of equations (5.9) and (7.2) the strain, velocity, electric field, stress and conduction current grow or decay as the pulse traverses the material according as $\overline{\xi}_0 > 0$ or $\overline{\xi}_0 < 0$. The case $\overline{\xi}_0 < 0$, which has no mechanical analogue, has been noted in treatments of small amplitude wave propagation in piezoelectric semiconductors with linear elastic response [4,24,34]. It is important to emphasize here that, in contrast to the situation which arises in some of the aforementioned analyses [4,34], ξ_0 is not an absolute constant for a given material. Its value depends on the properties of the material as well as the conditions prevailing therein. It is clear from the foregoing that the threshold condition, at which the amplitude of the pulse begins to grow, may be defined by

238

$$\xi_0 = 0. \tag{7.7}$$

Equation (7.7) with the definition (3.6) determines the threshold electric field $\mathcal{E}_T$ for the onset of amplification and acousto-electric domain formation.

In order to illustrate the influence which the biasing electric field $\mathcal{E}_0$ exerts on the propagation of pulses in semiconductors with linear elastic response, let us follow our own earlier work [14,16] and that of others and consider a semiconductor for which

$$\mathcal{G} = -um(\lambda + 1)\mathcal{E} + (\lambda + 1)D^C g \tag{7.8}$$

where m and D^C are the constant dimensionless mobility and diffusivity respectively. The dimensional[*] mobility $\overline{m}$ and the dimensional diffusivity $\overline{D}^C = -\overline{m}\partial_{\mu e}\overline{p}^e$ are related to m and D^C by the formulae

$$\overline{m} = \overline{m}_0 m, \quad \overline{D}^C = \overline{D}_0^C D^C, \quad \overline{m}_0 = \overline{c}_0/\overline{E}_0, \quad \overline{D}_0^C = -\overline{m}_0\overline{\rho}_0\overline{c}_0^2/\overline{\mu}_0^e. \tag{7.9}$$

It was shown in [14] that

$$u_0\alpha_2^{(0)} + \beta_1^{(0)}\alpha_3^{(0)}\gamma_3^{(0)}/\gamma_4^{(0)} = 0 \tag{7.10}$$

so that

$$\xi_0 = -\frac{u_0}{2c_0}\left(\mathcal{E}_0 + \frac{c_0}{m_0}\right) \tag{7.11}$$

and thus the <u>threshold field</u> is

$$\mathcal{E}_T = -\frac{c_0}{m_0}. \tag{7.12}$$

This result is the dimensionless form of the result originally derived by Hutson and White [4]. We may now rewrite (7.11) as

$$\xi_0 = -\frac{u_0}{2c_0}\left(\mathcal{E}_0 - \mathcal{E}_T\right) \tag{7.13}$$

and we note that ξ_0 will be negative whenerer $\mathcal{E}_0 > \mathcal{E}_T$ so that, in the pre-

----

[*] Here overbars are used to denote dimensional quantities.

sent approximation, the amplitudes of $\lambda^{(1)}$, $v^{(1)}$, $\mathcal{E}^{(1)}$ and $g^{(1)}$ will increase exponentially as the pulse traverses the material and acoustoelectric domains will be formed. Of course, the present analysis breaks down when $\lambda^{(1)}$ becomes sufficiently large, so that if $\mathcal{E}_0 > \mathcal{E}_T$ we can only expect the foregoing results to have a limited range of validity.

We now return to the study of small amplitude high frequency pulses in materials whose electroelastic response is nonlinear. The slowness of each wavelet will no longer be a constant and will vary in accordance with the formula (7.6). It is instructive to rewrite (7.6) in the form

$$s(X,\alpha) = 1/c_0 + \{s(0,\alpha) - 1/c_0\}e^{-\overline{\xi}_0 X} \tag{7.14}$$

where $s(0,\alpha) = 1/c_0 + aK_0 f(\alpha)$ is the slowness of the wavelet $\alpha = $ constant at the point $X = 0$. If the electric field ahead of the pulse is smaller than the threshold electric field so that $\overline{\xi}_0 > 0$, then the influence of the initial amplitude of the pulse on the slowness decreases as the pulse traverses the medium. When the field ahead of the pulse exceeds the threshold electric field, then $\overline{\xi}_0 < 0$ so that

$$s(X,\alpha) = 1/c_0 + \{s(0,\alpha) - 1/c_0\}e^{|\overline{\xi}_0| X} . \tag{7.15}$$

Clearly, if $s(0,\alpha) > 1/c_0$, i.e., the speed of propagation of the wavelet $\alpha = $ constant is less than the local speed of propagation of acceleration waves at $X = 0$, then the slowness of the wavelet increases as it traverses the material and as $X \to \infty$ the speed of propagation of the wavelet vanishes so that the wavelet becomes stationary. On the other hand if $s(0,\alpha) < 1/c_0$, the speed of propagation of the wavelet increases as the wavelet traverses the medium.

The acceleration at the wavelet α follows from (5.22) and (7.3) as

$$\begin{aligned}
a(X,\alpha) &= \frac{-ac_0 f'(\alpha)\exp(-\overline{\xi}_0 X)}{1 + aK_0 f'(\alpha)\{1 - \exp(-\overline{\xi}_0 X)\}/\overline{\xi}_0} \\[2mm]
&= \frac{a_c}{\{a_c/a(0,\alpha) - 1\}\exp(\overline{\xi}_0 X) + 1}
\end{aligned} \tag{7.16}$$

where $a_c = \xi_0/K_0$ is the critical acceleration and $a(0,\alpha) = -a c_0 f'(\alpha)$ is the acceleration at the wavelet α at $X = 0$. Of course, $a(X,\alpha)$ satisfies the Bernoulli equation (5.23) with $\overline{\beta}_0 = K_0/c$. If $\overline{\xi}_0 > 0$, the evolutionary behaviour of the acceleration at the wavelet will be qualitatively the same as that of an acceleration wave which propagates into a material with memory which is deformed ahead of the wavelet. Four distinct possibilities now arise:

(i) If $|a(0,\alpha)| < a_c$, then $a(X,\alpha) \to 0$ monotonically as $X \to \infty$,

(ii) If $a(0,\alpha) = a_c$, then $a(X,\alpha) = a_c$. Even though the acceleration at the wavelet α remains constant, the strain $\lambda^{(1)}(X,\alpha)$ will decrease monotonically as X increases,

(iii) If $\operatorname{sgn} a(0,\alpha) = -\operatorname{sgn} K_0$ and $|a(0,\alpha)| > |a_c|$, then $a(X,\alpha) \to 0$ monotonically as $X \to \infty$,

(iv) If $\operatorname{sgn} a(0,\alpha) = \operatorname{sgn} K_0$ and $|a(0,\alpha)| > |a_c|$, then the acceleration $a(X,\alpha)$ will become unbounded when the wavelet α has reached the point

$$X = -\overline{\xi}_0^{-1} \ell n\{1 - a_c/a(0,\alpha)\}. \tag{7.17}$$

On the other hand, if $\overline{\xi}_0 < 0$, four distinct possibilities again arise,

(i)* If $\operatorname{sgn} a(0,\alpha) = -\operatorname{sgn} K_0$ and $|a(0,\alpha)| < |a_c|$, then $a(X,\alpha) \to a_c$ monotonically as $X \to \infty$,

(ii)* If $a(0,\alpha) = a_c$, then $a(X,\alpha) = a_c$. Thus, even though the strain $\lambda^{(1)}(X,\alpha)$ increases as the wavelet α propagates through the material, the acceleration remains fixed,

(iii)* If $\operatorname{sgn} a(0,\alpha) = -\operatorname{sgn} K_0$ and $|a(0,\alpha)| > |a_c|$, then $a(X,\alpha) \to a_c$ monotonically as $X \to \infty$,

(iv)* If $\operatorname{sgn} a(0,\alpha) = \operatorname{sgn} K_0$, then the acceleration $a(X,\alpha)$ will become unbounded when the pulse reaches the point

$$X = |\overline{\xi}_0|^{-1} \ell n\{1 + |a_c/a(0,\alpha)|\} \tag{7.18}$$

It is important to appreciate that, since the acceleration at $X = 0$ will, in general, vary from one wavelet to another, the acceleration will behave differently at different points of the pulse. When $\xi_0 > 0$, an interesting situation arises in a pulse in which $a(0, \alpha_1) = a_c$ at one point in the pulse while $|a(0, \alpha)| < a_c$, $\alpha \neq \alpha_1$. Such a pulse will ultimately evolve into an acceleration wave which coincides with the wavelet α_1 and whose amplitude will have the constant value a_c. It is also interesting to note that when $\xi_0 < 0$ the acceleration will tend to the finite nonzero value a_c or else it will become unbounded. In particular, it never decays to zero.

The locations at which shocks form are determined by equations (7.17) or (7.18) according as $\xi_0 > 0$ or $\xi_0 < 0$. In either case, the shock which forms closest to $X = 0$ forms on the wavelet on which $|a_c/a(0, \alpha)|$ is a minimum or, equivalently, one which $|f'(\alpha)|$ is a maximum. Once the shock is formed, its amplitude will vary as it is fed by the faster wavelets which come from behind and by he slower wavelets ahead. If $\alpha^-(X)$ and $\alpha^+(X)$ are the wavelets just ahead of and immediately behind the shock, respectively, at the point X, then, when the shock has reached X, the portions of the signal contained on the wavelets α such that $\alpha^+(X) < \alpha < \alpha^-(X)$ will have been obliterated. Thus, the shock will cause part of the information contained in the pulse at $X = 0$ to be lost by the time the pulse reaches the point X. If $t = A(X)$ denotes the arrival time of a shock at X, then the jump conditions (2.6), (2.7) imply that the slowness S of the shock is determined by the formulae

$$S = S(X) = \frac{dA}{dX} = -\frac{[v]}{[\Sigma]} = -\frac{[\lambda]}{[v]} = \frac{[\mu]}{[\ ]} . \tag{7.19}$$

Furthermore, at the shock wave, we also have the conditions

$$[\phi] = [D] = [p^e] = 0 \tag{7.20}$$

and, since by assumption $(2.22)_5$ $\alpha_3 \neq 0$, it follows that $[u] = 0$.

A detailed analysis of shock waves has been presented elsewhere [16] and we content ourselves with presenting an outline of the situation which arises when the shocks are weak [25,35]. Thus, we assume that reflections from the shock may be neglected so that the formulae derived in Section 5 may be used to describe the fields on either side of the propagating shock wave. Since the shock and the wavelets $\alpha^+(X)$ and $\alpha^-(X)$ are simultaneously at the point X, it follows from (5.21) that, correct to order $1/\omega$, the time $A(X)$ at which the shock arrives at the point X is given by

$$\omega(A(X) - X/c_0) = \alpha^+ - a\phi(X)f(\alpha^+)$$
$$= a^- - a\phi(X)f(a^-), \tag{7.21}$$

where $\phi(X)$ is given by (7.3). Differentiation of (7.21) with respect to X implies that

$$\ell(a^-)\frac{d\alpha^-}{dX} = -\ell(\alpha^+)\frac{d\alpha^+}{dX}, \tag{7.22}$$

which in turn may be used to show that

$$\int_{\alpha^+}^{\alpha^-} f(\alpha)d\alpha = \frac{1}{2}\{f(\alpha^+) + f(\alpha^-)\}[\alpha]. \tag{7.23}$$

Once a, $f(\alpha)$ and $\phi(X)$ are prescribed, (7.21) and (7.22) determine the wavelets α^- and α^+ and the path of the shock then follows from $(7.21)_1$. It should also be noted that differentiation of (7.21) with respect to X also leads to the classical formula of 'weak' shock theory [25,35]

$$S = 1/c_0 + \frac{aK_0}{2\omega}\{\lambda^{(1)}(X,\alpha^+(X)) + \lambda^{(1)}(X,\alpha^-X))\}$$

$$= 1/c_0 + \frac{aK_0}{2\omega}\{f(\alpha^+) + f(\alpha^-)\}\exp(-\bar{\xi}_0 x). \tag{7.24}$$

Of course, the expression (7.24) for S must be compatible with the expression (7.19) and the conditions (7.20). It should be clear from (5.4) that in order to check the aforementioned compatibility to order $1/\omega$ we need to evaluate the jumps occuring in (7.19) to order $1/\omega^2$. When formulae (5.4), (5.9), (5.26), (7.2) and (7.3) are used to simplify expressions (5.28)-(5.34) and the definition $(2.4)_5$ is used, we have

$$v(X,\alpha) = -c_0\lambda(X,\alpha) + \frac{1}{\omega^2}\{\frac{1}{2}c_0^2 K_0 \lambda^{(1)2}(X,\alpha) + c_0 a\xi_0 M^{(1)}\mathcal{Y}(X,\alpha)\},$$

$$\Sigma(X,\alpha) = c_0^2\lambda(X,\alpha) - \frac{1}{\omega^2}\{c_0^3 K_0 \lambda^{(1)2}(X,\alpha) - 2c_0^2 a\xi_0 M^{(1)}\mathcal{Y}(X,\alpha)\}, \qquad (7.25)$$

$$\mathcal{P}(X,\alpha) = c_0 u_0 \lambda(X,\alpha) - \frac{1}{\omega^2}\{\frac{1}{2}c_0^2 u_0 K_0 \lambda^{(1)2}(X,\alpha) +$$

$$+ ac_0(u_0\xi_0 + c_0\pi_1 F_0)M^{(1)}\mathcal{Y}(X,\alpha)\},$$

$$\bar{\mu} = (u_0 F_0 + \bar{\mu}_r) + u_0\lambda - \frac{1}{\omega^2}aF_0 c_0\pi_1 M^{(1)}\mathcal{Y}(X,\alpha),$$

where $M(1) = \exp(-\bar{\xi}_0 X)$ and

$$\mathcal{Y}(X,\alpha) = \int_0^\alpha f(\xi)d\xi + \frac{1}{2}\frac{ac_0 K_0}{\xi_0}\{1 - \exp(-\bar{\xi}_0 X)\}f^2(\alpha). \qquad (7.26)$$

It follows from (7.26) that

$$[\mathcal{Y}] = \int_{\alpha^+}^{\alpha^-} f(\xi)d\xi + (ac_0 K_0/2\xi_0)\{1 - \exp(-\bar{\xi}_0 X)\}(f(\alpha^-) + f(\alpha^+)) \qquad (7.27)$$

and, when (7.21) and (7.23) are used, (7.27) implies that $[\mathcal{Y}] = 0$ i.e., the function $\mathcal{Y}(X,\alpha)$ is continuous at the shock. Since $\mathcal{Y}$ is continuous, it follows from formulae (5.26) that the conditions (7.20) are satisfied when terms of order $1/\omega^3$ are neglected. Furthermore, the continuity of $\mathcal{Y}$ also ensures the compatibility of (7.19) and (7.24).

Equations (7.21), (7.23) are well known in 'weak' shock analysis and have been studied in a variety of papers [22,23,31]. Explicit solutions of these equations have only been found in a number of particular situations, the details of which will be found in the aforementioned references. The situation becomes particularly simple at a shock which forms at the leading wavelet $\alpha = 0$. The position X_s at which this shock forms follows from either (7.17) or (7.18) on setting $\alpha = 0$. Of course $f(\alpha^+) = 0$ and it follows from (7.27) and the continuity of $\mathcal{Y}$ that

$$\int_0^{\alpha^-} f(\xi)d\xi = -\frac{ac_0 K_0}{2\xi_0}\{1 - \exp(-\bar{\xi}_0 X)\}f^2(\alpha^-). \qquad (7.28)$$

Next equation $(4.7)_1$ may be used to show that, correct to order $\frac{1}{\omega}$, the particle displacement at the reference point $X = 0$ is given by the formula

$$y(0,\alpha) = -c_0 a z(\alpha) \ , \quad z(\alpha) = \int_0^\alpha f(\xi)d\xi \tag{7.29}$$

and in terms of $z(\alpha)$ we may write (7.28) as

$$z'(\alpha^-) = -\frac{ac_0 K_0}{2\xi_0} \{1 - \exp(-\overline{\xi}_0 X)\} z^2(\alpha^-). \tag{7.30}$$

Equations (7.21) and (7.30) together imply that

$$\omega(A(X) - X/c_0) = \alpha^- - 2z(\alpha^-)/z'(\alpha^-). \tag{7.31}$$

Equations (7.29), (7.30) and (7.31) now determine the trajectory of the shock in terms of the particle displacement at the point $X = 0$. Notice in particular that the wavelet α can only coalesce into the shock if $\mathrm{sgn}(f(\alpha)) = -\mathrm{sgn}(K_0/\xi_0) = \mathrm{sgn}(\zeta_{11}^{(0)}/\xi_0)$. It follows from (7.29) and this observation that if $\xi_0 > 0$ and $\zeta_{11}^{(0)} > 0 \{\zeta_{11}^{(0)} < 0\}$ so that the electroelastic semiconductor hardens (softens) in extension at constant electric displacement and free electronic charge density, only displacements which left $X = 0$ when $y(0,t) < 0 \{y(0,t) > 0\}$ can catch up with the shock. If $\xi_0 < 0$, the situation just described is reversed. These results throw into sharp focus the critical role which the threshold field $\mathcal{E}_T$ plays in determining which wavelets may coalesce into a shock wave. In particular we note that those wavelets which may (may not) coalesce into a shock wave when $\mathcal{E} < \mathcal{E}_T$ cannot (can) do so when $\mathcal{E} > \mathcal{E}_T$.

To complete our analysis, we examine in detail the predictions of Section 6 for pulses of finite amplitude which propagate into electroelastic semiconductors which are in a uniform steady state in the region ahead of the pulse. The variation of the amplitude of the pulse as it traverses the material will now be governed by equation (6.5) and, since the material ahead of the pulse is assumed to be in a uniform steady state, the expression (6.6) for $\mathcal{J}$ reduces to

$$\mathcal{J} = \hat{\mathcal{J}}(\lambda) = \zeta_2(\hat{g}_0 - \hat{g})/2c^3 - \alpha_3 \hat{g}/2c^2 \tag{7.32}$$

where $\zeta_2 = \alpha_2/\beta_2$. In order to determine $\mathcal{J}$ as a function of λ we must first calculate $\hat{\mathcal{J}}$ as a function of λ by substituting from (6.1) into the constitutive relation (2.15) for $\mathcal{J}$. In the interest of simplicity we now assume that the response of the material may be adequately described by assuming that

$$\hat{\mathcal{E}}(\lambda) = \mathcal{E}_0 + \delta_1^{(0)}\lambda + \frac{1}{2}\delta_{11}^{(0)}\lambda^2, \quad \hat{\Sigma}(\lambda) = \Sigma_0 + \rho_0\zeta_1^{(0)}\lambda + \frac{1}{2}\rho_0\zeta_{11}^{(0)}\lambda^2,$$

$$\hat{\mathcal{J}}(\lambda) = \mathcal{J}_0 + c_0 u_0\lambda + \frac{\zeta_{11}^{(0)}}{2c_0}u_0\lambda^2, \quad \hat{g}(\lambda) = \pi_1^{(0)}\lambda + \frac{1}{2}\pi_{11}^{(0)}\lambda^2,$$

$$(7.33)$$

where, of course, the coefficients of λ and λ^2 are constant. For a material of this type $\zeta_2 = \zeta_2^{(0)} + \zeta_{21}^{(0)}\lambda$, where $\zeta_2^{(0)}$, $\zeta_{21}^{(0)}$ as well as $\alpha_3 = \alpha_3^{(0)}$ are constants. Substitution for ζ_2, $\hat{\mathcal{J}}$ and $\hat{g}$ in (7.32) now reduces the nonlinear transport equation (6.5) to the Bernoulli equation

$$\partial_x\hat{\lambda} = -\overline{\xi}_0\hat{\lambda}(1 - \xi_{11}\hat{\lambda}) \tag{7.34}$$

where

$$\overline{\xi}_0\xi_{11} = -\frac{\zeta_2^{(0)}u_0}{2c_0^3}\{c_0\zeta_{21}^{(0)}/\xi_2^{(0)} + \zeta_{11}^{(0)}/c_0\} \tag{7.35}$$

$$-\alpha_3^{(0)}\{\frac{1}{2}\pi_{11}^{(0)} - \zeta_{11}^{(0)}\pi_1^{(0)}/c_0\}/2c_0^3.$$

The solution of (7.34) corresponding to the boundary condition $\hat{\lambda}(\alpha,0) = k(\alpha)$ is

$$\hat{\lambda}(X,\alpha) = \frac{k(\alpha)\exp(-\overline{\xi}_0 X)}{1 - \xi_{11}k(\alpha)\{1 - \exp(-\overline{\xi}_0 X)\}}. \tag{7.36}$$

Notice the difference between expressions (7.2) and (7.36). In contrast to the situation which arises in the case of small amplitude finite rate pulses, the rate at which the strain on any wavelet varies as the wave traverses the material is now influenced by the strength of the signal carried by the wavelet. The signal is now said to be distorted by amplitude attenuation [19,31].

The asymptotic behavior of $\hat{\lambda}(X,\alpha)$ as X becomes large depends strongly on the sign of ξ_0. Thus, if $\mathcal{E}_0 < \mathcal{E}_T$, then $\xi_0 > 0$ and as X increases

$$\hat{\lambda}(X,\alpha)\exp(\overline{\xi}_0 X) \rightarrow \frac{k(\alpha)}{1 - \xi_{11}k(\alpha)} \tag{7.37}$$

so that the amplitude of the pulse ultimately decreases to zero more rapidly (slowly) than does the amplitude of a pulse of the type discussed in Section 5 if $\xi_{11}k(\alpha) > 0(\xi_{11}k(\alpha) < 0)$. On the other hand, if the steady electric field is larger than the threshold field, i.e. $\mathcal{E}_0 > \mathcal{E}_T$, then $\xi_0 < 0$ and as X increases

$$\hat{\lambda}(X,\alpha) \rightarrow - 1/\xi_{11}. \tag{7.38}$$

Thus, when $\mathcal{E}_0 > \mathcal{E}_T$, the strain on each wavelet of a pulse of finite amplitude will tend to a constant value. Of course, when this happens the stress, electric field, and conduction current will also tend to constant limiting values throughout the pulse. This is nothing more than the saturation phenomenon which has been observed experimentally by a number of authors [24]. Saturation occurs because of the nonlinear response exhibited by the material for pulses of finite amplitude. It has been explained in other work [12,24] by assuming that the electroelastic response is nonlinear. It should be clear from the foregoing that saturation may also occur in nonlinear electroelastic media even when the constitutive equation for the conduction current is linear.

When conditions are uniform in the region ahead of a pulse of finite amplitude, equations (6.10) and (6.11) reduce to

$$T(X,\alpha) = \alpha\tau_p + X/c_0 + \{\zeta_{11}^{(0)}/2c_0^2\xi_{11}\xi_0\}\ell n\{1 - k(\alpha)\xi_{11}[1 - \exp(-\overline{\xi}_0 X)]\}, \tag{7.39}$$

and

$$\ell(X,\alpha) = \tau_p - \{\zeta_{11}^{(0)}k'(\alpha)/2c_0^2\xi_0\}\{1-\exp(-\overline{\xi}_0 X)\}\{1-k(\alpha)\xi_{11}[1-\exp(-\overline{\xi}_0 X)]\}^{-1}, \tag{7.40}$$

respectively. A shock will form on the wavelet α when it reaches the point X_s provided X_s is a solution of the equation $\ell(X_s,\alpha) = 0$ i.e.,

$$X_s = -c_0/\xi_0 \ell n \{1 - [\xi_{11}k(\alpha) + Ak'(\alpha)]^{-1}\} \tag{7.41}$$

where $A = 2c_o^2 \xi_o \tau_\rho / \zeta_{11}^{(0)}$. Corresponding to (7.39), this is the criterion which must be satisfied for shock formation. Thus, if $\xi_o > 0$ then a shock will form if for some $\alpha \in [0,1]$

$$k'(\alpha) > A^{-1}\{1 - \xi_{11}k(\alpha)\}. \tag{7.42}$$

On the other hand, if $\xi_o < 0$ then a shock will form if for some $\alpha \in [0,1]$

$$-k'(\alpha) > A^{-1}\xi_{11}k(\alpha). \tag{7.43}$$

We close by noting that if the material under study is a linear electroelastic medium with nonlinear semiconduction characteristics then the amplitude of the pulse will no longer suffer amplitude dispersion since all the characteristic curves will be parallel straight lines. Shocks will never form in such a material. Of course, $\lambda(X,\alpha)$ will still be given by an expression of the type (7.36) with $\xi_{11} = -\alpha_3^{(0)} \pi_{11}^{(0)}/4c_o^3$. If $\xi_o < 0$ saturation will still occur but will now be caused solely as a result of the nonlinear semiconduction properties of the medium.

Appendix

When we substitute from (5.4) into equations (4.4), $(4.7)_2$, (4.8) and (4.12)-(4.14) and equate coefficients of ω^{-1} we obtain the differential equations

$$s_o \partial_x v^{(1)} + \partial_\alpha \lambda^{(1)} = 0, \tag{A.1}$$

$$\partial_\alpha D^{(1)} = \beta_1^{(0)} \partial_\alpha \lambda^{(1)} + \beta_2^{(0)} \partial_\alpha \mathcal{E}^{(1)} = 0, \tag{A.2}$$

$$(\gamma_1^{(0)} - u_o c_o)\partial_\alpha \lambda^{(1)} + \gamma_2^{(0)} \partial_\alpha \mathcal{E}^{(1)} + \gamma_4^{(0)} \partial_\alpha g^{(1)} = 0, \tag{A.3}$$

$$\partial_\alpha u^{(1)} = 0, \tag{A.4}$$

$$\partial_x \ell^{(1)} = \partial_\alpha s^{(1)}, \tag{A.5}$$

$$\partial_\alpha \phi^{(1)} = 0, \tag{A.6}$$

$$\partial_X \Sigma^{(1)} - c_0 \partial_X v^{(1)} + \frac{\alpha_2^{(0)}}{c_0 \beta_2^{(0)}} \mathcal{J}^{(1)} - \alpha_3^{(0)}(\partial_X u^{(1)} - g^{(1)}) = 0. \qquad (A.7)$$

In order to obtain a simple expression for $g(1)$, we note that equation $(2.3)_3$ may be rewritten in the form

$$s \partial_\alpha \bar{\mathcal{J}} - \bar{u} \partial_\alpha \bar{\lambda} - F \partial_\alpha \bar{u} = \ell \partial_X^{-} \qquad (A.8)$$

and it follows that

$$\partial_\alpha \mathcal{J}^{(1)} - F_0 c_0 \partial_\alpha u^{(1)} - u_0 c_0 \partial_\alpha \lambda^{(1)} = 0. \qquad (A.9)$$

Next we note that since $s = \dfrac{1}{c}$ then $s(1) = -s_0^2 c^{(1)}$. In equations (A.2), (A.3), (A.7) and (A.9) all coefficients are evaluated at $\lambda = \lambda_0(X)$, $\mathcal{E} = \mathcal{E}_0(X)$ and $u = u_0(X)$.

When terms of order ω^{-2} in the foregoing equations are equated, we arrive at the differential equations

$$\partial_\alpha v^{(2)} + c_0 \partial_\alpha \lambda^{(2)} = c_0 \ell^{(1)} \partial_X v^{(1)} - c^{(1)} \partial_\alpha \lambda^{(1)}, \qquad (A.10)$$

$$\partial_\alpha D^{(2)} = -\ell^{(1)}(u_0 \lambda^{(1)} + F_0 u^{(1)}), \qquad (A.11)$$

$$\beta_1^{(0)} \partial_\alpha \lambda^{(2)} + \beta_2^{(0)} \partial_\alpha \mathcal{E}^{(2)} = -\beta_1^{(1)} \partial_\alpha \lambda^{(1)} - \beta_2^{(1)} \partial_\alpha \mathcal{E}^{(1)}$$

$$+ c_0 \ell^{(1)}(\partial_X D^{(1)} - u^{(1)} F_0 - u_0 \lambda^{(1)}), \qquad (A.12)$$

$$(\gamma_1^{(0)} - u_0 c_0)\partial_\alpha \lambda^{(2)} + \gamma_2^{(0)} \partial_\alpha \mathcal{E}^{(2)} + (\gamma_3^{(0)} - c_0 F_0)\partial_\alpha u^{(2)} + \gamma_4^{(0)} \partial_\alpha g^{(2)}$$

$$= c_0 \ell^{(1)} \partial_X \mathcal{J}^{(1)} - (\gamma_1^{(1)} - c_0 u^{(1)} - u_0 c^{(1)})\partial_\alpha \lambda^{(1)} - \gamma_2^{(1)} \partial_\alpha \mathcal{E}^{(1)}$$

$$-(\gamma_3^{(1)} - c_0 \lambda^{(1)} - F_0 c^{(1)})\partial_\alpha u^{(1)} - \gamma_4^{(1)} \partial_\alpha g^{(1)}, \qquad (A.13)$$

$$\partial_\alpha u^{(2)} = c_0 \ell^{(1)}(\partial_X u^{(1)} - g^{(1)}), \qquad (A.14)$$

$$\partial_X \ell^{(2)} = \partial_\alpha s^{(2)}, \qquad (A.15)$$

$$\partial_\alpha \phi^{(2)} = c_0 \ell^{(1)}(\mathcal{E}^{(1)} + \partial_X \phi^{(1)}), \qquad (A.16)$$

$$c_o \partial_x \Sigma^{(2)} - c_o^2 \partial_x v^{(2)} + \frac{\alpha_2^{(0)}}{\beta_2^{(0)}} \mathcal{G}^{(2)} - c_o \alpha_3^{(0)} [\partial_x u^{(2)} - g^{(2)}]$$

$$= -c^{(1)} \partial_x \Sigma^{(1)} + 2c_o c^{(1)} \partial_x v^{(1)}$$

$$- (\beta_2^{(0)} \alpha_2^{(1)} + \alpha_2^{(0)} \beta_2^{(1)})/\beta_2^{(0)^2} \mathcal{G}^{(1)} \qquad \text{(A.17)}$$

$$+ (\alpha_3^{(0)} c^{(1)} + c_o \alpha_3^{(1)})(\partial_x u^{(1)} - g^{(1)}),$$

$$\partial_\alpha \mathcal{G}^{(2)} - c_o u_o \partial_\alpha \lambda^{(2)} - c_o F_o \partial_\alpha u^{(2)} = (c_o u^{(1)} + u_o c^{(1)}) \partial_\alpha \lambda^{(1)}$$

$$+ (c^{(1)} F_o + c_o \lambda^{(1)}) \partial_\alpha u^{(1)} + \ell^{(1)} \partial_x \mathcal{G}^{(1)}, \qquad \text{(A.18)}$$

where

$$\beta_1^{(1)} = \beta_{11}^{(0)} \lambda^{(1)} + \beta_{12}^{(0)} \mathcal{E}^{(1)}, \quad \beta_2^{(1)} = \beta_{12}^{(0)} \lambda^{(1)} + \beta_{22}^{(0)} {}^{1)},$$

$$\alpha_2^{(1)} = \alpha_{12}^{(0)} \beta^{(1)} + \alpha_{22}^{(0)} \mathcal{E}^{(1)}, \quad \alpha_3^{(1)} = \alpha_{33}^{(0)} u^{(1)},$$

$$\gamma_1^{(1)} = \gamma_{11}^{(0)} \lambda^{(1)} + \gamma_{12}^{(0)} \mathcal{E}^{(1)} + \gamma_{13}^{(0)} u^{(1)} + \gamma_{14}^{(0)} g^{(1)}, \qquad \text{(A.19)}$$

$$\gamma_2^{(1)} = \gamma_{12}^{(0)} \lambda^{(1)} + \gamma_{22}^{(0)} \mathcal{E}^{(1)} + \gamma_{23}^{(0)} u^{(1)} + \gamma_{24}^{(0)} g^{(1)},$$

$$\gamma_3^{(1)} = \gamma_{13}^{(0)} \lambda^{(1)} = \gamma_{23}^{(0)} \mathcal{E}^{(1)} + \gamma_{33}^{(0)} u^{(1)} + \gamma_{34}^{(0)} g^{(1)},$$

$$\gamma_4^{(1)} = \gamma_{14}^{(0)} \lambda^{(1)} + \gamma_{24}^{(0)} \mathcal{E}^{(1)} + \gamma_{34}^{(0)} u^{(1)} + \gamma_{44}^{(0)} g^{(1)},$$

with

$$\alpha_{33} = \partial_{e_\mu}^2 \hat{\Sigma}, \quad \gamma_{11} = \partial_F^2 \hat{\mathcal{G}}, \quad \gamma_{12} = \partial_{\mathcal{E}} \partial_F \hat{\mathcal{G}},$$

$$\gamma_{13} = \partial_{e_\mu} \partial_F \hat{\mathcal{G}}, \quad \gamma_{14} = \partial_g \partial_F \hat{\mathcal{G}}, \quad \gamma_{22} = \partial_{\mathcal{E}}^2 \hat{\mathcal{G}}, \quad \gamma_{23} = \partial_{e_\mu} \partial_{\mathcal{E}} \hat{\mathcal{G}}, \qquad \text{(A.20)}$$

$$\gamma_{24} = \partial_g \partial_{\mathcal{E}} \hat{\mathcal{G}}, \quad \gamma_{33} = \partial_{e_\mu}^2 \hat{\mathcal{G}}, \quad \gamma_{34} = \partial_g \partial_{e_\mu} \hat{\mathcal{G}}, \quad \gamma_{44} = \partial_g^2 \hat{\mathcal{G}}.$$

Finally, we have

$$s^{(2)} = -\frac{c^{(2)}}{c_0^2} + \frac{{c^{(1)}}^2}{c_0^2} \tag{A.21}$$

and

$$\Sigma^{(2)} = \alpha_1^{(0)}\lambda^{(2)} + \alpha_2^{(0)}\mathcal{E}^{(2)} + \alpha_3^{(0)}u^{(2)}$$

$$+ \frac{1}{2}\alpha_{11}^{(0)}{\lambda^{(1)}}^2 + \alpha_{12}^{(0)}\lambda^{(1)}\mathcal{E}^{(1)} + \frac{1}{2}\alpha_{22}^{(0)}{\mathcal{E}^{(1)}}^2 + \frac{1}{2}\alpha_{33}^{(0)}{u^{(1)}}^2$$

$$= \zeta_1^{(0)}\lambda^{(2)} + \alpha_3^{(0)}u^{(2)} + \frac{1}{2}\zeta_{11}^{(0)}{\lambda^{(1)}}^2 + \delta_1\alpha_2^{(0)}D^{(2)}. \tag{A.22}$$

REFERENCES

1. W.C. Wang, "Current Oscillations and Acoustic Flux Generation in CdS." Appl. Phys. Lett. <u>6</u>, 81 (1965).

2. J.H. McFee and P.K. Tien, "Acoustoelastic Current Distribution and Current Saturation in CdS." J. Appl. Phys. <u>37</u>, 2754 (1966).

3. W.H. Haydl, K. Harker and C.F. Quate, "Current Oscillations in Piezoelectric Semiconductors." J. Appl. Phys. <u>38</u>, 4295 (1967).

4. A.R. Hutson and D.L. White, "Elastic Wave Propagation in Piezoelectric Semiconductors." J. Appl. Phys. <u>33</u>, 40 (1962).

5. V.L. Gurevich, "On the theory of Sound Absorption in Piezoelectrics." Sov. Phys. Solid State <u>4</u>, 668 (1962).

6. B.K. Ridley and J. Wilkinson, "A Theory of Acoustoelectric Domains in Piezoelectric Semiconductors, 1. Linear Regime." J. Phys. C., Ser. 2, <u>2</u>, 1299 (1969).

7. V.L. Gurevich, "Limitation of the Sound Amplification in Piezoelectric Semiconductors." Sov. Phys. Solid State <u>5</u>, 892 (1963).

8. V.L. Gurevitch and B.D. Laikhtman, "Nonlinear Theory of Sound Instability in Piezoelectrics." Sov. Phys. J.E.T.P. <u>22</u>, 668 (1966).

9. B.K. Ridley and J. Wilkinson, "A Theory of Acoustoelectric Domains in Piezoelectric Semiconductors, 11 Nonlinear Interactions." J. Phys. C. Ser 2. <u>2</u>, 1307 (1969).

10. B.K. Ridley and J. Wilkinson, "A Theory of Acoustoelectric Domains in Piezoelectric Semiconductors, 111. Domain Solution." J. Phys. C. Ser. 2.2, 1321 (1969).

11. B.K. Ridley, "A Theory of Nonlinear Acoustic Waves in a Piezoelectric Semiconductor." J. Phys. C. <u>4</u>, 783 (1971).

12. M. Pawlik and G. Rowlands, "The Propagation of Solitary Waves in Piezoelectric Semiconductors." J. Phys. C. _8_, 1189 (1975).

13. H.G. de Lorenzi and H.F. Tiersten, "On the Interaction of the Electromagnetic Field with Heat Conducting Deformable Semiconductors." J. Math. Phys. _16_, 938 (1975).

14. M.F. McCarthy and H.F. Tiersten, "On Dimensional Acceleration Waves and Acoustoelectric Domains in Piezoelectric Semiconductors." J. Appl. Phys. _47_, 3389 (1976).

15. M.F. McCarthy and H.F. Tiersten, "The Growth of Wave Discontinuities and Piezoelectric Semiconductors." J. Math. Phys. _20_, 2682 (1979).

16. M.F. McCarthy and H.F. Tiersten, "Shock Waves and Acoustoelectric Domains in Piezoelectric Semiconductors." J. Appl. Phys. _48_, 159 (1977).

17. E. Varley and E. Cumberbatch, "Large Amplitude Waves in Stratified Media." J. Fluid Mech. _43_, 513 (1970).

18. D.F. Parker, "An Asymptotic Theory for Oscillatory Nonlinear Signals." J. Inst. Math. Applics., _7_, 92, 1971.

19. B.R. Seymour and M.P. Mortel, "Nonlinear Geometrical Acoustics." Mechanics Today, Vol. 2, 251, Pergamon, Oxford, 1975.

20. D.F. Parker, "Locally Simply Unsteady Deformations". Rheol. Acta, _16_, 134, 1977.

21. D.F. Parker and B.R. Seymour, "Finite Amplitude One-Dimensional Pulses in an Inhomogeneous Granular Material." Arch. Rational Mech. Anal., _72_, 265, 1980.

22. E. Varley and T.G. Rogers, "The Propagation of High Frequency, Finite Acceleration Pulses and Shocks in Viscoelastic Materials." Proc. Roy. Soc. A, _296_, 498, 1967.

23. B.R. Seymour and E. Varley, "High Frequency, Periodic Disturbances in Dissipative Systems. 1. Small Amplitude Finite Rate Theory." Proc. Roy. Soc. A, _314_, 387, 1970.

24. J.W. Tucker and V.W. Rampton, Microwave Ultrasonics in Solid State Physics, North Holland-Amsterdam (1972).

25. G. Whitham, Linear and Nonlinear Waves, Wiley-Interscience, New York (1974).

26. B.D. Coleman and M.E. Gurtin, "Waves in Materials with Memory, 11. On the Growth and Decay of One Dimensional Acceleration Waves", Arch. Ratl. Mech. Anal." _19_, 239 (1965).

27. W.A. Green, "Growth of Plane Discontinuities Propagating into a Homogeneously Deformed Elastic Material." Arch. Ratl. Mech. Anal. _16_, 79 (1964).

28. M.F. McCarthy, "Thermodynamic Influences on the Propagation of Waves in Electroelastic Materials." Int. J. Engng. Sci. _11_, 1301 (1973).

29. E. Varley, "Simple Waves in General Elastic Materials." Arch. Rational Mech. Anal. _20_, 309, 1965.

30. M.F. McCarthy and H.F. Tiersten, "On Integral Forms of the Balance Laws of Deformable Semiconductors." Arch. Ratl. Mech. Anal. _68_, 27 (1978).

31. D.F. Parker, "Propagation of a Rapid Pulse Through a Relaxing Gas." Phys. Fluids 15, 256, 1972.

32. P.B. Bailey and P.J. Chen, "On the Local and Global Behaviour of Acceleration Waves." Arch. Ratl. Mech. Anal. 41, 121 (1971); Addendum, Arch. Ratl. Mech. Anal. 44, 212 (1972).

33. M.F. McCarthy, "One Dimensional Pulse Propagation in Composite Materials Modelled as Interpenetrating Solid Continua." Wave Motion, 4, 221, 1982.

34. P.N. Butcher and N.R. Ogg, "A Nonlinear Theory of Acoustoelastic Gain and Current in Piezoelectric Semiconductors." J. Phys. D1, 1271 (1968).

35. A.H. Nayfeh and D.T. Mook, "Nonlinear Oscillations." Wiley Interscience, New York, 1979.

WEAK SOLUTIONS OF TRANSONIC FLOW BY COMPENSATED COMPACTNESS

Cathleen Synge Morawetz

Courant Institute of Mathematical Sciences
New York University
251 Mercer Street
New York, NY 10012

The existence of weak solutions for compressible inviscid potential transonic flow is an old problem going back to the 1930's when Busemann [1] demonstrated wind tunnel experiments showing that solutions were not likely to be smooth. Our objective is to establish an existence theory for such solutions. There are some missing links and we will point them out enroute. The main idea is to solve the problem with viscosity, to look at the limit as viscosity vanishes, and to establish that in the limit the equations of inviscid flow are satisfied.

1. The Problem with Viscosity

We consider two-dimensional flow past a bump on a wall on which the normal velocity is zero and we make the region finite by cutting it off by straight lines.

The viscous boundary value problem is a free boundary value problem. The velocity (u,v) is the gradient $\nabla\phi$ of a potential and satisfies

$$\text{div} (\rho \, \nabla\phi) = 0$$

where ρ is density. Viscosity is induced by the assumption that

$$\nu \, \nabla\rho\cdot\nabla\phi = |\nabla\phi|^2 - q_B^2(\rho),$$

where q_B is the speed as a function of density given by Bernoulli's law, see [2]. Note q_B has a maximum at $\rho = 0$.

The flow domain is $0 < x < a$, $Y(x) < y < b$, and on the line $x = 0$, $\phi = 0$, whereas on $y = b$, $\frac{\partial\phi}{\partial n} = 0$.

If $|\nabla\phi| < \max q_B$ on $x = a$ then $\phi = A$ there, while this condition on $y = Y$ implies $\frac{\partial\phi}{\partial n} = 0$. There is a free boundary where $|\nabla\phi| = \max q_B$ and $\phi = A$ a constant which confines the domain.

Since the system is of third order, we add the boundary condition $\rho = \rho_0$ on $x = 0$.

The first missing link is the existence of a smooth solution to this problem for $\nu > 0$. However there is a suitable variational procedure followed by a standard compactness argument which indicates that such a solution exists and that for A sufficiently small it has no free boundary.

The next step will be to proceed to the limit as $\nu \to 0$. For this there is a wealth of ν-independent estimates, the most important of which are:

(i) $|\nabla\phi(x,\nu)|$ is uniformly bounded by $\max q_B$.

(ii) $\nu^{-1}||\rho(x,\nu) - R(|\nabla\phi|)||^2$ is uniformly bounded.

(iii) $\nu(||\nabla v|| + ||\nabla n||)^2$ is uniformly bounded.

Here the norm is, roughly speaking, the L^2 norm and $R(|\nabla\phi|)$ is the density given by Bernoulli's law.

The first forces ϕ to converge and the other two allow us to bound independent of ν in H_{loc}^{-1}, that is,

(iv) $\mathrm{div}(fR\,\nabla\phi + gR^{-1}\nabla\psi)$ remains bounded in H_{loc}^{-1} , as $\nu\to0$,

where ψ is the stream function $(\psi_y = \rho\phi_x, \psi_x = -\rho\phi_y)$, provided that f and g satisfy certain equations in the u, v plane (the hodograph plane).

We are now in a position to apply the compensated compactness theories of Tartar and Murat [3,4]. If the hodograph equations for f, g were hyperbolic we would be able to apply DiPerna's theory [5]. But the equations are of mixed type and are determined locally as hyperbolic or elliptic according to whether the speed is above or below the speed of sound $c(R)$.

The problem is further complicated because the limit speed as $\nu \to 0$ could drop to zero in the interior or could increase to $\max q_B$. In fact as $\nu \to 0$ it could oscillate between these two extremes. The second missing link is the proof that this does not happen. Along with that we need to assume that the flow angle θ also remains bounded.

However, if we assume that at a given point (x,y) the flow angle is bounded and the speed as $\nu \to 0$ is bounded away from the two extremes, stagnation and cavitation, then we may use the theory of compensated compactness to prove that the weak limit ϕ satisfies weakly the boundary conditions and the equation

$$\text{div}(R\ \nabla\phi) = 0.$$

By Young's theory we have at (x,y)

$$h(\phi_x, \phi_y) \to \int h(u,v)\ d\mu(u,v;x,y)$$

where h is an arbitrary bounded function of its arguments and $d\mu$ is a measure with $\int d\mu = 1$. Our objective will be achieved if we can prove that $d\mu$ is a Dirac measure.

What we have to work with is the so-called div-curl lemma which implies that if f_1, g_1 and f_2, g_2 satisfy (iv), then

$$\int (f_1 g_2 - f_2 g_1)\ d\mu = \int f_1\ d\mu \int g_2\ d\mu - \int f_2\ d\mu \int g_1\ d\mu.$$

If the class of functions f,g (the entropy pairs) is rich enough, then this identity will rule out all measures that are not point measures. For the present problem one finds useful choices to be

$$f = \dot{H}_n e^{\pm in\theta}$$
$$g = \mp in\ H_n e^{\pm in\theta}$$

and

$$f = \dot{K}_n e^{\pm n\theta}$$
$$g = \mp n\ K_N e^{\pm n\theta},$$

where H_n, K_n satisfy a second order differential equation in the speed, θ is the flow angle and $\cdot = p(q)\dfrac{d}{dq}$.

The crucial identity, obtained by manipulation of the div curl identity, is

$$0 = \int [(H_n(q) - H_n(q'))(\dot{H}_n(q) - \dot{H}_n(q'))$$

$$+ 4 H_n(q)\dot{H}_n(q')\sin^2 \tfrac{1}{2}\, n(\theta - \theta')]d\mu\, d\mu'$$

Using the property that H_n behaves like q^n for low q, has other monotonicity properties in the elliptic region, and is bounded uniformly in the hyperbolic region, one can prove from this identity by taking n large that if the Young measure in the elliptic region is non-zero, then it is a point measure, and in any case the measure vanishes in the hyperbolic region. On the other hand if the Young measure has no elliptic support, one can use an analogous identity with K_n, $\sinh n\theta$ instead of H_n, $\sin n\theta$ to show that the measure if not entirely hyperbolic is confined between two characteristics $(p_1(q)dq \pm d\theta = 0)$ meeting at a point on the sonic line. Then one can adapt DiPerna's hyperbolic theory to show that the Young measure is a Dirac measure.

References

[1] Busemann, A. Widerstand bei geschunwinding keiten naher du schell geschwin-dijkeiten, Proc. Third Int. Congr. Appl. Mech. 1 (1930), 282-285.

[2] Synge, J.L., Motion of a viscous fluid conducting heat, Quart. of Appl. Math. Vol. 13, pp. 271-278, 1955.

[3] Tartar, L.C., Compensated compactness and applications to partial differential equations, Nonlinear Analysis and Mechanics, Heriot-Watt Symp. IV, (1979), 136-192. Research Notes in Mathematics, Pitman.

[4] Murat, F., Compacité par compensation, Ann. Scuola Norm. Sup. Pisa 5 (1978), 489-507.

[5] DiPerna, R.J., Convergence of approximate solutions to conservation laws, Arch. Rat. Mech. Anal. 82 (1983), 27-70.

EXTENDED THERMODYNAMICS OF IDEAL GASES

by

Ingo Müller

FB9-Hermann Föttinger Institute
Technical University, Berlin
West Germany

1. Introduction

Extended thermodynamics is a field theory with the primary objective of determining the fields of mass density, velocity, energy density, stress deviator and heat flux. The theory, as presented here, is applicable to monatonic gases, classical and degenerate, and it is strongly motivated by the kinetic theory of gases. Its main advantage over ordinary thermodynamics is that in a natural way it removes some inconsistencies of that older theory. In particular it predicts finite wave speeds of both thermal waves and shear waves. Also the theory removes discrepancies that exist between ordinary thermodynamics and the kinetic theory of gases. Thus it lets us see in proper perspective the violation of material frame indifference in the kinetic theory. Also the entropy flux in extended thermodynamics agrees with the entropy flux of the kinetic theory.

Extended thermodynamics goes back to an old paper by Müller [7] (see also [2]) and it was recently revisited and streamlined by Liu & Müller [3].

2. Juxtaposition of Ordinary and Extended Thermodynamics

The principal objective of

<table>
<tr><td>ordinary thermodynamics</td><td>extended thermodynamics</td></tr>
</table>

is the determination of the following

<table>
<tr><td>five fields:</td><td>thirteen fields:</td></tr>
<tr><td>mass density F</td><td>mass density F</td></tr>
<tr><td>momentum density F_i</td><td>momentum density F_i</td></tr>
<tr><td>energy density $\frac{1}{2} F_{ii}$</td><td>momentum flux density F_{ij}</td></tr>
<tr><td></td><td>energy flux density $\frac{1}{2} F_{ijj}$.</td></tr>
</table>

$$(2.1)$$

In order to achieve this we need field equations and these are based upon the equations of balance of

<table>
<tr><td>mass</td><td></td><td>mass</td></tr>
<tr><td>momentum</td><td></td><td>momentum</td></tr>
<tr><td>energy</td><td></td><td>momentum flux</td></tr>
<tr><td></td><td></td><td>energy flux</td></tr>
</table>

The form of these equations of balance is well known in ordinary thermodynamics. In extended thermodynamics the form of the last two balance equations is suggested by the form of the appropriate "equations of transfer" of the kinetic theory of gases. We write

$$\frac{\partial F}{\partial t} + \frac{\partial F_i}{\partial x_i} = 0 \qquad\qquad \frac{\partial F}{\partial t} + \frac{\partial F_i}{\partial x_i} = 0$$

$$\frac{\partial F_i}{\partial t} + \frac{\partial F_{ij}}{\partial x_j} = F(f_i + i_i^0) + 2W_{ik}F_k \qquad\qquad \frac{\partial F_i}{\partial t} + \frac{\partial F_{ij}}{\partial x_j} = F(f_i + i_i^0) + 2W_{ik}F_k$$

$$\frac{\partial F_{ii}}{\partial t} + \frac{\partial F_{iij}}{\partial x_j} = F_i(f_i + i_i^0) \qquad\qquad \frac{\partial F_{ij}}{\partial t} + \frac{\partial F_{ijk}}{\partial x_k} - P_{<ij>} = \qquad (2.2)$$

$$= 2F_{(i}(f_{j)} + i_{j)}^0) + 4F_{k(j}W_{i)k}$$

$$\frac{\partial F_{ieee}}{\partial t} + \frac{\partial F_{ieej}}{\partial x_j} - P_{ijj} =$$

$$= 3F_{(ij}(f_{j)} + i_{j)}^0) + 2W_{ik}F_{kee}$$

$P_{<ij>}$ and P_{ijj} are the production densities of momentum and energy flux respectively. f_i is the specific external body force, while i_i^0 is the part of the specific inertial force that is independent of momentum. W_{ik} is the matrix of angular velocity of the frame, so that the term $2W_{ik}F_k$ is recognized as part of the Coriolis force density. [+)]

Neither in ordinary nor in extended thermodynamics is the set (2.2) a set of field equations for the fields (2.1), because it contains additional quantities, viz.

[+)] All tensors in(2) are symmetric. Round brackets indicate a general symmetrization, i.e. the sum of all quantities with the N indices permuted divided by $N!$. Angular brackets <> indicate symmetric trace-less parts of tensors.

$$F_{\langle ij\rangle}, \; F_{iij} \qquad \qquad \Bigg\| \qquad \qquad F_{\langle ijk\rangle}, \; F_{ieej}, \; P_{\langle ij\rangle}, \; P_{ijj} \qquad (2.3)$$

Borrowing the jargon of the kinetic theory we may say that we have a closure problem. Continuum mechanics and thermodynamics solves that problem by saying that the quantities (2.3) are constitutive quantities which means that they are related to the fields (2.1) in a materially dependent manner by constitutive relations. In the case of viscous heat conducting fluids the constitutive relations have the following general form

$$F_{\langle ij\rangle} = \hat{F}_{\langle ij\rangle}\left(F, F_i, F_{ii}, \frac{\partial F}{\partial x_e}, \frac{\partial F_i}{\partial x_e}, \frac{\partial F_{ii}}{\partial x_e}\right)$$

$$F_{iij} = \hat{F}_{iij}\left(\text{—————————————}\right) \; \Bigg\| \; \begin{aligned} F_{\langle ijk\rangle} &= \hat{F}_{\langle ijk\rangle}(F, F_i, F_{ij}, F_{iee}) \\[6pt] F_{ieej} &= \hat{F}_{ieej}\left(\text{—————————}\right) \\ P_{\langle ij\rangle} &= \hat{P}_{\langle ij\rangle}\left(\text{—————————}\right) \\[6pt] P_{ijj} &= \hat{P}_{ijj}\left(\text{—————————}\right) \end{aligned} \qquad (2.4)$$

If we assume for the moment that we know all constitutive functions, we can eliminate the quantities (2.3) between the balance equations (2.2) and the constitutive relations (2.4) and come up with a set of

$$\text{five} \qquad \qquad \Bigg\| \qquad \qquad \text{thirteen}$$

explicit field equations. Every solution of those shall be called a <u>thermoelastic process</u>.

Note that the field equations of ordinary thermodynamics are of second order while the field equations of extended thermodynamics are of first order.

The above set of equations (2.2) are strongly motivated by the kinetic theory of gases. This is true in particular for the form of the inertial contributions in the last two equations on the right hand side of (2.2), but also in regard to the specific form of the left hand sides of both groups of equations in (2.2). Indeed, note that always the flux in one equation is repeated as the density in the following equation. The structure limits the applicability of this theory to the description of monatonic gases. The easiest way to see this is as follows. The trace F_{ii} of the momentum flux F_{ij} is equal to 3p in the rest frame of the fluid, where p is the pressure. On the other hand F_{ii} is twice the energy density which in the rest frame is the density $\rho\varepsilon$ of the internal energy. Thus we see that the theory based on (2) implies

$$p = \frac{2}{3}\, \rho\varepsilon \qquad (2.5)$$

and that relation is well known to be true only in monatonic ideal gases.

3. Constitutive Theory

3.1 Non Convective Quantities

The equations (2.2) have the neat simple form of equations of balance with supplies due to a body force and to inertial forces. That simple form was achieved by writing the equations for the full fluxes of momentum (say) or energy. The momentum flux F_{ij} contains, however, a convective part $\rho v_i v_j$, explicit in density ρ and velocity $\underset{\sim}{v}$ and, added to that, the non-convective momentum flux or pressure tensor m_{ij}. Similarly the energy flux can be decomposed in a convective term, the power of the pressure tensor and a non-convective contribution which we call the heat flux; the same is true for the fluxes F_{jk} and F_{ijjk} of momentum and energy flux. In fact, the kinetic theory of gases or an exercise in tensor representation shows us that the following decompositions are valid

$$
\begin{aligned}
F &= \rho \\
F_i &= \rho v_i \\
F_{ij} &= m_{ij} + \rho v_i v_j \\
F_{ijk} &= m_{ijk} + 3m_{(ij}v_{k)} + \rho v_i v_j v_k \\
F_{ijjk} &= m_{ijjk} + 4m_{(ijj}v_{k)} + 6m_{(ij}v_i v_j)} + \rho v_i v_j v_k v_e.
\end{aligned}
\qquad (3.1)
$$

Equation $(3.1)_3$ implies for the energy density $\frac{1}{2} F_{ii}$

$$\frac{1}{2} F_{ii} = \rho\varepsilon + \frac{\rho}{2} v^2 \tag{3.2}$$

which is the familiar decomposition of energy into a kinetic and an internal part. Similarly $(3.1)_4$ implies for the energy flux $\frac{1}{2} F_{iij}$

$$\frac{1}{2} F_{iij} = q_j + m_{ji} v_i + \rho(\varepsilon + \frac{1}{2} v^2)v_j \tag{3.3}$$

In the sequel it will be preferable to consider

$$\rho, v_i, \varepsilon, m_{\langle ij \rangle}, q_i \tag{3.4}$$

as the set of variables instead of $(2.1)_2$. By (3.1) obviously there is a one-to-one correspondence between the two sets. Similarly, instead of the constitutive quantities $(2.3)_2$ we shall consider

$$m_{\langle ijk \rangle}, \ m_{ieej}, \ P_{\langle ij \rangle}, \ P_i \equiv P_{iee} - 3v_{(j} P_{\langle ij \rangle)},$$

which are clearly equivalent but preferable, because they are all objective vectors. Thus we shall have constitutive equations of the form

$$
\begin{aligned}
m_{\langle ijk \rangle} &= \hat{m}_{\langle ijk \rangle}(\rho, v_i, \varepsilon, m_{\langle ij \rangle}, q_i) \\
m_{ieej} &= \hat{m}_{ieej}(\text{------------}) \\
P_{\langle ij \rangle} &= \hat{P}_{\langle ij \rangle}(\text{------------}) \\
P_{iee} - 3v_{(j} P_{\langle ij \rangle)} &= \hat{P}_i (\text{------------})
\end{aligned}
\tag{3.5}
$$

instead of $(2.4)_2$.

3.2 Restrictive Principles

In Chapter 2 we have assumed, for the purpose of discussion, that we know the constitutive functions. In reality, of course, we do not know these functions and

iii) The requirement of hyperbolicity

We shall proceed to formulate these principles and to list their consequences.

3.3 Material Frame Indifference

The principle of material frame indifference requires that the constitutive functions of nonconvective quantities be independent of frame. That is to say, the functions must have the same form in an inertial and a non-inertial frame.

One consequence of this principle is that the velocity $\underset{\sim}{v}$ cannot appear as a variable in the constitutive equations (3.5). Moreover the constitutive functions (3.5) must be isotropic functions of the remaining variables. There are representation theorems for isotropic functions and, if we restrict the attention to linear constitutive functions in $m_{<ij>}$ and q_i we obtain

$$
\begin{aligned}
m_{<ijk>} &= 0 \\
m_{ieej} &= \beta \delta_{ij} + \gamma\, m_{<ij>} \\
P_{<ij>} &= \sigma\, m_{<ij>} \\
P_{iee} - 3v_{(j}P_{<ij>)} &= \tau\, q_i ,
\end{aligned}
\tag{3.6}
$$

where all coefficients may be functions of ρ and ε. The reason for considering only linear terms in $m_{<ij>}$ and q_i is that there is particular interest in near-equilibrium phenomena, where products of $m_{<ij>}$ and q_i may be considered to be negligable.

3.4 Entropy Principle

The entropy principle states that the entropy inequality

$$
\frac{\partial h}{\partial t} + \frac{\partial}{\partial x_j}\,(hv_j + \phi_j) \geqslant 0
\tag{3.7}
$$

must hold for all thermodynamic processes. Here h and ϕ_k are the density and the non-convective flux of entropy respectively and they are both constitutive quantities that must satisfy the requirement of material frame indifference.

Including all terms quadratic in $m_{<ij>}$ and q_i we thus have [+)]

$$h = \rho n + h_1\, m_{<ij>} m_{<ij>} + h_2 q_i q_i$$
$$\Phi_k = \phi_1 q_k + \phi_2 m_{<ki>} q_i \tag{3.8}$$

Here again all coefficients may depend on ρ and ε.

The easiest way to exploit (3.7) is to write the new inequality

$$\frac{\partial h}{\partial t} + \frac{\partial}{\partial x_j}\,(hv_j + \Phi_j) - \Lambda\left(\frac{\partial F}{\partial t} + \frac{\partial F_i}{\partial x_i}\right) -$$
$$- \Lambda^i\left(\frac{\partial F_i}{\partial t} + \frac{\partial F_{ij}}{\partial x_j}\right) -$$
$$- \Lambda^{ij}\left(\frac{\partial F_{ij}}{\partial t} + \frac{\partial F_{ijk}}{\partial x_k} - P_{<ij>}\right) -$$
$$- \Lambda^{ijj}\left(\frac{\partial F_{iee}}{\partial t} + \frac{\partial F_{ieek}}{\partial x_k} - P_{iee}\right) \geq 0. \tag{3.9}$$

which must hold for all fields F, F_i, F_{ij}, F_{iee} or ρ, v_i, ε, $m_{<ij>}$, q_i rather than only for thermodynamic processes. That restriction has been taken care of in (3.9) by the Lagrange multipliers Λ which may be functions of the variables ρ, v_i, ε, $m_{<ij>}, q_i$. [++)]

The key to the exploitation of (3.9) is the observation that the left hand side of (3.9) is a linear function of all space and time derivatives of F, F_i, F_{ij} and F_{iee}. The factors of those derivatives must therefore vanish lest the inequality be violated by some choice of the derivatives. This condition gives many restrictions from which the Lagrange multipliers can be eliminated.

In this presentation I do not give a detailed account of the evaluating process. I merely list the results and comment on them.

One intermediate result is

[+)] Even in a linear, near-equilibrium theory the entropy density and entropy flux must include quadratic terms, since there is no linear entropy production.

[++)] In the evolution of the entropy principle it is irrelevant whether the frame is inertial or not. Therefore for simplicity we take it to be non-inertial.

$$d\eta = \Lambda^{ii}|_E (\rho,\varepsilon)(d\varepsilon + pd(\tfrac{1}{\rho})) \qquad (+) \qquad\qquad (3.10)$$

which shows that $\Lambda^{ii}|_E(\rho,\varepsilon)$ is <u>the</u> integrating factor of $d\varepsilon + pd(\tfrac{1}{\rho})$ that leads to the entropy as a potential. That integrating factor is defined as the reciprocal of the absolute temperature in thermodynamics. Therefore we may write

$$\Lambda^{ii}|_E(\rho,\varepsilon) = \frac{1}{T} \qquad\qquad (3.11)$$

which may be used - in principle - to calculate ε as a function of ρ and T so that T may replace ε as a variable. We shall use that new choice and henceforth consider ρ, v_i, T, $m_{\langle ij\rangle}$, q_i as variables. Eqn's (3.10) and (3.11) combine to give

$$d\eta = \frac{1}{T}(d\varepsilon(\rho,T) + p(\rho,T)d(\tfrac{1}{\rho})). \qquad\qquad (3.12)$$

This is the Gibbs equation. The integrability condition implied by it reads

$$p = T^{5/2}F(\frac{\rho}{T^{3/2}}), \qquad\qquad (3.13)$$

where (2.5) was used. F is an arbitrary function of the single variable $\frac{\rho}{T^{3/2}}$. Note that the thermal equation of state $p = p(\rho,T)$ of a monotomic ideal gas is thus reduced to a function of <u>one</u> variable.

The other definite results that come out of the inequality can be summarized as follows

$$m_{\langle ijk\rangle} = 0$$

$$m_{ieej} = \{5T^{7/2}(\int \frac{FF'}{z}\,dz + c) + C\}\delta_{ij} + 7\frac{T}{F}(\int \frac{FF'}{z}\,dz + c)m_{\langle ij\rangle}$$

$$P_{\langle ij\rangle} = \sigma m_{\langle ij\rangle}$$

$$P_{iee} - 3v_{(ij}P_{\langle ij\rangle)} = \tau q_i \qquad\qquad (3.14)$$

$$h = \frac{3}{2}\rho\int((\frac{F'}{z} - \frac{5}{3}\frac{F}{z^2})\,dz+K - \frac{m_{\langle ij\rangle}m_{\langle ij\rangle}}{4T^{7/2}F} - \frac{2q_iq_i}{35T^{9/2}\{c- \frac{5}{7}\frac{F^2}{z} +\int \frac{FF'}{z}\,dz\}}$$

$$\phi_i = \frac{q_i}{T} - \frac{2}{5}\frac{m_{\langle ij\rangle}q_j}{T^{7/2}F}$$

$(+)$ The index E denotes equilibrium.

Of course, $(3.14)_{1,3,4}$ have been taken over unchanged from $(3.6)_{1,3,4}$. They are listed again for completeness of the set of constitutive equations. The other equations, viz. $(3.14)_{2,5,6}$ have to be compared with $(3.6)_2$ and (3.8) and we see that the entropy principle has had a profound effect on them. Indeed, instead of the 7 functions of two variables each that were still permitted in $(3.6)_2$ and (3.8) we see that only one function of one variable remains in $(3.14)_{2,5,6}$ and in addition three constants of integration, namely c, C and K. The argument z in (3.14) stands for $\frac{\rho}{T^{3/2}}$.

There remains a residual entropy inequality of the form

$$ -\frac{1}{4pT}\, \sigma\, ^m{}_{<ij>}\, ^m{}_{<ij>} - \frac{2}{5pT}\, \frac{\tau q_i q_i}{7\, \frac{T}{F}\, \{c - \frac{5}{7}\frac{F^2}{z} + \int \frac{FF'}{z}\, dz\}} \geqslant 0 \qquad (3.15) $$

The left hand side represents the entropy production. Inspection shows that both summands in (15) must be positive whence we conclude

$$ \sigma \leqslant 0 \quad \text{and} \quad \frac{\tau}{\{c - \frac{5}{7}\frac{F^2}{z} + \int \frac{FF'}{z}\, dz\}} \leqslant 0. \qquad (3.16) $$

This exhausts the consequences of the entropy principle.

3.5 Hyperbolicity

In order to ensure that our system of field equations implies finite wave speeds and that regular Cauchy data provide a unique and regular solution locally we require that the system be symmetric hyperbolic in the sense of Friedrichs and Lax [4]. This requirement can be formally expressed by saying that the matrix of second derivatives of the entropy density with respect to the variables $V_\alpha = \{F, F_i, F_{ij}, F_{iee}\}$ must be negative definite:

$$ \frac{\partial^2 h}{\partial V_\alpha \partial V_\beta} \quad - \text{negative definite.} \qquad (3.17) $$

Exploitation of this requirement is again a rather laborious process of which I only list the results. It turns out that (3.17) implies

$$ 0 < F' < \frac{5}{3}\frac{F}{z} \quad \text{and} \quad c > \frac{5}{7}\frac{F^2}{z} - \int \frac{FF'}{z}\, dz. \qquad (3.18) $$

Thus hyperbolicity imposes a restriction upon the thermal equation of state (see (3.13)) and upon the constant of integration c that appeared before in (3.14), (3.15), and (3.16). If we combine (3.18) and the condition (3.16) that came from the entropy principle we see that both σ and τ must be non-negative:

$$\sigma \leq 0 \quad \text{and} \quad \tau \leq 0. \tag{3.19}$$

4. Field Equations

Insertion of the restricted form (3.14) of the constitutive equations into the balance equations $(2.2)_2$ and use of the decompositions (3.1) leads to the following set of field equations after a little calculation

$$\dot{\rho} + \rho \frac{\partial v_i}{\partial x_i} = 0.$$

$$\dot{\rho v_i} + \frac{\partial(p\,\delta_{ij} + m_{<ij>})}{\partial x_j} = \rho(f_i + i_i^0) + 2W_{ik}\rho v_k$$

$$\dot{\rho \varepsilon} + \frac{\partial q_i}{\partial x_i} + (p\,\delta_{ij} + m_{<ij>})\frac{\partial v_i}{\partial x_j} = 0$$

$$\dot{m}_{<ik>} + m_{<ik>}\frac{\partial v_e}{\partial x_e} + \frac{4}{5}\frac{\partial q_{<i}}{\partial x_{k>}} + 2m_{j<i}\frac{\partial v_{k>}}{\partial x_j} - 4m_{j<i}W_{k>j} = \sigma m_{<ik>}$$

$$\dot{q}_i + \frac{7}{2}q_i\frac{\partial v_e}{\partial x_e} + \frac{1}{2}\frac{\partial(\frac{5}{7}\rho\gamma\delta_{ij} + \gamma m_{<ij>})}{\partial x_j} + \frac{2}{5}q_k\frac{\partial v_k}{\partial x_i} + q_k\frac{\partial v_i}{\partial x_k} -$$

$$- \frac{1}{\rho}(m_{<ik>}\frac{\partial m_{<kj>}}{\partial x_j} + m_{<ij>}\frac{\partial p}{\partial x_j} + \frac{5}{2}p\frac{\partial m_{<ij>}}{\partial x_j} + \frac{5}{2}p\frac{\partial p}{\partial x_i}) -$$

$$- 2W_{ik}q_k = \tau q_i$$

$$\tag{4.1}$$

where we have

$$p = \frac{2}{3}\rho\varepsilon = T^{5/2}F\left(\frac{\rho}{T^{3/2}}\right) \quad \text{and} \quad \gamma = 7\frac{T}{F}\left(\int \frac{FF'}{z}\,dz + c\right). \tag{4.2}$$

We conclude that this set of field equations contains but one unknown constant c, one unknown function of one variable, viz. $F\left(\frac{\rho}{T^{3/2}}\right)$ and two negative-valued functions of two variables each, namely $\sigma = \sigma(\rho,T)$ and $\tau = \tau(\rho,T)$.

We shall now proceed to show how these unknowns can be determined.

5. Determination of the Unknown Constitutive Functions

5.1 Thermal Equations of State in a Monatomic Gas

The most conspicuous unknown in the set (4.1) is the function $F(\frac{\rho}{T^{3/2}})$ which occurs in the thermal equation of state (see (3.13)). But it is exactly this fact that makes it possible to determine F. Indeed the thermal equation of state is an equilibrium equation whose form has been derived from first principles in statistical mechanics. Any book on statistical mechanics will give the result which I list here in implicit form

$$p(\rho,T) = \frac{2}{3}\frac{k}{\mu}\frac{1}{A}T^{5/2}I_4^{\pm}(\alpha) \quad \text{with} \quad I_2^{\pm}(\alpha) = A\frac{\rho}{T^{3/2}} \tag{5.1}$$

k is the Boltzmann constant and μ is the atomic mass. The factor A is defined as

$$A = \frac{h^3}{w}\frac{1}{8\pi}\frac{1}{\sqrt{2}}\frac{1}{k^{3/2}}\frac{1}{\mu^{5/2}} \tag{5.2}$$

where h is the Planck's constant and $w = 2s + 1$ for atoms with spin $s\frac{h}{2\tilde{h}}$.

the expressions $I_4^{\pm}(\alpha)$ and $I_2^{\pm}(\alpha)$ in (1) are integrals defined as

$$I_n^{\pm}(\alpha) = \int_0^{\infty}\frac{x^n dx}{e^{x^2+\alpha}\pm 1}. \tag{5.3}$$

The + sign holds for atoms that are Bose particles while the - sign holds for Fermi particles.

Comparison of (5.1) with (3.13) shows that we have

$$F(z) = \frac{2}{3}\frac{k}{m}\frac{1}{A}I_4^{\pm}(\alpha) \quad \text{with} \quad I_2^{\pm}(\alpha) = Az. \tag{5.4}$$

In order to obtain $p(\rho,T)$ we must solve $(5.1)_2$ for α and insert the result into $(5.1)_1$. Unfortunately this cannot be done explicitly in general, because of the complexity of the integrals (5.3). However there are three limiting cases in which the evaluation is possible and they refer to the classical ideal gas for which $\alpha \gg 1$ holds and to the strongly degenerate gases of Bosons $\alpha \gtrsim 0$ and Fermions $\alpha \ll -1$. In those cases it is easy to solve (5.3) and we obtain from (5.1) by a little calculation

$$p = \rho \frac{k}{\mu} T \qquad \text{classical ideal gas}$$

$$p = \nu_F \rho^{5/3} \qquad \text{strongly degenerate Fermi gas} \qquad (5.5)$$

$$p = \nu_B T^{5/2} \qquad \text{strongly degenerate Bose gas}$$

ν_F and ν_B are constants which are of no interest here. They are listed in the original paper [10] along with the special forms which the constitutive quantities $(14)_{2,5,6}$ take in the cases (5.5).

It may be worthwhile noting that in the cases $(5.5)_{2,3}$ the hyperbolicity condition $(18)_1$ is violated.

5.2 Entropy and Entropy Flux in a Classical Ideal Gas

The only explicit constitutive relations which I list here are those for the entropy density and for the entropy flux in a classical ideal gas. These are particularly instructive, since they may be compared with the corresponding formulae of the kinetic theory of gases.

Insertion of $p = \rho \frac{k}{\mu} T$ into $(3.14)_{5,6}$ gives

$$\frac{h}{\rho} = \frac{k}{\mu} \ln \frac{T^{3/2}}{\rho} + K - \frac{{}^m\!\!<ke>\,{}^m\!\!<ke>}{4\rho^2 \frac{k}{\mu} T^2} - \frac{2}{5\rho^2 \frac{k}{\mu} T^2} \frac{q_i q_i}{2 \frac{k}{\mu} T + \frac{7}{k/\mu} \frac{T^{5/2}}{\rho} c} , \qquad (5.6)$$

$$\Phi_i = \frac{q_i}{T} - \frac{2}{5\rho \frac{k}{\mu} T^2} \, {}^m\!\!<ie>\, q_e .$$

We note that the entropy flux is entirely specific and it is <u>not</u> just equal to q_i/T as is often assumed in non-equilibrium thermodynamics. Rather it has an additional term that is quadratic in ${}^m\!\!<ie>$ and q_e.

The specific entropy h/ρ has two non-equilibrium contributions, one is quadratic in ${}^m\!\!<ik>$ and the other one is quadratic in q_i. Also the entropy is not quite specific since $(5.6)_1$ contains two constants K and c. The constant K is additive, it is the entropy constant of classical thermodynamics which can be determined by making use of the third law of thermodynamics; but this does not concern us here. The other constant c is not additive and therefore potentially more serious. However, that constant can be seen to be equal to zero by the following argument. The kinetic theory offers the possibility to approximate the distribution function by Grad's thirteen moment method (see [5]) and one can use that function to calculate h/ρ and ϕ_i. It turns out that the functions thus calculated coincide completely with (5.6), down to the numerical coefficients, _provided_ that c is set equal to zero.

Of course, once c is seen to be equal to zero in a classical gas, it is also zero in degenerate gases, because upon cooling and compression a classical gas can be made into a degenerate one - unless phase transitions intervene. [+)]

Thus we see that now, with $F(z)$ known by (5.4) and with $c = 0$ the left hand side of the field equations (4.1) is entirely specific. There remain only two unknown functions in those field equations, viz. $\sigma(\rho,T)$ and $\tau(\rho,T)$ on their right hand sides. We proceed to determine these.

5.3 Viscosity and Heat Conductivity

The most natural question to ask of extended thermodynamics is how to get back to the set of balance laws and constititutive equations of ordinary thermodynamics. The answer to this question is provided by an iterative procedure which I proceed to describe.

First of all the balance equations of ordinary thermodynamics are still valid in extended thermodynamics and in (4.1) they are represented by the first three equations, the balance of mass, momentum and internal energy. The remaining two partial differential equations in (4.1) will now be converted into constitutive relations for $m_{<ik>}$ and q_i as follows. On the left hand sides of equations $(4.1)_{4,5}$ we replace $m_{<ij>}$ and q_i by their equilibrium values, namely zero and thus calculate first iterates $\pi_{<ik>}^{(1)}$ and $q_i^{(1)}$. Then these are introduced into the left hand sides and we calculate second iterates $m_{<ik>}^{(2)}$ and $q_i^{(2)}$, etc. This scheme is an adaptation of an iterative scheme that was invented by Maxwell in the

---

[+)] This of course is what generally happens, and this is why it is difficult to observe a degenerate gas.

kinetic theory of gases.[+)]

The form of the first iterate can easily be read off from $(4.1)_{4,5}$ and we obtain

$$m_{<ij>}^{(1)} = \left\{ \frac{2p}{\sigma} \right\} \frac{\partial v_{<i}}{\partial x_{j>}}$$

$$q_i^{(1)} = \left\{ \frac{5}{4\tau} \left(7T^{5/2} \int \frac{FF'}{z} \, dz - 5T^4 \frac{F^2}{\rho} \right) \right\} \frac{\partial T}{\partial x_i} \tag{5.7}$$

These equations are easily recognized as the constitutive equations of Navier-Stokes and Fourier for the pressure deviator and the heat flux respectively. With this interpretation we must identify the coefficients as follows

$$-\frac{p}{\sigma} = \mu \qquad \qquad \text{- viscosity}$$

$$-\frac{5}{4\tau} \left(7T^{5/2} \int \frac{FF'}{z} \, dz - 5T \frac{4F^2}{\rho} \right) = \kappa \qquad \text{- heat conducitivity} \tag{5.8}$$

These identifications provide the possibility to determine the two functions $\sigma(\rho,T)$, $\tau(\rho,T)$ that were the only unknowns left in the field equations (4.1). Indeed the viscosity $\mu(\rho,T)$ and the heat conductivity have been measured for many substances, in particular, for gases with great care and the results are listed in tables and diagrams. Thus, since $p = T^{5/2}F(z)$ holds and since $F(z)$ is known, see (4), we can calculate σ and τ for each pair ρ,T.

5.4 A Wave Solution

Now that the set of field equations (4.1) is fully explicit, we can begin to inquire about its solutions. It is of course quite difficult to solve this system of coupled non-linear partial differential equations for any non-trivial set of boundary and initial conditions. Therefore here we limit ourselves to the simplest kind of solution, and we investigate plane harmonic waves of small amplitude in an infinite gas. The assumed form of solution is given by

---

[+)] In Maxwell's method (e.g. see [6]) more and more equations of transfer were drawn into the iterative scheme as it proceeded. This cannot be done here, since we have only 13 equations.

$$\rho = \tilde{\rho} + \overline{\rho}e^{i(\omega t - kx)}, \quad v_i = \begin{pmatrix} \overline{v} \\ 0 \\ 0 \end{pmatrix} e^{i(\omega t - kx)}, \quad T = \tilde{T} + \overline{T}e^{i(\omega t - kx)}$$

$$m_{\langle ij \rangle} = \begin{pmatrix} \overline{m} & & 0 \\ & -\dfrac{\overline{m}}{2} & \\ 0 & & -\dfrac{\overline{m}}{2} \end{pmatrix} e^{i(\omega t - kx)}, \quad q_i = \begin{pmatrix} \overline{q} \\ 0 \\ 0 \end{pmatrix} e^{i(\omega t - kx)} \tag{5.9}$$

where $\tilde{\rho}$ and $\tilde{T}$ are the constant mean values of the density and the temperature and the barred quantities are small amplitudes whose products we neglect. ω and k are frequency and wave number of the waves and

$$v_{Ph} = \frac{\omega}{Re(k)} \quad \text{and} \quad \alpha = -Im(k) \tag{5.10}$$

are their phase speed and absorption coefficient.

For (9) to be a solution, ω and k must satisfy the dispersion relation

$$\frac{\partial p}{\partial \rho} \frac{p}{T} \frac{1}{2} \left(\gamma - \frac{5}{2} \frac{p}{\rho} \right)\left(\frac{9}{2} + \frac{5}{2} \sigma \frac{i}{\omega} \right)\left[\frac{k}{\omega} \right]^4 -$$

$$- \left\{ \frac{1}{2} \left(\gamma - \frac{5}{2} \frac{p}{\rho} \right)\left(\frac{4}{5} \frac{\partial p}{\partial T} + \frac{5}{2} \frac{p}{T} \right) + \frac{9}{2} \frac{p}{\rho} \frac{\partial p}{\partial T} + \right.$$

$$\left. + \frac{i}{\omega} \left[\frac{5}{2} \frac{p}{T} \frac{1}{2} \left(\gamma - \frac{5}{2} \frac{p}{\rho} \right)\sigma + \frac{\partial p}{\partial T} \frac{p}{\rho} \left(\frac{9}{2} \tau + \frac{5}{2} \sigma \right) \right] - \frac{5}{2} \frac{1}{\omega^2} \frac{\partial p}{\partial T} \frac{p}{\rho} + \sigma\tau \right\}\left[\frac{k}{\omega} \right]^2 +$$

$$+ \frac{3}{2} \frac{\partial p}{\partial T} \left[1 + \frac{i}{\omega} (\sigma + \tau) - \frac{1}{\omega^2} \sigma\tau \right] = 0 \tag{5.11}$$

This is a quartic equation with two physically relevant solutions representing two waves of which one is the ordinary sound wave while the other one has never been observed, since it is heavily damped.

It is instructive to note that, by (5.5), the first term in (5.11) vanishes in a strongly degenerate Bose gas so that there is only one wave in this limit. [+] In the strongly degenerate Fermi gas the last term vanishes, so that one phase speed

---

[+] The case $\alpha = 0$ in the case of Bosons must be treated separately. This case corresponds to a gas with Bose-Einstein condensation which has again two wave speeds, the first and second sound (see [7]).

goes to infinity in that case. This is in line with the observation that the criterion $(3.18)_1$ of hyperbolicity is violated in the limit of strong degeneracy.

6. Remarks on Material Frame Indifference and Temperature

6.1 Material Frame Dependence in Ordinary Thermodynamics

The first iterative step in the procedure described in Section 5.3 was sufficient for the identification of the two functions $\sigma(\rho,T)$ and $\tau(\rho,T)$ in terms of the viscosity and the heat conductivity and that identification was our goal. However, it is quite instructive to make one more step in that scheme so as to be able to illustrate the break-down of the principle of material frame indifference in ordinary thermodynamics.

If the first iterates (5.7) are introduced into the left hand side of (4.7) we obtain second iterates for $m_{\langle ij \rangle}$ and q_i in a classical gas of the form [+)]

$$m^{(2)}_{\langle ij \rangle} = \frac{2\mu}{\sigma} \frac{\partial v_{\langle i}}{\partial x_{j \rangle}} + \frac{2p}{\sigma^2} \Big[\frac{d}{dt} \Big(\frac{\partial v_{\langle i}}{\partial x_{j \rangle}} \Big) + \frac{\partial v_{\langle k}}{\partial x_{i \rangle}} \Big(\frac{\partial v_{[j}}{\partial x_{k]}} - \underline{\underline{2W_{jk}}} \Big) +$$

$$\frac{\partial v_{\langle k}}{\partial x_{j \rangle}} \Big(\frac{\partial v_{[i}}{\partial x_{k]}} - \underline{\underline{2W_{ik}}} \Big) \Big] + \frac{5}{3} \frac{2p}{\sigma^2} \frac{\partial v_e}{\partial x_e} \frac{\partial v_{\langle i}}{\partial x_{j \rangle}} +$$

$$4 \frac{k}{\mu} \frac{1}{2\tau\sigma} \Big(\frac{\partial^2 pT}{\partial x_{\langle i} \partial x_{j \rangle}} - T \frac{\partial^2 p}{\partial x_{\langle i} \partial x_{j \rangle}} \Big), \tag{6.1}_1$$

$$q^{(2)}_i = \frac{5}{2} \frac{k}{\mu} \frac{p}{\tau} \frac{\partial T}{\partial x_i} + \frac{1}{2\tau^2} \Big[\frac{d}{dt} \Big(5p \frac{k}{\mu} \frac{\partial T}{\partial x_i} \Big) + \frac{5}{2} p \frac{k}{\mu} \frac{\partial T}{\partial x_j} \Big(\frac{\partial v_{[i}}{\partial x_{j]}} - \underline{\underline{2W_{ij}}} \Big) \Big] +$$

$$+ \frac{5}{3} \frac{1}{2\tau^2} \frac{\partial v_e}{\partial x_e} - \frac{p}{\rho} \frac{1}{\sigma\tau} \frac{\partial v_{\langle i}}{\partial x_{j \rangle}} \frac{\partial p}{\partial x_j} + 7 \frac{k}{\mu} T \frac{1}{\sigma\tau} \frac{\partial}{\partial x_j} \Big(\frac{\partial v_{\langle i}}{\partial x_{j \rangle}} \Big) \tag{6.1}_2$$

These are equations of the type of the constitutive equations of ordinary thermodynamics in that $m_{\langle ij \rangle}$ and q_i are related to ρ, v_i, T and the spatial and time derivatives of ρ, v_i, T. However, the equations (1) do not satisfy the principle of material frame indifference, because the angular velocity of the frame appears explicitly in them.

[+)] Square bracket [] indicate antisymmetric tensors.

In order to emphasize that point let us focus on the heat flux and let us consider a process in which $\dot{T} = 0$ holds and in which the velocity field is a rigid rotation. In that case $(1)_2$ reduces to the form

$$q_i = -\kappa \left(\frac{\partial T}{\partial x_i} + \frac{1}{\tau_1} \left(\frac{\partial v_{[i}}{\partial x_{j]}} - W_{ij} \right) \frac{\partial T}{\partial x_j} \right) \tag{6.2}$$

Here we see the frame dependence of the relation between q_i and the temperature gradient $\frac{\partial T}{\partial x_i}$ in its clearest form. Indeed, if the gas is at rest in an inertial frame, so that $\frac{\partial v_{[i}}{\partial x_{j]}} - W_{ij}$ is equal to zero, the heat flux is parallel to the temperature gradient. But if the gas rotates with respect to an inertial frame, for instance if it is at rest in a particular non-inertial frame, the heat flux has a component perpendicular to the temperature gradient.

The frame dependence implied by (6.2) has been suggestively interpreted in [8] as the effect of the Coriolis force on atoms in "free" flight. We refer to that paper for an intuitive appreciation of the effect.

Frame indifference is no problem in extended thermodynamics, because in that theory the inertial contributions are put into the balance equations where they belong, see equations (2.2). The constitutive relations in extended thermodynamics could then be made frame independent.

6.2 Discontinuity of Temperature

The reason why one prefers temperature over internal energy as a variable is easy measurability of the former. And the reason for that easy measurability is the fact that the temperature is continuous at a thermometric (sic!) wall. Let us consider the situation in the context of extended thermodynamics.

A thermometric wall may be considered as a wall which does not itself participate in the thermodynamic processes in the adjacent bodies. In particular, there shall not be an entropy production in such a wall and this implies that the normal component of the entropy flux is continuous at the wall. We write this condition as

$$[| \phi_i n_i |] = 0 \tag{6.3}$$

Where n_i is a unit normal vector to the wall. The brackets denote the difference of the bracketed quantity on the different sides of the wall.

Insertion of Φ_i from $(5.6)_2$ into (6.3) gives

$$\left[\left| \left(\frac{q_i}{T} - \frac{2}{5\rho \frac{k}{\mu} T^2} m_{<ij>} q_j \right) n_i \right|\right] = 0 \qquad (6.4)$$

Let us first ignore the non-linear second term in (6.4). If we do that we have

$$\left[\left| \frac{q_i}{T} n_i \right|\right] = 0 \qquad (6.5)$$

and since $q_i n_i$ is itself continuous under very general conditions, we obtain

$$\left[\left| T \right|\right] = 0. \qquad (6.6)$$

Thus the temperature is continuous and the wall is properly called <u>thermometric</u>.

However, when we do not ignore the non-linear temr in the entropy flux, we get from (6.4) by a little calculation

$$\frac{T^+}{T^-} = \frac{1 - \frac{2}{5} \frac{m^+_{<ij>} q^+_j n_i}{p_+ q_i n_i}}{\frac{2}{5} \frac{m^-_{<ij>} q^-_j n_i}{p_- q_i n_i}} \qquad (6.7)$$

where + and - refer to the different sides of the wall.

We conclude that the temperature is not continuous at a wall without entropy production. This fact should considerably detract from the usefulness of the temperature as a variable in non-equilibrium conditions.

7. Outlook

Extended thermodynamics has been presented here as a theory for monotomic ideal gases. And while the theory is phenomenological in character it quite obviously has considerable similarity to the kinetic theory of gases which has in fact served as motivation.

However, even as a theory of monatomic gases it offers more than standard kinetic theory does, because it can easily be applied to degenerate gases as we have seen.

Extensions of the theory which are either already completed or in preparation point in three directions:

i) The restriction of the theory to monatomic gases is kept, but relativistic thermal and caloric equations of state are considered. This diretion is pursued by Müller and Ruggeri [9] who deal with non-degenerate gases. Work on relativistic degenerate gases is under way.

ii) The theory is generalized so as to apply to materials other than monatomic gases. Liu [10] and Kremer [11] have investigated vapours and Kremer [11] has also developed a theory of molecular gases. More recently Müller and Wilmanski [12] have formulated an extended thermodynamics of rheological fluids.

iii) More than thirteen fields are taken into account. In this effort one must exercise good judgement so as not to develop a physically meaningless theory. Kremer [11] has used 14 fields, adding a non-equilibrium pressure to the 13 fields of this paper. The choice of 14 fields is indeed tempting, because it is suggested by the relativistic theory [9].

REFERENCES

[1] Müller, I. Zur Ausbreitungsgeschwindigkeit von Störungen in kontinuierlichen Medien. Dissertation TH Aachen (1966).

[2] Müller, I. On the Paradox of Heat Conduction. Z. für Physik 198 (1967).

[3] Liu, I-Shih, Müller, I. Extended Thermodynamics of Classical and Degenerate Gases. Arch. Rational Mech. Anal. 83 (1983).

[4] Friedrichs, K.O., Lax, P.D. Systems of Conservation Equations with a Convex Extension. Proc. Nat. Acad. Sci. 68 (1971).

[5] Grad, H. Principles of the Kinetic Theory of Gases. Handbuch der Physik XII Springer Berlin. Heidelberg, Göttingen (1958).

[6] Ikenberry, E., Truesdell, C.A. On the Pressures and the Flux of Energy in a Gas according to Maxwell's Kinetic Theory I. J. Rational Mech. Anal. 5 (1956).

[7] Greco, A., Müller, I., Extended Thermodynanics and Superfluidity. Arch. Rational Mech. Anal. 85 (1984).

[8] Müller, I., On the Frame Dependence of Stress and Heat Flux. Arch. Rational Mech. Anal. <u>45</u> (1972).

[9] Müller, I., Ruggeri, T., Relativistic Thermodynamics of Non-Degenerate Ideal Gases. Submitted to Annals of Physics.

[10] Liu, I-Shih, Extended Thermodynamics of Fluids and Virial Equations of State. Arch. Rational Mech. Anal. <u>88</u> (1985).

[11] Kremer, G.M. Zur erweiterten Thermodynamik idealer und dichter Gase. Dissertation TU Berlin (1985).

[12] Müller, I., Wilmanski, K. Extended Thermodynamics of Second Order Fluids. In preparation.

PHASE TRANSITIONS IN ONE DIMENSIONAL NONLINEAR VISCOELASTICITY:

ADMISSIBILITY AND STABILITY [#]

Robert L. Pego[*]

Department of Mathematics
University of Michigan
Ann Arbor, Michigan   48109

1. Introduction

One scientific approach toward comprehending phase changes in material
systems has been through study of continuum theories. In an attempt to describe
phase transitions in elastic bars, Ericksen [3] considered a one space dimensional
situation in which the stress σ is taken to be a nonmonotone function of the
strain, or displacement gradient, u_x, where $u(x,t)$ measures the displacement of
the material point x. Phases of the bar were interpreted to correspond to maxi-
mal intervals of monotonicity of $\sigma(u_x)$. The graph of a typical σ is indicated
in fig. 1. The equation of motion of a homogeneous elastic bar in this context
may be written

$$u_{tt} = \sigma(u_x)_x \, , \quad 0 < x < 1. \tag{1}$$

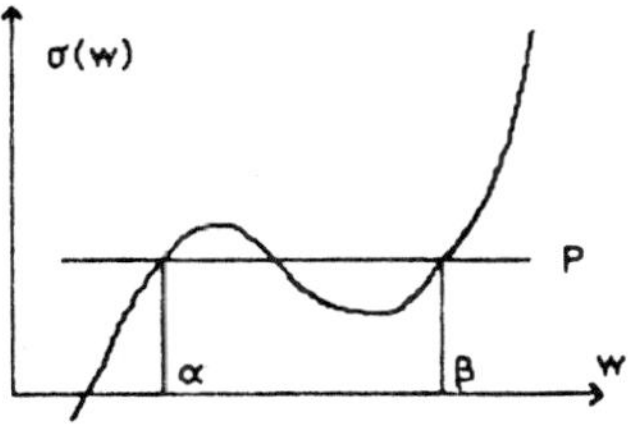

Fig. 1 Stress-strain relation

[#] This material is based on work done partly while the author was on leave at the
Institute for Mathematics and its Applications, University of Minnesota.

[*] Partially supported by the National Science Foundation under Grant DMS-84-01614.

One also thinks of (1) rewritten as a system of conservation laws for mass and momentum in the form,

$$w_t = v_x$$
$$v_t = \sigma(w)_x ,$$

(2)

where $w = u_x$ is the strain and $v = u_t$ is the velocity.

The bar is said to be in a "soft loading device" if one end is held fixed and a given force P is applied at the other end. This situation is governed by the boundary conditions

$$u(0,t) = 0, \qquad \sigma(u_x)(1,t) = P.$$

For the bar to be at rest, the stress must be constant,

$$\sigma(u_x) = P, \qquad\qquad 0 < x < 1.$$

(3)

Nonmonotonicity of $\sigma(w)$ permits the existence of more than one state of strain corresponding to some values of the given load P . This allows one to imagine modeling hysteresis phenomena that might occur upon raising and lowering the load P. One might also describe coexisting phases, by equilibrium solutions of the form

$$(w,v)(x) = \begin{cases} (\alpha,0) & \text{for } 0 < x < a, \\ (\beta,0) & \text{for } a < x < 1, \end{cases}$$

wherein the strain is allowed to be discontinuous (see again, figure 1).

In the above context, one hopes to use the dynamic model (1) to identify the stable equilibria and to describe the dynamic process of transition between phases. Experimentally in metal bars, this transition is associated with slowly moving, sharply defined waves [9]. But (1) is difficult to deal with mathematically. For values of strain in ranges where the stress $\sigma(w)$ is decreasing, (1) is of elliptic type, and the initial value problem is ill posed. Ideally one would like to work with weak solutions having discontinuous strain confined outside the elliptic range. But here too is a quandary. Uniqueness fails for the simple Riemann problem of seeking a centered wave solution $(w,v)(x/t)$ of (2) with initial data

$$(w,v)(x,0) = \begin{cases} (w_-,v_-) & \text{for } x < 0, \\ (w_+,v_+) & \text{for } x > 0. \end{cases} \tag{4}$$

With σ of the form shown in fig. 1, James [9] has demonstrated that the Riemann problem can have a two-parameter family of centered-wave weak solutions. One is faced with the problem of identifying a (unique, one hopes) physically admissible solution from the various possibilities.

To address the issues of stability and well-posedness, I follow an approach which includes the effects of a physically relevant dissipative mechanism that regularizes the initial value problem. While other mechanisms may also be relevant, depending on the application, consider here only the additional effect of a viscosity term of rate type, replacing (1) by

$$u_{tt} = (\sigma(u_x) + \mu u_{xt})_x, \quad \mu > 0. \tag{5}$$

The issues of stability of equilibria and admissibility of waves will occupy our attention. These issues are not unrelated. The stability results reinforce the status of stationary waves with regard to the admissibility criterion of §3.

2. Stability of Coexistent Phases

We will study the dynamic approach to equilibrium of a viscoelastic bar in a soft loading device (with $\mu = 1$) by considering the initial boundary value problem

$$u_{tt} = (\sigma(u_x) + u_{xt})_x, \quad 0 < x < 1, \ t > 0,$$

$$\begin{aligned} u(0,t) &= 0, \\ (\sigma(u_x) + u_{xt})(1,t) &= P, \end{aligned} \qquad t > 0 \tag{6}$$

$$\begin{aligned} u_x(x,0) &= u_0(x), \\ u_t(x,0) &= u_1(x), \qquad 0 < x < 1. \end{aligned}$$

Details of the analysis of (6) will be found in [10]. Global existence of smooth solutions with $u(\cdot,t)$ and $u_t(\cdot,t)$ in $C^{2+\alpha}(0,1)$ was established by Dafermos [2], for an equation of the general form

$$u_{tt} = \sigma(u_x, u_{xt})x,$$

(Under a physically rather restrictive assumption on the growth of σ for large u_x). Dafermos showed that the velocity u_t and the total stress, here $\sigma(u_x) + u_{xt}$, decay uniformly to zero as $t \to \infty$, and argued that in general the strain u_x may not converge asymptotically to a continuous limit. Andrews and Ball [1] found that under the quite weak condition that $(\sigma(w) - P)w > 0$ for large w, the strain u_x is uniformly bounded and converges in the weak-* sense as $t \to \infty$. However, they could not determine whether this weak limit was an equilibrium solution. They also extended the existence theory to cover situations where the solution might have discontinuous strain, taking $u_0 \in L^\infty$, $u_1 \in L^2$. In that case, they obtain weak solutions globally in time.

In [10] the present author shows that for the solutions constructed by Andrews and Ball, the strain u_x does in fact converge pointwise almost everywhere to a limit as $t \to \infty$, so that solutions of (6) do approach some equilibrium state asymptotically. Furthermore, the existence theory for $u_0 \in L^\infty$, $u_1 \in L^2$ is simplified, and the smoothing effect of the viscosity is shown to be limited. For example, for $t > 0$ one finds that $u(\cdot,t) \in W^{1,\infty}(0,1)$, $u_t(\cdot,t) \in W^{1,\infty}(0,1)$, and that discontinuities present in the initial strain $u_0(x)$ persist without moving for all time. (Hoff and Smoller [6] have shown that discontinuities in density persist in isothermal gas dynamics with viscosity.) However, the total stress satisfies $\sigma(u_x) + u_{xt} \in W^{2,\infty}(0,1)$ for $t > 0$, which means equation (6) holds in an almost classical sense.

A dynamic stability criterion is established in [10] for equilibrium solutions $u(x) \in W^{1,\infty}(0,1)$, which satisfy $\sigma(u_x) = P$ a.e. Such a solution is <u>stable</u> provided that

$$\sigma'(u_x(x)) \ge c > 0 \quad \text{for a.e. } x \text{ in } (0,1). \tag{7}$$

It is <u>unstable</u> if $\sigma'(u_x) < 0$ on a set of positive measure. The stability property asserted here is to be taken in a rather strong sense, to be made clear pre-
sently. But it is important to relate the results to traditional energy
approaches to stability. The total stored energy of the bar in the soft loading
device has the form

$$I(u) = \int_0^1 (W(u_x(x)) - Pu_x(x))dx, \quad \text{where} \quad W'(w) = \sigma(w). \tag{8}$$

A typical graph for the stored-energy density $W(w) - Pw$ appears in fig. 2.

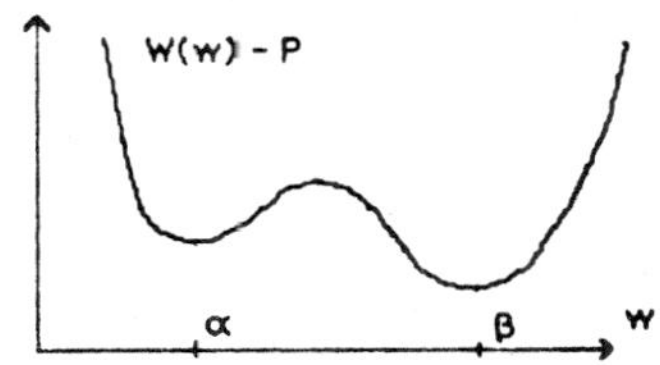

Fig. 2 Stored-energy density

Critical points of $W(w) - Pw$ correspond to possible equilibrium values of the
stability criteria focus on states u for which $I(u)$ is at an absolute minimum.
(See, for example, [3] and [8].) This can only occur if u is a "strong minimizer" of $I(u)$, in the language of the calculus of variations. Here this implies
that the stored-energy density is everywhere at an absolute minimum. In the
situation in fig. 2, where the local minima of $W(w) - Pw$ are global minima,
a unique strong minimizer exists having $u_x = \beta$. Coexistence of phases in a
strong minimizer can occur only if the local minima of $W(w) - Pw$ are global
minima which occurs for only one value of P. Observations of hysteresis and
coexistent phases in metals, however, indicate that the stored energy density
need not be homogeneous at rest, nor at an absolute minimum everywhere, suggesting
that "metastable" states taking strain values partly at merely local minima of the
stored energy density are significant. Such states are "weak relative minimizers"
of $I(u)$. In the situation depicted in fig. 2, such a state may assume values of
strain at both α and β. It is precisely these "metastable" states which are
found to be stable by the criterion (7).

The stability of equilibria satisfying (7) may be asserted in a number of senses (see [10]). First, a homogeneous metastable state ($u_x \equiv \alpha$ in fig. 2) is asymptotically stable to perturbations which are sufficiently small in a uniform sense. Furthermore, such perturbations decay at an exponential rate. Since perturbations remain small, however, solutions are not affected by non-monotonicity of $\sigma(w)$, so this result is not surprising. (Exponential decay to equilibrium was established for smooth solutions when $\sigma(w)$ is monotone by Greenberg and MacCamy [4].) More interesting is that asymptotic stability and exponential decay continue to hold whenever the equilbrium state satisfies (7) and the (strain, velocity) perturbations are small in (L^{∞},L^2). In this situation the equilibrium state can contain coexistent "metastable" and "stable" phases, but all the discontinuities in the asymptotic state of strain are present in the initial data. What can be said for smooth data, which we know yield smooth solutions?

It is found that if the velocity perturbation is small in L^2 (small kinetic energy) and the strain perturbation is bounded and small except on a set of small measure ε, then the asymptotic strain is equal to the unperturbed strain except on the same set of measure ε. In particular, this stability property guarantees that for any smooth initial data near a "metastable" equilibrium having discontinuous strain and satisfying (7), the (smooth) solution of (6) approaches an asymptotic state nearby which also has discontinuous strain.

The proofs of this last stability property and the convergence to equilibrium depend on the use of a modified strain q defined by Andrews as:

$$q(x,t) = u_x(x,t) - p(x,t)$$

where

$$p(x,t) = \int_1^x u_t(y,t)dy.$$

For each x, the quantity q satisfies an ODE in t, namely

$$q_t = -\sigma(p + q) + P. \tag{9}$$

It follows from the definition of p that

$$\sup|p(x,t)| \;\leqslant\; \left(\int_0^1 u_t^2(x,t)dx\right)^{1/2} ,$$

and the kinetic energy can be shown to be bounded and to decay to zero as $t \to \infty$. Provided that

$$\|p(\cdot,t)\|_\infty \;\leqslant\; \varepsilon , \quad \text{for} \quad t > t_0,$$

it is easy to see that an interval $[q_-,q_+]$ is <u>positively invariant</u> for $q(x,\cdot)$ in (9) for $t > t_0$ if

$$\sigma(q) - P > 0 \quad \text{when} \quad |q - q_+| < \varepsilon, \text{ and}$$
$$\sigma(q) - P < 0 \quad \text{when} \quad |q - q_-| < \varepsilon.$$

Three such invariant intervals are indicated in fig. 3, corresponding to the σ illustrated in fig. 1. The presence of such invariant intervals for the modified strain means that the strain in (6) cannot change phase <u>at any point</u>, provided only that the <u>total</u> kinetic energy of the bar is small. The stability result in view thus depends on showing that the total kinetic energy cannot grow if the perturbation from a "metastable" equilibrium has a small enough effect on the energy, even if the solution contains smooth transition layers joining the phases.

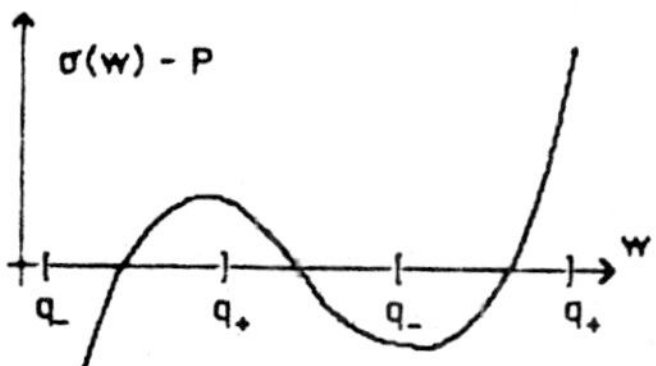

Fig. 3   Invariant intervals

An interesting observation is that in terms of the quantities p and q the problem (6) can be transformed to become a system consisting of an ODE coupled to a parabolic PDE

$$p_t = p_{xx} + \sigma(p + q) - P,$$
$$q_t = -\sigma(p + q) + P,$$
$$p_x(0,t) = 0,$$
$$p(1,t) = 0,$$
$$p(x,0) = p_0(x),$$
$$q(x,0) = q_0(x).$$

$$(10)$$

Thus, (6) may be viewed as an abstract semilinear parabolic equation, so allowing a significant simplification and extension of the existence and regularity theory of [1] by making use of standard theory as found in Henry [5]. The form of (10) also exhibits explicitly a zero-speed characteristic in the ODE for q. The persistence of discontinuities in the strain follows easily from continuous dependence on parameters reversing time in the ODE.

3. Admissibility of Waves

Nonuniqueness for the Riemann problem for (2) appears to turn on the question of which discontinuous wave solutions of the form

$$(w,v)(x,t) = \begin{cases} (w_-,v_-) & \text{for } x < st, \\ (w_+,v_+) & \text{for } x > st, \end{cases}$$

$$(11)$$

are to be regarded as physically relevant, or admissible. The states in (11) are related by the Rankine-Hugoniot jump conditions, which may be (re)written as

$$s^2(w_+ - w_-) = \sigma(w_+) - \sigma(w_-),$$
$$v_+ - v_- = -s(w_+ - w_-).$$

$$(12)$$

Our approach to this question is to identify those waves (11) for (2) which correspond to traveling waves of a class of viscoelastic systems,

$$w_t = v_x,$$
$$v_t = (\sigma(w) + \Lambda(w,w_t))_x,$$

$$(13)$$

derived from the viscoelastic equation

$$u_{tt} = (\sigma(u_x) + \Lambda(u_x, u_{xt}))_x$$

in which the stress dependence on the strain rate has the general form considered by Dafermos [2]. If a traveling wave $(w,v)(x - st)$ exists for (13) satisfying

$$(w,v)(\xi) \rightarrow \begin{cases} (w^-, v^-) & \text{as } \xi \rightarrow -\infty \\ (w^+, v^+) & \text{as } \xi \rightarrow +\infty \end{cases} \tag{14}$$

then the wave (11) will be regarded as <u>admissible</u>.

If a traveling wave exists for (13) satisfying (14), then by subsituting the form $(w,v)(x - st)$ into (13), eliminating v, and integrating once, using (14), we find that $w(\xi)$ must satisfy

$$-\Lambda(w, -sw') = \sigma(w) - \sigma(w_-) - s^2(w_+ - w_-). \tag{15}$$

The class of viscoelastic systems considered here are given by those Λ that are smooth and invertible on all of R in their second argument, with $\Lambda(a,0) = 0$, $\Lambda_b(a,b) > 0$ for all a,b. Write $b = \Lambda^{-1}(a,c)$ if $\Lambda(a,b) = c$, for any real c. (Note that $\text{sgn } b = \text{sgn } c$.) Under the assumptions, Equation (15) becomes

$$-sw' = \Lambda^{-1}(w, -(\sigma(w) - \sigma(w_-) - s^2(w_+ - w_-))). \tag{16}$$

There are now two cases. Consider first the case $s \neq 0$. Then it is easy to see that the ODE (16) has a solution with the required limits from (14) if and only if

$$\text{sgn } s(w_+ - w_-) = \text{sgn}(\sigma(w) - \sigma(w_-) - s^2(w - w_-)), \tag{17}$$
$$\text{for } w \text{ between } w_- \text{ and } w_+.$$

This requirement is called the <u>chord condition</u> (it appears in [13]), since it means that the chord from $(w_-, \sigma(w_-))$ to $(w_+, \sigma(w_+))$ lies above the graph of $\sigma(w)$ if $s(w_+ - w_-) > 0$, below if $s(w_+ - w_-) < 0$. Note from (12) that the slope of this chord is the squared wave speed s^2.

In the second case, $s = 0$, the stationary wave (11) for (2) satisfies

$$\sigma(w_+) = \sigma(w_-),$$
$$v_+ = v_-.$$

But the wave (11) is also a stationary-wave solution of (13); it corresponds to itself. Equation (16) degenerates to the algebraic condition $\sigma(w) = \sigma(w_-)$. Here the corresponding viscous wave is not continuous, but the discussion of §1 indicates that this is irrelevant because at least when the viscosity has the linear form, $\Lambda(u_x, u_{xt}) = \mu u_{xt}$, such stationary waves can be stable limits of smooth solutions.

To summarize, then, we have the following <u>viscoelastic admissibility criterion</u> for waves of the form (11).

> If $s \neq 0$, the wave is admissible if the chord condition (17) holds.
>
> If $s = 0$, the wave is always admissible.

$$(18)$$

Remarkably, this criterion was introduced partly on an <u>ad hoc</u> basis by Shearer [12], who showed that with $\sigma(w)$ as in fig. 4, whose graph changes convexity at just one point in the decreasing portion, the Riemann problem for (2) always has a unique solution if discontinuities are required to satisfy condition (18). So the viscoelastic admissibility criterion yields a self-consistent selection principle for the solution of the Riemann initial value problem for (2).

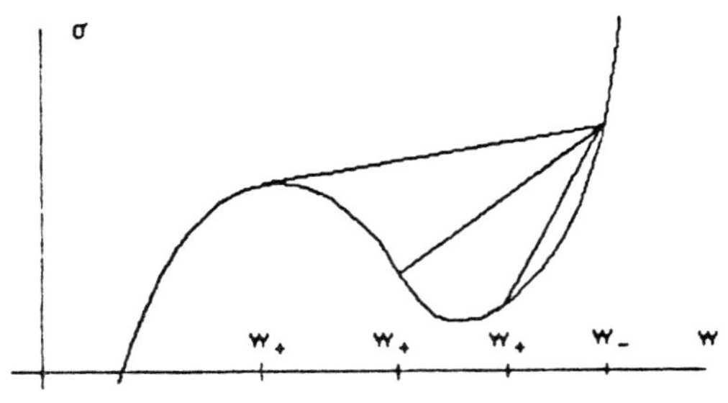

Fig. 4 Admissible waves for s > 0

4. Dynamics of Phase Transition

We discuss briefly an initial-boundary value problem for an elastic bar governed by equation (2) which illustrates how a phase transition can proceed dynamically via a slow, propagating wave. The particular mechanism we describe has been examined by Pence [11] and suggested by Hunter and Slemrod [7] in the context of shear flow in a polymeric fluid. The admissibility criterion of §3

allows us to assert uniqueness property for the solution.

In order to apply the results of Shearer [12], suppose $\sigma(w)$ is as in fig. 1,

changing convexity at just one point w_0 with $\sigma'(w_0) < 0$. Let

$P_\beta = \min\{\sigma(w)|w > w_0\}$. Suppose that at $t = 0$, the elastic bar lies at rest

in a soft loading device at a stress level P_- slightly larger than P_β, with

homogeneous strain $w(x) = w_-$, where w_- is the rightmost root of $\sigma(w) = P_-$

(lying in the same phase as $w = \beta$ in fig. 1). Imagine that the stress at the

boundary $x = 1$ is suddenly lowered to a level $P_+ < P_\beta$, and held constant.

This now implies that the strain at the boundary must satisfy $w(1,t) = w_+$, where

$\sigma(w_+) = P_+$ (w_+ lies in the same phase as $w = \alpha$ in fig. 1). Physically, one

expects the bar to change phase, and ultimately equilibrate with uniform strain

$w = w_+$. Dynamically, this process is described by the solution of the following

initial-boundary value problem. Find a weak solution of the elastic system (2)

having initial conditions

$$w(x,0) = w_-, \quad v(x,0) = 0 \quad \text{for} \quad 0 \leqslant x \leqslant 1,$$

and boundary conditions $\hspace{8cm}$ (19)

$$w(1,t) = w_+, \quad v(0,t) = 0 \quad \text{for} \quad t > 0.$$

For a certain time we can ignore the boundary condition at $x = 0$, and construct

a solution to problem (19)   by finding a <u>centered 1-wave</u> solution to the ordinary

Riemann problem for (2), connecting the state $(w_-,0)$ on the left to the state

(w_+,v_+) on the right, for some v_+ to be determined. That is, we seek v_+ so

that the solution of the Riemann problem (4) obtained by Shearer under the

admissibility criterion (18) contains only waves with speed less than zero so that

the boundary condition in (19) at $x = 1$ will be satisfied. The solution

obtained from this Riemann problem by shifting the x variable and restricting to

the slab $0 < x < 1$ is then valid until the leading wave traveling to the left

impinges on the boundary $x = 0$. By invoking the results of Shearer, one finds

that a <u>unique</u> v_+ does always exist with the property specified above (for

details, see [10]). If P_+ is sufficiently close to P_β, the solutions so considered consist of a 1-rarefaction wave connecting $(w_-,0)$ to an intermediate state (w_0,v_0), which is then connected immediately to (w_+,v_+) by a 1-shock. The value w_0 lies in the same phase as w_- and is determined by the requirement that the chord from $(w_0,\sigma(w_0))$ to $(w_+,\sigma(w_+))$ lies below the graph of $\sigma(w)$ and is tangent to it at w_0. The slope of this chord is the squared speed of the shock, which clearly can be arbitrarily close to zero, depending on P_+. The speed of the leading edge of the rarefaction wave is $\sqrt{\sigma'(w_-)}$.

References

1. G. Andrews and J.M. Ball, Asymptotic behaviour and changes of phase in one-dimensional nonlinear viscoelasticity, J. Diff. Eqns. 44 (1982) 306-341.

2. C.M. Dafermos, The mixed initial-boundary value problem for the equations of nonlinear one dimensional viscoelasticity, J. Diff. Eqns. 6 (1969), 71-86.

3. J.L. Ericksen, Equilibrium of bars, J. Elasticity 5 (1975), 191-201.

4. J.M. Greenberg and R.C. MacCamy, On the exponential stability of solutions of $E(u_x)u_{xx} + \lambda u_{xtx} = \rho u_{tt}$, J. Math. Anal. Appl. 31 (1970) 406-417.

5. D. Henry, Geometric theory of semilinear parabolic equations, Lec. Notes in Math. v. 840, Springer, New York, 1981.

6. D. Hoff and J. Smoller, Solutions in the large for certain nonlinear parabolic systems, Analyse Nonlineaire (to appear).

7. J. Hunter and M. Slemrod, Viscoelastic fluid flow exhibiting hysteretic phase changes, Phys. Fluids 26 (1983), 2345-2351.

8. R.D. James, Coexistence phases in the one dimensional static theory of elastic bars, Arch. Rat. Mech. Anal. 72 (1980), 99-140.

9. R.D. James, The propagation of phase boundaries in elastic bars, Arch. Rat. Mech. Anal. 73 (1980), 125-158.

10. R.L. Pego, Phase transitions in one dimensional nonlinear viscoelasticity: admissibility and stability, Arch. Rat. Mech. Anal. (to appear).

11. T.J. Pence, On the emergence and propagation of a phase boundary in an elastic bar with a suddenly applied end load, J. Elasticity (to appear).

12. M. Shearer, The Riemann problem for a class of conservation laws of mixed type, J. Diff. Eqns. 46 (1982), 426-443.

13. B. Wendroff, The Riemann problem for materials with nonconvex equations of state I; Isentropic flow, J. Math. Anal. Appl. 38 (1972), 454-466.

DYNAMIC PHASE TRANSITIONS AND COMPENSATED COMPACTNESS

V. Roytburd[*]

M. Slemrod[#]

Department of Mathematical Sciences
Rensselaer Polytechnic Institute
Troy, New York   12180

0. Introduction

In the paper [1] R. DiPerna showed how Luc Tartar's method of compensated
compactness [2] may by applied to prove existence of weak solutions of the Cauchy
initial value problem

$$v_t + F(w)_x = 0,$$
$$w_t - v_x = 0, \qquad -\infty < x < \infty, \quad t > 0 \qquad (0.1)$$

$$w(x,0) = w_0(x), \quad v(x,0) = v_0(x), \qquad (0.2)$$

when $p' < 0$ and $p'' \neq 0$. These conditions on p make the system (0.1) strictly
hyperbolic and genuinely nonlinear (in the sense of Lax). In this lecture we are
concerned with p of the form shown in Figure 1.

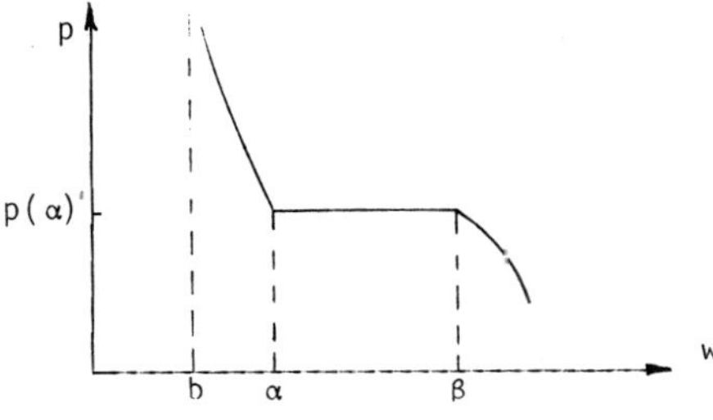

Figure 1

[*] Supported in part by NSF Grant No. DMS-8408260.

[#] This research was sponsored in part by the Air Force Office of Scientific
Research, Air Force Systems Command, USAF, Contract/Grant No. AFOSR-81-0172.
The U.S. Government's right to retain a nonexclusive royalty free license in
and to copyright this paper for governmental purposes is acknowledged.

Here p is continuous and

$$p \text{ is smooth for } b < w < \alpha, \quad w > \beta,$$
$$p' < 0 \quad \text{on} \quad 0 < b < w < \alpha, \quad w > \beta,$$
$$p'' > 0 \quad \text{on} \quad b < w < \alpha,$$
$$p'' < 0 \quad \text{on} \quad \beta < w$$
$$p = \text{const} = p(\alpha) \quad \text{on} \quad \alpha \leqslant w \leqslant \beta.$$

Notice that since $P' = 0$ on $\alpha < w < \beta$ (0.1) is not strictly hyperbolic.

We shall sketch how under certain restrictions we can prove weak solutions of (0.1), (0.2) exist as $\varepsilon \to 0+$ as limits of weak solutions of the regularized parabolic problem

$$v_t^\varepsilon + p(w^\varepsilon)_x = \varepsilon w_{xx}^\varepsilon,$$
$$w_t^\varepsilon - v_x^\varepsilon = \varepsilon v_{xx}^\varepsilon, \tag{0.3}$$

$\varepsilon > 0$ a small parameter.

The motivation for this choice of p lies in the theory of phase transitions where p has the shape of a slightly subcritical van der Waals isotherm with the Maxwell construction. In this context w represents the specific volume of a fluid, v the velocity and p the pressure. System (0.1) is the Lagrangian form of the balance of linear momentum.

It is known that one may observe such a p vs. w diagram as shown in Figure 1 in the _equilibrium_ configurations of materials undergoing first order phase transitions. Classical equilibrium theories of phase transitions have reconciled such observed p-w graphs with non-monotonoe constitutive relations such as the van der Waals fluid (without the Maxwell construction). A typical slightly subcritical isotherm is sketched in Figure 2.

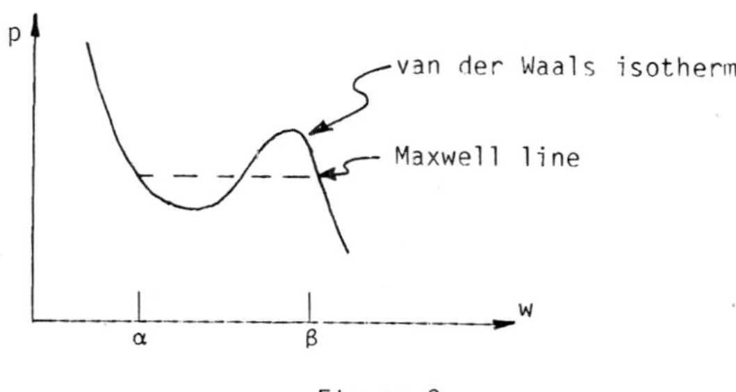

Figure 2

It is our goal to study non-equilibrium fluid mechanics. Hence it may lack
physical consistency to choose as a constitutive relation the equilibrium p-w
graph of Figure 1 as opposed to the underlying van der Waals constitutive relation
of Figure 2. Nevertheless our mathematical abilities at the moment force this
choice upon us. It is our hope that understanding the dynamics of (0.1) with
p given by Figure 1 will prepare us for the analysis of the case where p is
given by Figure 2.

As in DiPerna's paper [1] we attempt to show that compensated compactness
will yield as weak $* L^\infty$ limits of w^ε, v^ε of (0.3) a weak solution of (0.1).

1. An L^∞ a Priori Estimate

Consider the system

$$v_t + p_\delta(w)_x = \varepsilon v_{xx},$$
$$w_t - v_x = \varepsilon w_{xx},$$

(1.1)

where p_δ has the shape shown in Figure 3.

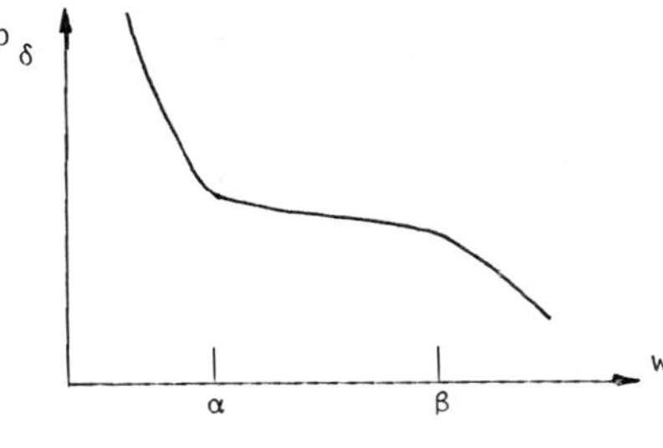

Figure 3

Assume p_δ is C^∞ and has precisely one inflection point at $w_i = \frac{1}{2}(\alpha + \beta)$, and $p_\delta \to p$ uniformly in w as $\delta \to 0$. For the system (1.1) we may define Riemann invariants

$$\left\{ \begin{matrix} r \\ s \end{matrix} \right\} = v \mp \int_\alpha^w (-p'_\delta(\xi))^{1/2} d\xi. \tag{1.2}$$

From the theory of [3], [4] we know (1.1) determines a positively invariant domain D_δ defined by the inequalities

$$\begin{aligned} r &< r_a, \quad s > s_a \quad \text{for} \quad w_a \leq w \leq w_i, \\ r &> r_b, \quad s < s_b \quad \text{for} \quad w_i \leq w \leq w_b. \end{aligned} \tag{1.3}$$

By positively invariant region we mean that if initial data $(w_0(x), v_0(x))$ lie in D_δ for all $x \in R$, then $(w(x,t), v(x,t))$ lie in D_δ for all $x \in R$, $t > 0$. A sketch of D_δ is shown in Figure 4.

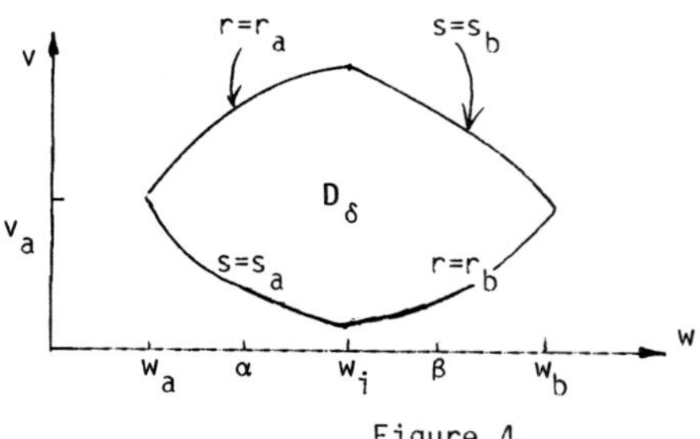

Figure 4

Formally we see that as $\delta \to 0$, $p'_\delta \to 0$ for $\alpha < w < \beta$ and $\frac{dv}{dw} \to 0$ on the r=const., s = const. lines. Hence one might expect that the limit as $\delta \to 0$ for the positively invariant region D_δ will yield a positive invariant region D for (0.3) of the shape shown in Figure 5.

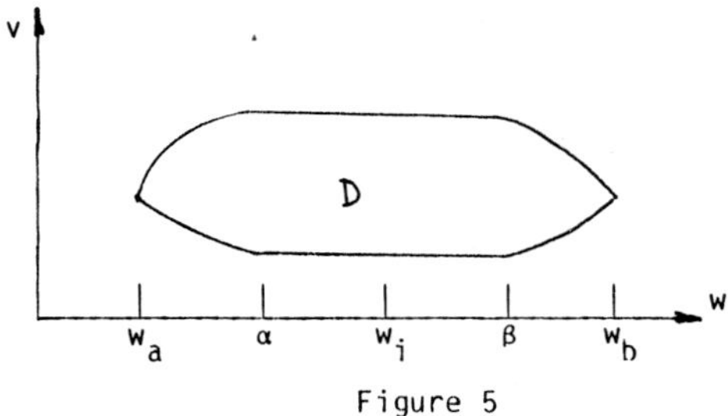

Figure 5

In fact in [5] we have proven this result. We then note that for initial data $(w_0(x), v_0(x))$ in D for all $x \in R$ we have the estimate

$$\|(w_\varepsilon, v_\varepsilon)\|_{L^\infty} \le \text{const.} \tag{1.4}$$

where $\|\ \|_{L^\infty}$ denotes the $L^\infty((-\infty,\infty) \times [0,T])$ norm; $T > 0$ arbitrary, and the const. is independent of $\varepsilon > 0$.

2. The Young Measure and Weak Convergence

We know from (1.4) that the sequence $\{w^\varepsilon(x,t), v^\varepsilon(x,t)\}$ takes values in a compact set $K \subset R^2$ and hence we may extract a subsequence also denoted by $\{w^\varepsilon(x,t), v^\varepsilon(x,t)\}$ so that

$$\begin{aligned} w^\varepsilon &\to \overline{w} \\ v^\varepsilon &\to \overline{v} \end{aligned} \quad \text{weak} * L^\infty((-\infty,\infty) \times [0,T]) \tag{2.1}$$

as $\varepsilon \to 0+$. If in addition we can prove

$$p(w^\varepsilon) \to p(\overline{w}) \quad \text{weak} * L^\infty((-\infty,\infty) \times [0,T]) \tag{2.2}$$

as $\varepsilon \to 0+$ it is easy to see that $(\overline{w}, \overline{v})$ will be a solution of (0.1) in the sense of distributions. The proof of this weak continuity of p is our goal.

Our method of analysis is based on the representation of weak convergence in terms of the Young measure. Roughly the idea is as follows: Let f be a continuous function from R^2 to R. If (2.1) holds

$$\|f(w^\varepsilon, v^\varepsilon)\|_{L^\infty} \le \text{const.} \tag{2.3}$$

and hence by possibly extracting an additional subsequence of $\{w^\varepsilon, v^\varepsilon\}$ we have

$$f(w^\varepsilon, v^\varepsilon) \to \overline{f} \quad \text{weak} * L^\infty((-\infty,\infty) \times [0,T]) \tag{2.4}$$

as $\varepsilon \to 0+$ as well. We can rewrite (2.4) as

$$\iint \phi(x,t) \, f(w^\varepsilon(x,t), v^\varepsilon(x,t)) \, dx \, dt \to \iint \phi(x,t) \, \overline{f}(x,t) dx \, dt$$

as $\varepsilon \to 0+$ for all $\phi \in L^1(R \times [0,T])$ or

$$\iint \phi(x,t) < \delta_{w^\epsilon(x,t),v^\epsilon(x,t)} , f > dx\ dt \rightarrow \iint \phi(x,t)\overline{f}(x,t)dx\ dt \qquad (2.5)$$

as $\epsilon \rightarrow 0+$. Here $< , >$ denotes the duality between the bounded measures and the continuous functions on K. Also $\delta_{a,b}(\lambda_1,\lambda_2)$ denotes the Dirac measure with mass at $\lambda_1 = a$, $\lambda_2 = b$. Hence the equality

$$f(w^\epsilon(x,t),v^\epsilon(x,t)) = \int_{R^2}\int \delta_{w^\epsilon(x,t),v^\epsilon(x,t)}(\lambda_1,\lambda_2)f(\lambda_1,\lambda_2)d\lambda_1\ d\lambda_2 \qquad (2.6)$$

which has been used to show (2.5) is apparent.

Consider the sequence $<\delta_{w^\epsilon(x,t),v^\epsilon(x,t)},f>$. Since $|<\delta_{w^\epsilon,v^\epsilon},f>|<\text{const.}\|f\|_{L^\infty}$ when f is continuous we see $\delta_{w^\epsilon,v^\epsilon}$ is a uniformly bounded sequence in X^*. Here X^* is the dual via $< , >$ of $C(K)$. Hence via the weak $*$ compactness of the unit ball in X^* $\delta_{w^\epsilon,v^\epsilon}$ possesses a subsequence also denoted by $\delta_{w^\epsilon,v^\epsilon}$ so that as $\epsilon \rightarrow 0+$

$$\delta_{w^\epsilon(x,t),v^\epsilon(x,t)} \rightarrow \nu_{x,t} \quad \text{weak} * X^* \qquad (2.7)$$

where $\nu_{x,t}$ lies in X^* (the space of bounded measures) or

$$<\delta_{w^\epsilon(x,t),v^\epsilon(x,t)},f> \rightarrow <\nu_{x,t},f> \qquad (2.8)$$

as $\epsilon \rightarrow 0+$ for all $f \in C(K)$ where the support of $\nu_{x,t}\subset K$. Inserting (2.8) into the left hand side of (2.5) we see

$$\iint \phi(x,t) <\nu_{x,t},f>dx\ dt = \iint \phi(x,t)\ \overline{f}(x,t)dx\ dt \qquad (2.9)$$

and hence

$$\overline{f}(x,t) = <\nu_{x,t},f> \qquad (2.10)$$

for almost all x,t. This is the fundamental representation formula for weak limits in terms of the Young measure $\nu_{x,t}$. In particular if $\nu_{x,t}$ reduces to the Dirac measure with mass at $\overline{w},\overline{v}$ we have

$$\overline{f}(x,t) = f(\overline{w}(x,t),\overline{v}(x,t))$$

and hence we have the weak continuity of f. In our problem of course the case of interest is $f(w,v) = p(w)$.

3. Compensated Compactness

We say a pair of functions $(\eta(w,v), q(w,v))$ is an entropy-flux pair for
(0.1) if η, q satisfy

$$\eta_w + q_v = 0,$$
$$-p'(w)\eta_v + q_w = 0. \tag{3.1}$$

In this it is easy to see that η satisfies the linear wave equation

$$\eta_{ww} + p'(w)\eta_{vv} = 0. \tag{3.2}$$

In addition we note that (3.1) implies that for all smooth solutions of (0.1)
$\eta(w,v)$, $q(w,v)$ will satisfy the additional conservation law

$$\eta_t + q_x = 0. \tag{3.3}$$

For our purposes the importance of entropy-flux pairs is their weak sequential
continuity as reflected in the following lemma of Murat [6] and Tartar [2].

<u>Compensated compactness lemma</u>: Let (η_1, q_1), (η_2, q_2) be two entropy-flux pairs
which satisfy

$$\begin{array}{c}\eta_{1_t} + q_{1_x} \\ \eta_{2_t} + q_{2_x}\end{array} \quad \text{lie in a compact subset of } H^{-1}_{loc} \tag{3.4}$$

along solutions of (0.3). Then for almost all (x,t) we have

$$\langle \nu_{x,t}, \eta_1 q_2 - \eta_2 q_1 \rangle = \langle \nu_{x,t}, \eta_1 \rangle \langle \nu_{x,t}, q_2 \rangle - \langle \nu_{x,t}, \eta_2 \rangle \langle \nu_{x,t}, q_1 \rangle. \tag{3.5}$$

The compensated compactness lemma shows the commutativity property of the
Young measure over the form $\eta_1 q_2 - \eta_2 q_1$ for entropy-flux pairs. Our goal now is
to show that by judicious choice of entropy-flux pairs satisfying (3.4), equality
(3.5) implies $\nu_{x,t} = \delta_{\overline{w}(x,t), \overline{v}(x,t)}$.

4. An a Priori Energy Estimate

Formally multiply (0.3_a) by v and (0.3_b) by $-p(w)$ and add. If we denote P as the primitive of p we have

$$(\frac{v^{\varepsilon 2}}{2} - P(w^\varepsilon))_t + (v^\varepsilon p(w^\varepsilon))_x =$$
$$= \varepsilon(v^\varepsilon v_x^\varepsilon)_x - \varepsilon(p(w^\varepsilon)w_x^\varepsilon)_x - \varepsilon v_x^{\varepsilon 2} + \varepsilon p'(w^\varepsilon)w_x^{\varepsilon 2}. \qquad (4.1)$$

Integration of (4.1) yields

$$\int_{-\infty}^{\infty} \frac{v^{\varepsilon 2}}{2} - P(w^\varepsilon)\big|_{t=T} \, dx + \varepsilon \int_{-\infty}^{\infty} \int_0^T v_x^{\varepsilon 2} - p'(w^\varepsilon)w_x^{\varepsilon 2} \, dx \, dt =$$
$$= \int_{-\infty}^{\infty} \frac{v_0^2}{2} - P(w_0) dx. \qquad (4.2)$$

While p' is not continuous and w^ε, v^ε are not smooth (4.2) can be derived by approximating p by p_δ as in Section 1. Also we note that we have assumed v^ε, w^ε approach constant states as $|x| \to \infty$ and hence by trivially choosing new variables also denoted by $v^\varepsilon, w^\varepsilon$ that the constant states are zero.

From (4.2) we have

$$\varepsilon \int_{-\infty}^{\infty} \int_0^T v^{\varepsilon 2} \, dxdt < \text{const.}$$
$$\varepsilon \int\int_H w_x^{\varepsilon 2} \, dxdt < \text{const.} \qquad (4.3)$$

where $H = \{(x,t); w^\varepsilon(x,t) < \alpha \text{ or } w^\varepsilon(x,t) > \beta\}$.

Note (4.3_b) follows from (4.2) since $p' < -\text{const.} < 0$ in H. Here all constants are independent of ε.

5. Entropy Construction - Part 1

Let n, q be an arbitrary C^2 entropy-flux pair for (0.1). If $w^\varepsilon, v^\varepsilon$ is a solution of (0.3) a formal calculation shows

$$n(w^\varepsilon, v^\varepsilon)_t + q(w^\varepsilon, v^\varepsilon)_x = \qquad (5.1)$$
$$= \varepsilon(n_v v_x^\varepsilon)_x + \varepsilon(n_w w_x^\varepsilon)_x - \varepsilon(n_{vv} v_x^{\varepsilon 2} + 2n_{wv} v_x^\varepsilon w_x^\varepsilon + n_{ww} w_x^{\varepsilon 2}).$$

(Some care should be taken in a rigorous derivation since $(w^\varepsilon, v^\varepsilon)$ is not smooth. Then (5.1) is taken in the sense of distributions.)

Set $I_1^\varepsilon(x,t) = \varepsilon(\eta_v v_x^\varepsilon + \eta_w w_x^\varepsilon),$

$$I_2^\varepsilon(x,t) = -\varepsilon(\eta_{vv} v_x^{\varepsilon^2} + 2\eta_{wv} v_x^\varepsilon w_x^\varepsilon + \eta_{ww} w_x^{\varepsilon^2}).$$

As before let $H = \{(x,t)$ in $R \times [0,T], w^\varepsilon(x,t) < \alpha$ or $w^\varepsilon(x,t) > \beta\}$ and set $E =$ complement of H in $R \times [0,T]$.

We wish to estimate the L^2 norm of I_1^ε and the L^1 norm of I_2^ε. There is some difficulty in doing this since we <u>don't</u> have a nice estimate of the form

$$\varepsilon \int_0^T \int_{-\infty}^\infty w_x^{\varepsilon^2}(x,t)dx\, dt < \text{const.}$$

but only (4.3_b). However if we assume $\eta = \phi(v)$, $q = 0$, $\alpha < w < \beta$ for some $\phi \in C^2$ things sufficiently simplify to allow us to do our estimates. Specifically we see

$$\int_0^T \int_{-\infty}^\infty I_1^\varepsilon(x,t)^2 dx\, dt = \varepsilon^2 \int_0^T \int_{-\infty}^\infty (\eta_v v_x^\varepsilon + \eta_w w_x^\varepsilon)^2\, dx\, dt$$

$$\leq \varepsilon^2 \text{ const } \{ \iint_H w_x^{\varepsilon^2}\, dxdt + \iint v_x^{\varepsilon^2}\, dxdt\}$$

$$\leq \varepsilon \cdot \text{const.}$$

So $I_1^\varepsilon \to 0$ as $\varepsilon \to 0+$ in $L^2(R \times [0,T])$ and hence $I_{1x}^\varepsilon \to 0$ as $\varepsilon \to 0+$ in $H^{-1}(R \times [0,T])$ and I_{1x}^ε is in a compact subset of $H^{-1}(R \times [0,T])$.

Next note

$$\int_0^T \int_{-\infty}^\infty |I_2^\varepsilon(x,t)|dx\, dt$$

$$= \varepsilon \iint_H |\eta_{vv} v_x^{\varepsilon^2} + 2\eta_{wv} v_x^\varepsilon w_x^\varepsilon + \eta_{ww} w_x^{\varepsilon^2}|dx\, dt$$

$$+ \varepsilon \iint_E |\eta_{vv} v_x^{\varepsilon^2}|dx\, dt \leq \text{const.}$$

by use of (4.3). Also we note that from the L^∞ estimate on $(w^\varepsilon, v^\varepsilon)$ that $\eta(w^\varepsilon, v^\varepsilon)_t + q(w^\varepsilon, v^\varepsilon)_x$ and hence the right hand side of (5.1) lies in a bounded set of $W^{-1,\infty}$. Thus from the above integral estimates we have that the right hand side of (5.1) is of the form

$$I_1^\varepsilon(x,t)_x + I_2^\varepsilon(x,t)$$

where $I_1^\varepsilon(x,t)_x$ lies in a compact subset of H_{loc}^{-1}, $I_2(x,t)$ lies in a bounded subset of L^1, $I_1^\varepsilon(x,t)_x + I_2^\varepsilon(x,t)$ lies in a bounded subset of $W^{-1,\infty}$. A lemma of Murat [2] implies the right hand side of (5.1) lies in a compact subset of H_{loc}^{-1}.

So for families of C^2 entropy-flux pairs (η,q) satisfying $\eta = \phi(v)$, $\alpha < w < \beta$ the compensated compactness lemma applies and we can attempt to show $\nu_{x,t}$ reduces to the Dirac mass centered at $\overline{w},\overline{v}$. Actually, since p' suffers a jump across $w = \alpha$ and $w = \beta$ (3.2) and (3.1b) show η_{ww} and q_w also possess jump discontinuities at these points. However this degree of regularity in η and q is sufficient to allow us to do the above estimates.

6. Entropy Construction - Part 2

As noted in Section 5 to apply the compensated compactness lemma we shall need entropies $\eta(w,v)$ satisfying

$$\eta_{ww} + p'(w)\eta_{vv} = 0 \tag{6.1}$$

with

$$\eta = \phi(v) \quad \text{for} \quad \alpha < w < \beta. \tag{6.2}$$

To do this we make the following construction. Consider the extended problem of solving

$$\eta_{e_{ww}} + p_e'(w)\eta_{e_{vv}} = 0 \tag{6.3}$$

where p_e is the odd extension of p about $w = \alpha$ as shown in Figure 6.

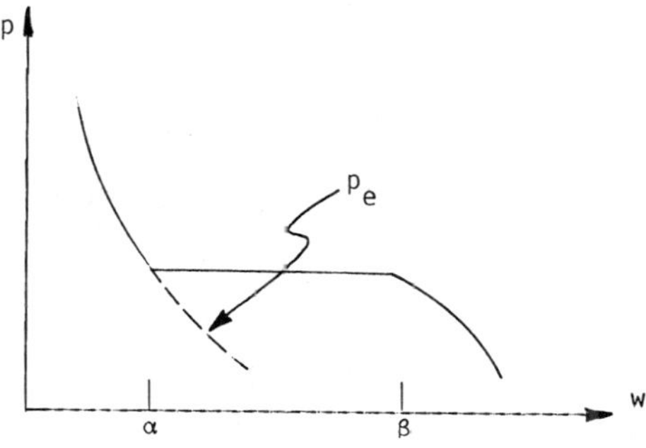

Figure 6

Solve this equation subject to the boundary condition

$$n_e = \phi(v) \quad \text{on} \quad w = \alpha. \tag{6.4}$$

Set $\quad n(w,v) = \frac{1}{2} [n_e(w,v) + n_e(2\alpha - w,v)]$ for $w < \alpha$. Then $\quad n \quad$ satisfies (6.1) on $w < \alpha$ and $\frac{\partial n}{\partial w} = 0$ on $w = \alpha$. Hence

$$n(w,v) = \frac{1}{2} [n_e(w,v) + n_e(2\alpha - w,v)], \quad w < \alpha,$$

$$= \phi(v) \quad , \quad \alpha < w < \beta,$$

is a piecewise smooth solution of (6.1), (6.4) for $w < \beta$. In fact n, n_v, n_w, n_{wv}, n_{vv} are continuous for $w < \beta$ and also n_w is Lipschitz continuous in w for fixed v. If we do a similar reflection argument about $w = \beta$ we have constructed a piecewise smooth entropy in that n, n_v, n_w, n_{wv}, n_{vv} are continuous and n_{ww} is continuous except at $w = \alpha$, $w = \beta$ where it experiences a jump discontinuity. We note that no convexity is either needed or expected for these entropies.

An interesting special case is obtained by this construction when $\phi(v) = e^{kv} V_0(v)$, $k > 0$ a parameter, $V_0(v)$ a C_0^∞ function with small compact support. In this case a graphical representation of the support of the solution n_k is shown in Figure 7.

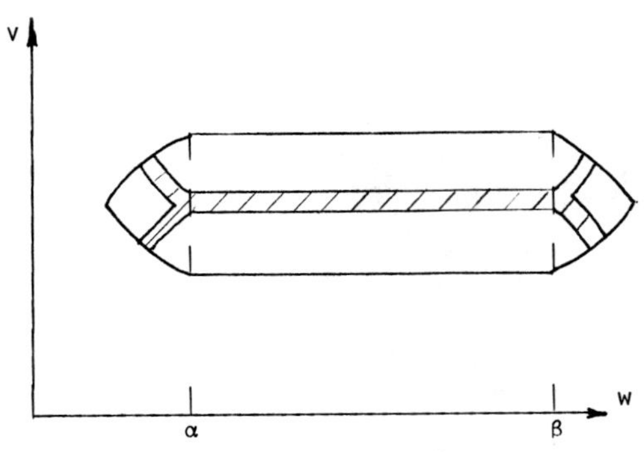

Figure 7

If we denote the Riemann invariants

$$\left\{ {r \atop s} \right. = v \mp \int_\infty^w \sqrt{-p'(\xi)}\, d\xi$$

an asymptotic (in large k) solution of (6.3) in the manner of Lax [7] yields the formulas

$$\eta_k = e^{kv}\{V_0^r + \frac{V_1^r}{k} + \ldots + 0(\frac{1}{k^N})\} + e^{ks}\{V_0^s + \frac{V_1^s}{k} + \ldots + 0(\frac{1}{k^N})\}$$

$$q_k = \lambda\eta_k + 0(\frac{1}{k})\eta_k, \quad \lambda = (-p'(w))^{1/2},$$

$$\text{for } w < \alpha,\ w > \beta, \tag{6.5}$$

$$\eta_k = e^{kv}V_0(v),$$

$$q_k = 0, \quad \text{for } \alpha < w < \beta.$$

Here $V_0^r(w,v)$, $V_1^r(w,v),\ldots,V_0^s(w,v),V_1^s(w,v),\ldots$ etc. satisfy first order hyperbolic equations with the data

$$V_0^r(\alpha,v) = V_0^r(s,v) = V_0(v),$$

$$V_0^s(\alpha,v) = V_0^s(\beta,v) = V_0(v),$$

$$V_n^r(\alpha,v) = V_n^r(\beta,v) = V_n^s(v,v) = V_n^s(\beta,v) = 0,\ n = 2,\ldots,N$$

7. Boundary Measures

We show in this section how the Young measure $\nu_{x,t}$ can be reduced to the Dirac mass in a simple case. Assume the Young measure has support completely inside the left hand triangle of our invariant region D and such that the support is enclosed completely by the smallest characteristic rectangle $R = \{(w,v);\ r_- \leqslant r(w,v) \leqslant r_+,\ s_- \leqslant s(w,v) \leqslant s_+,\ r_-,r_+,s_-,s_+$ constants$\}$. In addition assume R is strictly to the left of $w = \alpha$. This case is illustrated in Figure 8

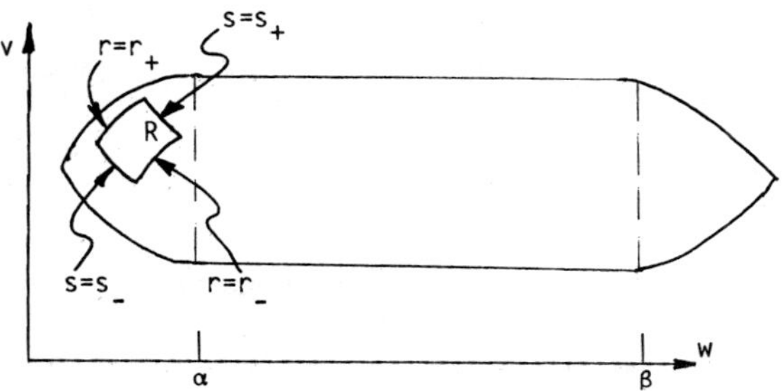

Figure 8

Consider data for (6.2) $\phi(v) = e^{kv}V_0^+(v)$, $k > 0$ where $V_0^+(v) > 0$ on $(r_+ - \rho, r_+)$ and is zero elsewhere. Here $\rho > 0$ is small. Then on a strip S_+ running parallel to $r = r_+$ namely $r_- - \rho \leqslant r \leqslant r_+$, $s_- \leqslant s \leqslant s_+$, (6.5) implies

$$\eta_k^+ = e^{kr}\{V_{0+}^r + \frac{V_{1+}^r}{k} + \ldots + O(\frac{1}{k^N})\} \tag{7.1}$$

$$q_k^+ - \lambda \eta_{k+}^+ = O(\frac{1}{k})\eta_k^+ \tag{7.2}$$

since the terms V_0^s, $V_1^s,\ldots$ and the error term from s part of the expansion for η_k are identically zero on S_+ if ρ is sufficiently small. (Indeed this is why we required R strictly inside the triangle).

Set

$$<\mu^+,h> = \lim_{k \to \infty} \frac{<\eta_k^+,\nu h>}{<\nu_k^+,h>} \tag{7.3}$$

for continuous functions $h(w,v)$. Then μ^+ is a measure with support on $r = r_+$,

$s_- \leqslant s \leqslant s_+.$

Do a similar construction near $r = r_-$, i.e. that $\phi(v) = e^{kv}V_0^-(v)$, $k > 0$ where $V_0^-(v) > 0$ on $(r_-, r_- + \rho)$ where $\rho > 0$ is small. Then on a strip S_- running parallel to $r = r_-$ namely $r_- \leqslant r \leqslant r_- + \rho$, $s_- \leqslant s \leqslant s_+$ (6.5) implies

$$\eta_k^- = e^{kr}\{V_{0-}^r + \frac{V_{1-}^r}{k} + \ldots + O(\frac{1}{k^N})\}, \tag{7.4}$$

$$q_k^- - \lambda\eta_k^- = O(\frac{1}{k})\eta_k^-, \tag{7.5}$$

and since V_{0-}^s, V_{1-}^s, .. and the error term in s part of the asymptotic expansion for η_k^- are identically zero in S- if ρ is sufficiently small. Set

$$\langle \mu^-, h \rangle = \lim_{k \to \infty} \frac{\langle \eta_k^-, h \rangle}{\langle \eta_k^-, h \rangle}$$

for continuous function $h(w,v)$. Then μ^- is a measure with support on $r = r_- + \rho$, $s_- \leqslant s \leqslant s_+.$

The rest of the argument parallels the proof given by DiPerna in Section 5 of [1]. We know from the compensated compactness lemma that

$$\langle \nu, q \rangle - \frac{\langle \nu, \eta \rangle \langle \nu, q_k^+ \rangle}{\langle \nu, \eta_k^+ \rangle} = \frac{\langle \nu, \eta_k^+ q - \eta q_k^+ \rangle}{\langle \nu, \eta_k^+ \rangle} \tag{7.7}$$

for the entropy-flux pair (η_k^+, q_k^+) and an arbitrary entropy-flux pair (η, q). Hence since by (7.2) $q_k^+ = \lambda\eta_k^+ + O(\frac{1}{k})\eta_k^+$ we see that (7.7) implies

$$\langle \nu, q \rangle - \frac{\langle \nu, \eta \rangle \langle \nu, \lambda\eta_k^+ + O(1/k)\eta_k^+ \rangle}{\langle \nu, \eta_k^+ \rangle} =$$
$$\frac{\langle \nu, \eta_k^+ q - \eta(\lambda\eta_k^+ + O(1/k)\eta_k^+) \rangle}{\langle \nu, \eta_k^+ \rangle}. \tag{7.8}$$

Let $k \to \infty$ in (7.8). We see

$$\langle \nu, q \rangle - \langle \nu, \eta \rangle \langle \mu^+, \lambda \rangle = \langle \mu^+, q - \lambda\eta \rangle. \tag{7.9}$$

A similar argument shows

$$\langle \nu, q \rangle - \langle \nu, \eta \rangle \langle \mu^-, \lambda \rangle = \langle \mu^-, q - \lambda\eta \rangle \tag{7.10}$$

or rewriting (7.9), (7.10) we have

$$\langle \nu, q - \langle \mu^+, \lambda \rangle \eta) = \langle \mu^+, q - \lambda \eta \rangle,$$
$$\langle \nu, q - \langle \mu^-, \lambda \rangle \eta \rangle = \langle \mu^-, q - \lambda \eta \rangle. \tag{7.11}$$

Now apply the compensated compactness lemma to (η_k^+, q_k^+), (η_k^-, q_k^-). We have

$$\langle \nu, \eta_k^+ \rangle \langle \nu, q_k^- \rangle - \langle \nu, \eta_k^- \rangle \langle \nu, q_k^+ \rangle = \langle \nu, \eta_k^+ q_k^- - \eta_k^- q_k^+ \rangle. \tag{7.12}$$

Now note by (7.1), (7.2), (7.4), (7.5)

$$\eta_k^+ q_k^- = 0, \quad \eta_k^- q_k^+ = 0 \quad \text{on} \quad R$$

since the supports of η_k^+ and q_k^-, η_k^- and q_k^+ don't intersect. Hence the right hand side of (7.12) is identically zero. Thus (7.12) implies

$$\frac{\langle \nu, q_k^- \rangle}{\langle \nu, \eta_k^- \rangle} - \frac{\langle \nu, c_k^+ \rangle}{\langle \nu, \eta_k^+ \rangle} = 0. \tag{7.13}$$

Again use (7.2), (7.5) and let $k \to \infty$. Then (7.13) gives

$$\langle \mu^-, \lambda \rangle = \langle \mu^+, \lambda \rangle \tag{7.14}$$

which by (7.11) yields

$$\langle \mu^+, q - \lambda \eta \rangle = \langle \mu^-, q - \lambda \eta \rangle \tag{7.15}$$

for any entropy-flux pair (η, q). Now we make the judicious choice $\eta = \eta_k^+$, $q = q_k^+$ and substitute into (7.15). An analysis of q_k^+, η_k^+ shows

$$q_k^+ - \lambda \eta_k^+ = \frac{e^{kr}}{k} ([\text{non-zero term}] + 0(\tfrac{1}{k})) \tag{7.16}$$

on the strip S^+ where its support lies. Hence $|\langle \mu^+, q_k^+ - \lambda \eta_k^+ \rangle| \geq \dfrac{\text{const. } e^{kr_+}}{k}$ for k large. On the other hand

$$\langle \mu^-, q_k^+ - \lambda \eta_k^+ \rangle = 0 \quad \text{for all} \quad k$$

since μ^- has support at $r = r_- + \rho$ where $q_k^+ - \lambda \eta_k^+ = 0$. This contradicts (7.15) and hence we conclude $r_- = r_+$. A similar argument gives $s_- = s_+$. So if we know $(w_\varepsilon(x,t), v_\varepsilon(x,t)) \subset R$ we have support $\nu_{x,t} \subset R$ and $p(w) = \text{weak } * \lim p(w_\varepsilon)$ as

$\varepsilon \to 0^+$. A similar result holds for R strictly to the right of $w = \beta$.

Of course if $(w_\varepsilon(x,t), v_\varepsilon(x,t))$ lies in $\alpha \leqslant w \leqslant \beta$ then $p(w_\varepsilon(x,t)) = p(\alpha)$ and trivially $p(w_\varepsilon(x,t)) \to p(\alpha)$, while $w_\varepsilon \to \overline{w}$ weak $* \ L^\infty$ as $\varepsilon \to 0+$. Since $\overline{w}$ must lie in $[\alpha, \beta]$ $p(\overline{w}) = p(\alpha)$ and we have $p(w_\varepsilon(x,t)) \to p(\overline{w})$ as $\varepsilon \to 0+$.

Thus in any of these three cases we know $(\overline{w}, \overline{v})$ is a distributional solution of (0.1), (0.2).

Of course the above arguments apply to cases where special assumptions have been made on the location of the support of the Young measure. Further results with complete proofs will be published elsewhere.

References

1. R. DiPerna, Convergence of approximate solutions to conservation laws, Archive for Rational Mechanics and Analysis, 82 (1983), 27-70.

2. L. Tartar, Compensated compactness and applications to partial differential equations, in Research Notes in Mathematics, Nonlinear analysis and mechanics: Heriot-Watt Symposium Vol. 4 ed. R.J. Knops, Pitman Press (1979).

3. H. Weinberger, Invariant sets for weakly coupled parabolic and elliptic systems, Rendiconti di Matematica 8 (1975), 295-310.

4. K.N. Chueh, C.C. Conley, and J.A. Smoller, Postively invariant regions for systems of nonlinear diffusion equations, Indiana Univ. Math. J. 26 (1977), 372-411.

5. V. Roytburd, M. Slemrod, Positively invariant regions for a problem in phase transitions, to appear in Archive for Rational Mechanics and Analysis.

6. F. Murat, Compacité par compensation, Ann. Scuola Norm. Sup. Pisa Sci. Fis. Mat. 5 (1978), 489-507.

7. P.D. Lax, Shock waves and entropy, in Contributions to Nonlinear Functional Analysis, ed. E.A. Zarantonello, Academic Press (1971), 603-634.

RECENT ADVANCES IN NONLINEAR WAVE EQUATIONS

Jalal Shatah[*]

Courant Institute
251 Mercer Street
New York University
New York, New York 10012

Various phenomena in physics are described by nonlinear wave equations of
the form

$$\Box\, u + m^2 u + F(u, \partial u, \partial^2 u) = 0 \tag{1}$$

where $u = u(x_1, x_2, x_3, t)$, $\Box = \partial_t^2 - \partial_1^2 - \partial_2^2 - \partial_3^2$, ∂u and $\partial^2 u$ are the first and
second partial derivatives of u, and F is a smooth function with $F(0) = F'(0) = 0$.
Associated with the above equation is the initial value problem to determine u
such that

$$u(x,0) = f(x) , \qquad \partial_t u(x,0) = g(x) ,$$

where f and g are given smooth functions.

An essential question about these equations is the following: do solutions
exist for all time or do they develop singularities in finite time? Here we will
give a short survey of results that address this question for small solutions,
i.e. solutions such that $u \sim 0$. Both the case $m^2 > 0$ (referred to as NLKG
equation), and the case $m^2 = 0$ (referred to as a NLW equation) will be discussed.
Then we will discuss a new technique for studying small solutions of nonlinear
evolution equations.

The local existence problem for small initial data is easy because the
equation is hyperbolic for u near zero. Thus standard energy methods give
local existence of small smooth solutions [6], [11]. In contrast, the global
existence problem is extremely difficult and is not true in this generality. That
is, there are nonlinearities where solutions with arbitrarily small initial data
develop singularities in finite time. Indeed a result of P. Lax [10] states that
solutions of the 1 + 1 dimensional genuinely nonlinear wave equation blow up in
finite time. In 3 + 1 dimensions F. John [3] proved that there are quasilinear

* This work was partially supported by ARO Contract DAAG29-81-K-0043.

wave equations whose solutions blow up in finite time. Moreover he showed that for any k, the solutions exist up to a time of order $1/\varepsilon^k$, where ε is some norm of the initial data. The order was later improved to $\exp(1/\varepsilon)$ for the semilinear case by F. John [5] and T. Sideris [13], and for the nonlinear case by S. Klainerman [5]. As demonstrated by F. John, this result is sharp, when $F = (\partial_t u)(\partial_t^2 u)$.

Although singularities develop in $3 + 1$ dimensions there is a large class of NLW equations whose small solutions remain smooth for all time. The first results on global existence are due to S. Klainerman [7], and later J. Shatah [11] and S. Klainerman and G. Ponce [8]. They state that if $F(y) = O(|y|^p)$, then small solutions of the $n + 1$ dimensional NLW equation exist for all time provided $(n - 1)(p - 1)^2/(2p) > 1$ $(n(p - 1)^2/(2p) > 1$ for NLKG equation). In particular for $n = 3$ the smallest integer allowed is $p \geqslant 3$. This is not very restrictive for the NLW equation since the results of F. John exclude $p = 2$ for general nonlinearities [3]. However there is a large class of NLW equations with $p = 2$ and whose solutions exist for all time. The most significant example is the Yang-Mills equations where D. Eardley and V. Moncrief [2] showed that all solutions exist for all time. For small solutions of the NLW equations in $3 + 1$ dimensions we have the following theorem.

Theorem A. Suppose that the nonlinearity F satisfies the "null condition":

$$F(u, s\,n, t\,nn) \text{ is at least cubic in } s \text{ and } t \text{ near zero}$$

whenever $n = (n_0, n_1, n_2, n_3)$ is a null vector i.e. $n_0^2 - n_1^2 - n_2^2 - n_3^2 = 0$. Then small smooth solutions of the NLW equation exist for all time in $3 + 1$ dimensions.

The proof of the above theorem was given by S. Klainerman [16] using weighted $L^\infty - L^1$ estimates and by D. Christodoulou [15] using a conformal mapping method. At the end of this paper we'll mention a new method for deriving this condition.

For the NLKG equation in $3 + 1$ dimensions the condition $p \geqslant 3$ is unnatural and due to the limitation of the method used. Indeed a recent result states

Theorem B. In 3 + 1 dimensions the NLKG equation has global smooth solutions for small initial data.

This theorem was proved, independently, by J. Shatah [12] and S. Klainerman [9]. The proof given in [9] is based on weighted L^∞ - L^1 estimates, while the proof in [12] is based on Poincare's theory of normal forms, which is used to study solutions of ordinary differential equations [1]. We will illustrate this technique for ODE's and then explain how to adapt it to PDE's.

Given a differential equation

$$\dot{x} = Ax + f(x) \qquad f(x) = 0(|x|^2),$$

can we reduce it to the linear equation

$$\dot{y} = Ay$$

by a change of variables of the form $x = y + h(y)$? A quick way of checking this is to try to eliminate the quadratic term in f. Since y is a solution of the linear equation, a quadratic term in y involves frequencies $\exp(\lambda_i + \lambda_j)t$ where (λ_k) are the eigenvalues of the matrix A. If the eigenvalues of A are second order resonant, that is, $\lambda_\ell = \lambda_i + \lambda_j$, then it is obvious that this cannot be done. However, if the eigenvalues are not second order resonant then the quadratic term can be eliminated. A simple calculation shows that the lowest degree term of h should satisfy

$$h'A - Ah = f.$$

Now we'll explain how to apply this technique to nonlinear PDE's. Consider the NLKG equation as a model. Solutions of the linear equation consist of plane waves, $\exp[i(wt + p\cdot x)]$, $w = \pm\,(|p|^2 + m^2)^{1/2}$. Two plane-wave solutions interact and the result is again a plane wave $\exp\{2[(w_1 + w_2)t + (p_1 + p_2)\cdot x]\}$. The condition for resonance is that the resulting plane wave is a solution of the linear equation

$$w_1 + w_2 = \pm\,(|p_1 + p_2|^2 + m^2)^{1/2},$$

which is impossible if $m^2 > 0$. Therefore we conclude that quadratic terms of the

NLKG equation are nonresonant. This implies that we can find a transformation that will eliminate the quadratic term. Once this is done we will obtain an equation with a nonlinear term of degree at least = 3, and this can be solved for all time by the technique of [11]. For more details on this procedure we refer the reader to [12].

A very interesting situation occurs if one tries to apply this method of the NLW equation. In this case there are resonant frequencies. This implies that the transformation will be singular and may not exist. However if the nonlinearity satisfies the null condition then the singularity will cancel and the transformation will exist.

Finally for the case when the nonlinearity depends only on u there are many results that deal with existence and scattering. For this case we refer the reader to W. Strauss [14].

REFERENCES

1. V.I. Arnold, Generic Methods in the Theory of Ordinary Differential Equations, Springer-Verlag, 1983.

2. D. Eardley and V. Moncrief, The Global Existence of Yang-Mills in 4-dimensional space time, Comm. Math. Phys., 83, 1982, p. 193-212.

3. F. John, Blow-up for Quasilinear Wave Equations in Three Space Dimension, Comm. Pure. Appl. Math., 34, 1981, p. 29-51.

4. F. John, Lower Bounds for the life-spac of Solutions of Nonlinear Wave Equations in Three Space Dimensions. Comm. Pure Appl. Math, 36, 1983, p. 1-35.

5. F. John and S. Klainerman, Almost Global Existence to Nonlinear Wave Equations in Three Space Dimensions, Comm. Pure Appl. Math. 37, 1984, p. 443-455.

6. T. Kato, Linear and Quasilinear Equations of Hyperbolic Type, C.I.M.E. II Ciclo, 1976.

7. S. Klainerman, Global Existence for Nonlinear Wave Equations, Comm. Pure Appl. Math., 33, 1980, p. 43-101.

8. S. Klainerman and G. Ponce, Global Small Amplitude Solutions to Nonlinear Evolution Equations, Comm. Pure Appl. Math., 36, 1983, p. 133-141.

9. S. Klainerman, Global Existence of Small Amplitude Solutions to Nonlinear Klein-Gordon Equations in 4-Space-Time Dimensions, 38, 1985.

10. P. Lax, Development of Singularities of Solutions of Nonlinear Hyperbolic Partial Differential Equations, J. Math. Phys. 5, 1964, p. 611-613.

11. J. Shatah, Global Existence of Small Solutions to Nonlinear Evolution
    Equations, J.D.E., 46, 1982, p. 409-425.

12. J. Shatah, Normal Forms and Quadratic Nonlinear Klein-Gordon Equations, Comm.
    Pure Appl. Math., 38, 1985.

13. T. Sideris, Global Behavior of Solutions to Nonlinear Wave Equations in Three
    Dimensions, Comm. P.D.E., 8, 1983, p. 1291-1323.

14. W. Strauss, Stable and Unstable States of Nonlinear Wave Equations,
    Contemporary Math., 17, 1983, p. 429-441.

15. D. Christodoulou, Global existence Nonlinear Wave Equations, in preparation.

16. S. Klainerman, The Null condition and Global existence to Nonlinear Wave
    Equations, preprint.

SOME EXISTENCE, UNIQUENESS AND NON-UNIQUENESS RESULTS FOR WEAKLY

HYPERBOLIC EQUATIONS IN THE GEVREY CLASSES

Sergio Spagnolo

Dipartimento di Matematica
Universitá di Pisa
Via F. Guonavoth 2
56100 Pisa, Italy

1. Introduction

In this lecture we review some results (obtained in collaboration with F. Colombini, E. DeGiorgi, and E. Jannelli) concerning the well-posedness of <u>initial value problems</u> like

$$
\begin{aligned}
u_{tt} - \sum (a^{ij}(x,t)u_{x_i})_{x_j} + c(x,t) &= 0 \\
u(x,0) = \phi(x), \quad u_t(x,0) &= \psi(x),
\end{aligned}
\qquad (x \in R^n,\ t \in [0,T]) \qquad (1)
$$

where the coefficients a^{ij} are real functions such that $a^{ij} = a^{ji}$ $(i,j = 1,\ldots,n)$ and

$$
0 < \lambda < \sum a^{ij} \xi_i \xi_j |\xi|^{-2} < \Lambda \qquad (\xi \in R^n). \qquad (2)
$$

Moreover, we shall always assume that all the coefficients and the initial data of Pb. (1) are C^∞ functions in the variables $x = (x_1,\ldots,x_n)$, whereas we shall allow the coefficients to be non-regular in the variable t.

We recall that Pb. (1) is said to be <u>well-posed</u> in some topological space $F \_ C^\infty(R_x^n)$ when it has a unique solution

$$
u: [0,T] \to F
$$

for every ϕ, ψ in F. The best known case of well-posedness occurs when the a^{ij}'s satisfy (2) with $\lambda > 0$ (<u>strict hyperbolicity</u>) and, in addition,

$$
| \frac{\partial}{\partial t} a^{ij} | < M < \infty \quad \text{on} \quad R_x^n \times [0,T]. \qquad (3)
$$

Under such hypotheses, Pb. (1) is well-posed in $C^\infty(R_x^n)$.

We are interested in the cases where $\lambda = 0$ or $M = \infty$. A motivation for this are the problems of stability of the solutions under some perturbations of the coefficients (see §2 below).

In §3 and §4, we shall exhibit some (complementary) results of well-posedness and non-uniqueness for Pb. (1).

2. Perturbations of the Coefficients

Let us consider the family of I.V.P.'s

$$u_{tt} - \sum (a_k^{ij}(x,t)u_{x_i})_{x_j} + c(x,t)u = 0$$
$$u(x,0) = \phi(x), \quad u_t(x,0) = \phi(x),$$

$$(4)$$

which depend on the parameter $k \in N$, and let us assume that

$$0 < \lambda_k \leq \sum a_k^{ij} \xi_i \xi_j |\xi|^{-2} \leq \Lambda_k \qquad (5)$$

and

$$(6) \qquad \left| \frac{\partial}{\partial t} a_k^{ij} \right| \leq M_k < \infty. \qquad (6)$$

Hence, for every ϕ, ψ in $C^\infty(R_x^n)$, there exists a unique solution $u_k(x,t)$ of Pb. $(4)_k$, so that we can pose the following question.

Question 1

Does the convergence

$$a_k^{ij} \to a^{ij} \quad \underline{in} \ L_{loc}^1 \qquad (k \to \infty) \qquad (7)$$

implies the convergence

$$u_k \to u \quad \underline{in} \ L_{loc}^1 \qquad (k \to \infty) \qquad (8)$$

for every ϕ, ψ in $C^\infty(R_x^n)$?

More generally, one could try to characterize the weakest topology τ (on the coefficients) for which

$$[a_k^{ij}] \overset{\tau}{\to} [a^{ij}] \quad \text{implies (8)} \qquad (9)$$

<u>Remark 1</u>

Similar questions for the <u>elliptic</u> boundary problems were at the origin of the mathematical theory of <u>G-convergence</u> (cf. [S], [BLP]). In fact, given a sequence of Dirichlet problems such as

$$-\sum (a_k^{ij}(x)u_{k,x_i})_{x_j} = -\sum (a^{ij}(x)u_{x_i})_{x_j} = f(x), \quad \text{on} \quad \Omega,$$

$$u_k = u = 0, \quad \text{on} \quad \partial\Omega,$$

where Ω is a bounded open subset of R^n and the a_k^{ij}'s satisfy (5) with $\lambda_k \equiv \lambda > 0$ and $\Lambda_k \equiv \Lambda < \infty$, the G-convergence

$$[a_k^{ij}] \overset{G}{\to} [a^{ij}] \quad \text{on} \quad \Omega$$

is just the weakest topology τ for which the inference (9) holds (for every $f \in H^{-1}(\Omega)$).

We emphasize that G-convergence is a convergence on the <u>matrices</u> $[a_k^{ij}]$ of the coefficients, and not on the single entries of these matrices.

The more relevant properties of G-convergence are the <u>compactness</u> and the <u>locality</u>, i.e. the independence on the type of boundary conditions considered.

Let us go back to Pbs. $(4)_k$: if the conditions (5), (6) are fulfilled uniformly in k, Question 1 has an affirmative answer. More precisely we have:

<u>Theorem 1</u> $([CS]_1)$

<u>Assume that the coefficients</u> $a^{ij}(x,t)$ <u>of Pb.</u> (4) <u>satisfy</u> (5) <u>and</u> (6) <u>with</u>

$$\lambda_k \equiv \lambda > 0, \; \Lambda_k \equiv \Lambda < \infty, \; M_k \equiv M < \infty$$

<u>and that, as</u> $k \to \infty$,

$$[a_k^{ij}(\cdot,t)] \overset{G}{\to} [a^{ij}(\cdot,t)] \quad \text{in} \quad R_x^n, \quad \text{for all t} \tag{10}$$

<u>Then</u> for all ϕ, ψ <u>in</u> $C^\infty(R_x^n)$

$$u_k \to u \; \underline{\text{in}} \; C([0,T],H_{loc}^1) \cap C^1([0,T],L_{loc}^2).$$

<u>Remark 2</u>

A special case of (10) is that in which

$$a_k^{ij} \to a^{ij} \quad \text{in} \quad L_{loc}^1(R_x^n \times [0,T]), \quad \forall i,j.$$

Another special case in which (10) holds (for some a^{ij} independent of x) is that of the <u>homogenization</u> in x, where

$$a_k^{ij}(x,t) = \alpha^{ij}(kx,t)$$

for some $\alpha^{ij}(y,t)$, <u>periodic</u> in y and satisfying (2) with $\lambda > 0$.

If one of the conditions (5), (6) is not uniform with respect to k, i.e. $\lambda_k \to 0^+$ or $M_k \to +\infty$ for $k \to \infty$, the answer to Question 1 may be negative, even for problems of the particular form:

$$\begin{aligned} u_{tt} - a_k(t)u_{xx} &= 0 \\ u(x,0) = \phi(x), \quad u_t(x,0) &= \phi(x). \end{aligned} \qquad (x \in R^1, \ t \in [0,T]). \qquad (11)$$

Indeed:

<u>Theorem 2</u> ($[CS]_1$)

i) <u>There exist a sequence</u> $\{a_k(t)\}$ <u>of</u> C^∞ <u>functions such that</u>

$$a_k(t) \geqslant \frac{1}{2}, \quad a_k(t) \to 1 \quad \underline{in} \quad C([0,T])$$

(<u>as</u> $k \to \infty$) <u>and a pair</u> (ϕ,ψ) <u>of initial data, for which the sequence</u> $\{u_k\}$ <u>of the solutions to Pbs. (11)</u> <u>is not bounded in the sense of distributions.</u>

ii) <u>There exist</u> $\{a_k(t)\}$ in C^∞ <u>and such that</u>

$$a_k(t) \geqslant 0, \ a_k(t) \to 0 \quad \underline{in} \quad C^\infty([0,T])$$

<u>and</u> ϕ,ψ <u>as above, for which the same conclusions as in</u> (i) <u>holds.</u>

The proof of Thm. 2 is based on the existence of <u>ordinary equations</u> of Sturmian type, which have periodic coefficients and a solution with <u>exponential growth</u> at the infinity. For instance, if we consider the equations

315

$$(\frac{d}{d\tau})^2 \psi + \alpha(\epsilon;\tau)\psi = 0 \quad \text{on} \quad R^1_\tau, \tag{12}$$

where $0 < \epsilon < \frac{1}{10}$ and

$$\alpha(\epsilon;\tau) = 1 - 4\epsilon \sin 2\tau - \epsilon^2(1 - \cos 2\tau)^2,$$

we see that the function

$$\psi(\tau) = \sin \tau \cdot \exp[\epsilon(\tau - \frac{1}{2} \sin 2\tau)]$$

is the solution of (12) such that $\psi(0) = 0$ and $\psi'(0) = 1$.

By taking

$$\phi(x) = 0, \quad \psi(x) = \int_0^\infty e^{-\sqrt{h}} \sin(hx)$$

and

$$a_k(t) = \alpha(\frac{1}{\sqrt{k}} ; kt) \qquad \text{in case (i)}$$

$$a_k(t) = e^{-\sqrt{k}} \cdot \alpha(\epsilon; kt) \quad \text{in case (ii)},$$

we get Thm. 2.

In the special case of Pbs. (11), the inference $[(7) \Rightarrow (8)]$ is still true, provided that we limit ourselves to <u>analytic</u> initial data. More generally we have:

<u>Theorem 3</u> ($[CS]_2$, $[J]$)

<u>Assume that the coefficients</u> $a^{ij}(x,t)$ <u>of Pbs</u> (4) <u>satisfy, besides</u> (5) <u>and</u> (6), <u>the analyticity conditions</u>

$$|D_x^\alpha a_k^{ij}(x,t)| \leqslant C_r \Lambda^{|\alpha|} \alpha! , \quad \text{for} \quad |x| < r,$$

<u>for every</u> $r > 0$, $\alpha \in N^n$, <u>and that also</u> $c(x,t)$ <u>satisfies similar conditions.</u>

<u>Then, from</u>

$$a_k^{ij} \to a^{ij} \quad \underline{in} \quad L^1_{loc}(R^n_x \times [0,T])$$

<u>we can deduce that</u>

$$u_k \to u \quad \underline{in} \quad C^1([0,T], C^\infty(R^n_x)),$$

<u>for every</u> ϕ, ψ <u>in the space</u> $A(R^n_x)$ <u>of the real analytic functions.</u>

Remark 3

By duality, we also get some (weak) stability results for non-regular initial data.

For example, if ϕ, ψ are C^∞-functions with compact support, then in the hypotheses of Thm. 3 we have

$$u_k \to u \quad \text{in} \quad C^1([0,T], A'(R^n_x))$$

where $A'(R^n)$ is the space of real analytic functionals.

3. Some Results of Well-Posedness

The continuity (or not) of the map

$$[\text{coefficients}] \to [\text{solutions}]$$

for fixed initial data, is strictly related to the well-posedness (or not) of the individual problems.

Indeed, the stability of the solution under a perturbation of the coefficients is a consequence of an appropriate estimate on the solution.

Conversely, a starting point to constructing an example of I.V.P. which is not well-posed, is a sequence of problems for which the inference $[(7) \Rightarrow (8)]$ is false.

For instance, Thm. 3 is based on the following existence result:

Theorem 4 ($[CS]_2$, $[J]_1$)

Assume that the coefficients $a^{ij}(x,t)$ and $c(x,t)$ of Pb (1) are analytic in x, more precisely that they belong to $L^1([0,T], A(R^n_x))$, and that (2) holds with $\lambda = 0$.

Then Pb. (1) is well-posed in $A(R^n_x)$.

In Thm. 4 no regularity in the variable t (apart from measurability) is required of the a^{ij}'s; for this reason we can only solve our Problem for ϕ, ψ analytic. If the a^{ij}'s are somehow regular in t, it is possible to solve Pb.

(1) for ϕ, ψ in wider classes than $A(R_x^n)$, as the Gevrey classes $E^1(R_x^n)(*)$.

In order to give a short exposition of some results in this direction (and also in view of the examples of §4) we shall confine ourselves to the special case in which the coefficients a^{ij} are independent of x, i.e.

$$a^{ij} \equiv a^{ij}(t).$$

Theorem 5 ([CDS], [CJS]$_1$)

Assume that $c(x,t)$ belongs to $L^1([0,T],E^s(R_x^n))$ and that either

$$\left|\begin{array}{l} (2)\ \text{holds with}\ \lambda = 0 \\[4pt] a^{ij}(t)\ \epsilon\ C^{0,\theta}([0,T]) \qquad\qquad (0 < \theta < 1) \\[4pt] 1 < s < 1 + \dfrac{\theta}{1 - \theta} \end{array}\right. \qquad (13)$$

or

$$\left|\begin{array}{l} (2)\ \text{holds with}\ \lambda > 0 \\[4pt] a^{ij}(t)\ \epsilon\ C^{k,\theta}([0,T]) \qquad (k\ \epsilon\ N,\ 0 < \theta < 1) \\[4pt] 1 < s < \dfrac{k + \theta}{2}. \end{array}\right. \qquad (14)$$

Then Pb. (1) is well-posed in $E^s(R_x^n)$.

Remark 4

Thm. 5 has been extended to the general case of coefficients $a^{ij}(x,t)$ which depend also on x, by T. Nishitani ([N]) and E. Jannelli ([J]$_2$). For instance, in (13) we can assume that $a^{ij}(x,t)$ belong to $C^{0,\theta}([0,T],E^1(R_x^n))$.

$(*)$ A C^∞-function $w(x)$ is said to belong to $E^s(R_x^n)$, where s is a real number ≥ 1, when

$$|D_x^\alpha w(x)| < C_r \Lambda_r^{|\alpha|} |\alpha|^s, \qquad \text{for}\ |x| < r,$$

$\forall r > 0$, $\alpha\ \epsilon\ N^n$.

4. Non-Uniqueness Results

We shall now prove that the results of §3 are _optimal_, in the sense that one cannot enlarge the class of well-posedness without strengthening the regularity of the coefficients.

To this end we must prove that Pb. (1) is not well-posed in certain classes F; this may be done in two different ways: by disproving the existence of solutions or by disproving the uniqueness.

In the first case, we must find a pair of initial data $\overline{\phi}, \overline{\psi}$ in F for which there is no solution in F, after a certain time T_* $(0 < T_* < T)$: in practice we find another class $\tilde{F} \supset F$, in which the Problem is well-posed, such that the solution $\overline{u}(\cdot,t)$ with initial data $(\overline{\phi},\overline{\psi})$ belongs to $\tilde{F} \setminus F$ for $t > T_*$.

In the second case, we must find a non-trivial solution $u: [0,T] \to F$ with initial data $(\phi,\psi) \equiv (0,0)$.

Getting non well-posedness through nonuniqueness is, in some sense, stronger than through nonexistence. Indeed, if there is non-uniqueness in the class F, then the Problem cannot be well-posed in any class $\tilde{F} \supset F$. Moreover, from an example of non-uniqueness we can get an example of non-existence by the _Holmgren's duality method_.

Let us now give some results of non-uniqueness for the single equation

$$(15) \qquad u_{tt} - a(t)u_{xx} + c(x,t)u = 0 , \quad \text{on} \quad R^2.$$

Theorem 6 ([CJS]$_2$)

i) _There exist_ $a(t)$ _and_ $c(x,t)$ _such that_

$$a(t) > \frac{1}{2} , \ a(t) \in C^{0,\theta}(R) \quad \text{for each} \quad \theta < 1,$$

$$c(x,t) \in C^{\infty}(R^2) \cap L^{\infty}(R^2),$$

for which Eq. (15) _has a nontrivial_ C^{∞} _solution on_ R^2 _with_

$$u \equiv 0 \quad \underline{if} \quad t < 0.$$

ii) <u>There exist</u> a(t) <u>and</u> c(x,t) <u>such that</u>

$$a(t) > 0 \qquad \underline{if} \ t > 0$$
$$a(t) \equiv c(x,t) \equiv 0 \quad \underline{if} \ t < 0,$$
$$a(t) \ \varepsilon \ C^{\infty}(R), \qquad c(x,t) \ \varepsilon \ C^{\infty}(R^2) \cap L^{\infty}(R^2),$$

<u>for which the same conclusion as in</u> (i) <u>holds.</u>

iii) <u>More generally, for each</u> θ, s <u>s.t.</u> $s > 1 + \theta/(1 - \theta)$ (<u>resp. for each</u> k, θ, s <u>s.t.</u> $s > 1 + (k + \theta)/2$) <u>there exist</u> $c \ \varepsilon \ C^{\infty}([0,T],E^s(R_x^n))$ <u>and</u> $a(t) \geqslant \frac{1}{2}$, $a(t) \ \varepsilon \ C^{0,\theta}$ (<u>resp.</u> $a(t) \geqslant 0$, $a(t) \ \varepsilon \ C^{k,\theta}$) <u>for which the same conclusion holds.</u>

Remark 5

The results of Thm. 6 are complementary to those of Thm. 5, hence they are optimal. For example, if a(t) belongs to $C^{0,\theta}$ for every $\theta < 1$, Thm. 5 ensures the uniqueness as soon as c(x,t) is not merely C^{∞} but is a Gevrey function in x.

We also observe that there is uniqueness (for c(x,t) merely C^{∞}) whenever $a(t) \geqslant \lambda > 0$ and $a(t) \ \varepsilon \ C^{0,1}$ or when $a(t) \geqslant 0$ and a(t) is analytic (or, more generally, a(t) has only a finite number of oscillations near its value zero). //

The first example of non-uniqueness for a linear <u>Kovalewskian</u> problem with C^{∞} coefficients, was given by A. Plis in 1954 ([P]).

In 1955 E. DeGiorgi ([D]) constructed a <u>weakly hyperbolic</u> equation of the form

$$D_t^8 u + a(x,t)D_x^4 u + b(x,t)D_x^2 u + c(x,t)u = 0 \tag{16}$$

with coefficients real and C^{∞} (more precisely belonging to every Gevrey class E^s with $s > 2$), for which there is no uniqueness for the I.V.P. This example was generalized by J. Leray ([L]) and Y. Ohya ([O]).

Other examples of non uniqueness, concerning the weakly hyperbolic equations

320

$$D_t^m u + c(x,t)D_x u = 0 \qquad (m \geqslant 2) \qquad\qquad (17)$$

were given by P. Cohen ([C]) and L. Hörmander ([H]).

Recently, S. Nakane ([Na]) constructed several examples of non-uniqueness for equations such as

$$u_{tt} - e^{-2/t}u_{xx} + it^{-4}e^{-1/t}u_x + c(x,t)u = 0. \qquad\qquad (18)$$

<u>Remark 6</u>

By contrast to Eq. (16) or (17), Eq. (18) is (pointwise) <u>hyperbolic</u>, i.e. the imaginary parts of its characteristic roots are bounded for each fixed (x,t). However, the l.u.b. of these imaginary parts goes to the infinity as $t \to 0^+$; in other words, the hyperbolicity of Eq. (18) is not uniform with respect to t.

In this sense, Eq. (15) of Thm. 6 is <u>uniformly hyperbolic</u>.

<u>Remark 7</u>

The coefficients $c(x,t)$ of Eq. (15), in Thm. 6, are complex valued functions. It would be very interesting to give an example with $c(x,t)$ real valued.

Another (related) question is to give an example of non-uniqueness for equations like

$$u_{tt} - (a(x,t)u_x)_x = 0$$

with $a(x,t) \geqslant 0$.

References

[BLP]   A. Bensoussan, J.L. Lions, G. Papanicolaou, Asymptotic methods in periodic structures, North Holland, Amsterdam, 1978.

[C]   P. Cohen, The non-uniqueness of the Cauchy problem, O.N.R. Technical Report <u>93</u>, Stanford University (1960).

[CDS]   F. Colombini, E. DeGiorgi, S. Spagnolo, Sur les équations hyperboliques...., Ann. Scu. Norm. Pisa, <u>6</u> (1979), p. 511-559.

[CJS]$_1$ F. Colombini, E. Jannelli, S. Spagnolo, Well-posedness in the Gevrey classes..., Ann. Scu. Norm. Pisa, <u>10</u> (1983), p. 291-312.

[CJS]$_2$ F. Colombini, E. Jannelli, S. Spagnolo, Non-uniqueness in hyperbolic Cauchy problems, in preparation.

[CS]$_1$ F. Colombini, S. Spagnolo, On the convergence of solutions of hyperbolic equations, Comm. Part. Diff. Ec., $\underline{3}$ (1978), p. 77-103.

[CS]2   .Second order hyperbolic equations with coef- fficients real analytic in space..., J. Analyse Math., $\underline{38}$ (1980), p. 1-33.

[D]   E. DeGiorgi, Un esempio di non unicita della soluzione ..., Rend. Mat., $\underline{14}$ (1955), p. 382-387.

[H]   L. Hormander, Non-uniqueness for the Cauchy problem , in Lect. Notes Math, $\underline{459}$, Springer (1975), p. 36-72.

[J]$_1$ E. Jannelli, Weakly hyperbolic equations of second order ..., Comm. Part. Diff. Eq., $\underline{7}$ (1982), p. 537-558.

[J]$_2$ E. Jannelli, Gevrey well-posedness for a class of weakly hyperbolic equations, J. Math. Kyoto Un. (1985).

[L]   J. Leray, Equations hyperboliques non-strictes ..., Math. Ann., $\underline{162}$ (1966), p. 228-236.

[Na]   S. Nakane, Uniqueness and non-uniqueness ... Proc. Japan Acad., $\underline{59}$ Ser. A, 7 (1983), p. 318-320.

[N]   T. Nishitani, Sur les equations hyperboliques a coefficients qui sont hölderiens en t et de classe de Gevrey en x , Bull. Sci. Math., $\underline{107}$ (1983), p. 113-138.

[O]   Y. Ohya, Non unicité de solutions ..., in "Equations aux derivées par- tielles hyperboliques et holomorphes", Seminaire J. Vaillaint, Hermann, Paris, 1984, p. 166-182.

[P]   A. Plis, The problem of uniqueness for the solutions of a system ..., Bull. Acad. Pol. Sc. $\underline{2}$ 1954, p. 55-57.

[S]   S. Spagnolo, Convergence in energy for elliptic operators , in "Numerical solutions of partial differential equations-III, Synspade 1975", B. Hubbart Ed., Academic Press, New York-San Francisco-London 1976, p. 469-499.

Table of Contents from Other Volumes
from the Program in
Continuum Physics and Partial Differential Equations

Homogenization and effective moduli of materials

October 22 – October 26, 1984

J. L. Ericksen
D. Kinderlehrer
R. Kohn
J.-L. Lions
Conference Committee

Theory and applications of liquid crystals

January 21 – 25, 1985

J. L. Ericksen
D. Kinderlehrer
Conference committee

Berry, G.	Rheological and rheo-optical studies with nematogenic solutions of a rodlike polymer: a review of data on poly (phenylene benzobisthiazole)
Brezis, H.	Singularities of liquid crystals
Capriz, G. and Giovine, P.	On virtual inertia effects during diffusion of a dispersed medium in a suspension
Choi, H. I.	Degenerate harmonic maps and liquid crystals
Cladis, P.	A review of cholesteric blue phases
Cohen, R., Hardt, R., Kinderlehrer, D., Lin, S.-Y., and Luskin, M	Minimum energy configurations for liquid crystals: computational results
Di Benedetto, E.	The flow of two immiscible liquids through a porous medium: regularity of the saturation
Doi, M.	Molecular theory for the nonlinear viscoelasticity of polymeric liquid crystals
Hardt, R. and Kinderlehrer, D.	Mathematical questions in liquid crystal theory
Huang, C. C.	The effect of the magnitude of the disordered phase temperature range on the given phase transition in liquid crystals
Leslie, F.	Some topics in equilibrium theory of liquid crystals
Leslie, F.	Theory of flow phenomena in nematic liquid crystals
Maddocks, J.	A model for disclinations in nematic liquid crystals

Miranda, M.	Some remarks about a free boundary type problem
Ryskin, G.	Computer simulation of flow of liquid crystal polymers
Sethna, J.	Theory of the blue phases of chiral nematic liquid crystals
Spruck, J.	On the global structure of solutions to some semilinear elliptic problems

(tentative contents)

Amorphous polymers and non-newtonian fluids

March 4 - March 8, 1985

J. L. Ericksen
D. Kinderlehrer
M. Tirrell
S. Prager
Conference Committee

Tentative contributors: Bird, R., Caswell, B., Dafermos, C., Hrusa, W. and Renardy, M., Joseph, D. D., Kearsley, E., Marcus, M. and Mizel, V., Nohel, J. and Renardy, M., Rabin, M., Wool, R. P.

Oscillation theory, computation, and methods of compensated compactness

April 1 – April 4, 1985

C. Dafermos
J. L. Ericksen
D. Kinderlehrer
M. Slemrod
Conference Committee

Chacon, T. and Pironneau, O.	Convection of microstructures by incompressible and slightly compressible flows
DiPerna, R.	Oscillations in solutions to nonlinear differential equations
Forest, M.G. and Lee, J.-L.	Geometry and modulation theory for the periodic nonlinear Schrödinger equation
Harten, A.	On high-order accurate interpolation for non-oscillatory shock capturing schemes
Lax, P.	On the weak convergence of dispersive difference schemes
Majda, A.	Nonlinear geometric optics for hyperbolic systems of conservation laws
McLaughlin, D.	On the construction of a modulating multiphase wavetrain for a perturbed KdV equation
Nunziato, J., Gartling, D., and Kipp, M.	Evidence of nonuniqueness and oscillatory solutions in in computational fluid mechanics
Osher, S. and Chakravarthy, S	Very high order accurate T V D schemes
Rascle, M.	Convergence of approximate solutions to some systems of conservation laws: a conjecture on the product of the Riemann invariants

Schonbek, M.   Applications of the theory of compensated compactness

Serre, D.   A general study of the commutation relation given by
                     L. Tartar

Slemrod, M.   Interrelationships among mechanics, numerical analysis,
                     compensated compactness, and oscillation theory

Venakides, S.   The solution of completely integrable systems in the
                     continuum limit of the spectral data

Warming, R.   Stability of finite-difference approximations for
and Beam, R.   hyperbolic initial boundary value problems

Yee, H.   Construction of a class of symmetric T V D schemes

Metastability and incompletely posed problems

May 6 - May 10, 1985

S. Antman
J. L. Ericksen
D. Kinderlehrer
I. Müller
Conference Committee

Antman, S. and Malek-Madani, R.	Dissipative mechanisms
Ball, J.	Does rank-one convexity imply quasiconvexity?
Brezis, H.	Metastable harmonic maps
Calderer, M.	Bifurcation of constrained problems in thermoelasticity
Chipot, M. and Luskin,M.	The compressible Reynolds lubrication equation
Ericksen, J.	Twinning of crystals I
Evans, L. C.	Quasiconvexity and partial regularity in the calculus of variations
Goldenfeld, N.	Introduction to pattern selection in dendritic solidification
Gurtin, M.	Some results and conjectures in the gradient theory of phase transitions
James, R.	The stability and metastability of quartz
Kenig, C.	Continuation theorems for Schrodinger operators
Kinderlehrer, D.	Twinning of crystals II

Kitsche, W., Müller, I., and Strehlow, P.   Simulation of pseudoelastic behaviour in a system of rubber balloons

Lions, J.-L.   Asymptotic problems in distributed systems

Liu, T.P.   Stability of nonlinear waves

Maderna, C., Pagani, C., and Salsa, S.   The Nash-Moser technique for an inverse problem in potential theory related to geodesy

Mosco, U.   Variational stability and relaxed Dirichlet problems

Pitteri, M.   A contribution to the description of natural states for elastic crystalline solids

Rogers, R.   Nonlocal problems in electromagnetism

Vazquez, J.   Hyperbolic aspects in the theory of the porous medium equation

Vergara-Caffarelli, G.   Green's formulas for linearized problems with live loads

Wright, T.   Some aspects of adiabatic shear bands